Postbiotics in Metabolic Disorders

This book discusses the role of postbiotics in human health. The initial chapter highlights the potential of postbiotics beyond traditional prebiotics and probiotics. Subsequent chapters present the beneficial effects on various health conditions, including cardiovascular disorders, cancer, diabetes, gestational diabetes, obesity, and mitochondrial disorders.

The book covers the role of postbiotics in immune system modulation and inflammation-mediated metabolic disorders. It also explores the interaction of postbiotics with gut microbiota, detailing molecular mechanisms and signaling pathways. In addition, it offers current and future perspectives on postbiotics in human health, addressing their role in managing insulin resistance, glucose metabolism, metabolic endotoxemia, and inflammation in type 2 diabetes. The text provides an in-depth analysis of the impact of postbiotics on nonalcoholic fatty liver disease (NAFLD), lipid profiles, and hypertension. In addition, the chapters:

1. Explore the promising concept of postbiotics and their potential beyond prebiotics and probiotics.
2. Examine the beneficial effects of postbiotics on cardiovascular health, cancer, diabetes, and other conditions.
3. Discuss the role of postbiotics in immune system modulation and inflammation-mediated metabolic disorders.
4. Cover the interaction of postbiotics with gut microbiota, highlighting molecular mechanisms and signaling pathways.
5. Analyze the current and future perspectives on postbiotics, including their impact on various metabolic disorders.

This book is a useful resource for researchers and students of pharmacology and biotechnology.

Postbiotics in Metabolic Disorders

Edited by
Sumel Ashique
Shakira Ghazanfar
Naheed Mojgani
Fernanda Maria Policarpo Tonelli

CRC Press
Taylor & Francis Group
Boca Raton London New York

CRC Press is an imprint of the
Taylor & Francis Group, an **informa** business

Designed cover image: Shutterstock

First edition published 2026
by CRC Press
2385 NW Executive Center Drive, Suite 320, Boca Raton FL 33431

and by CRC Press
4 Park Square, Milton Park, Abingdon, Oxon, OX14 4RN

CRC Press is an imprint of Taylor & Francis Group, LLC

ISBN: 978-1-041-00008-2 (hbk)
ISBN: 978-1-041-00010-5 (pbk)
ISBN: 978-1-003-60777-9 (ebk)

DOI: 10.1201/9781003607779

Typeset in Times
by SPi Technologies India Pvt Ltd (Straive)

Contents

About the Editors

Sumel Ashique has been working as an assistant professor at the Department of Pharmaceutical Technology, Bharat Technology, West Bengal, India. He has 4 years of teaching experience. He has achieved 80 publications of international and national accredited and reputed journals. He has knowledge in drug delivery, nanotechnology, and targeted treatment strategy. He has also five granted patents from IP and Australia, with expertise in the fields of targeted drug delivery, nanotechnology, gut microbiome, and postbiotics. Currently he is pursuing his PhD from Lovely Professional University, Punjab, India (NAAC A++) in the Pharmaceutical Science department.

Dr. Shakira Ghazanfar is a prominent microbiologist with a focus on probiotics. Since 2007, she has been a senior scientific officer at the Pakistan Agricultural Research Council (PARC) and founded the Probiotic Development Lab in Islamabad. Dr. Ghazanfar earned her PhD in microbiology from Quaid-i-Azam University, specializing in gut microbial genomics. Her research has led to AI-driven probiotic products for humans and animals and the invention of a bioinformatics tool named Probio Predict. Currently, she is working on AI development in Pakistan and has published over 100 articles. She has also won several awards, including the Young Biotechnology Award in 2021, continuing to impact the industry with her groundbreaking research in microbial genomics and probiotic advancement.

Professor Dr. Naheed Mojgani, with a PhD in microbiology (1997), heads the Human and Animal Probiotic Research Lab at the Razi Vaccine and Serum Research Institute in Iran. With over 70 peer-reviewed publications and extensive experience in infectious diseases, bacterial vaccines, and probiotics, she has made significant contributions to her field. Dr. Mojgani has received multiple awards, including the Outstanding Researcher Award and recognition from the Iranian Ministry of Education. She serves on several key committees and acts as a senior consultant, facilitating technology transfer for probiotic supplements. Dr. Mojgani has also supervised numerous students and organized workshops, enhancing education in her area of expertise.

Dr. Fernanda Maria Policarpo Tonelli, PhD, is an assistant professor at Federal University of São João del Rei, Divinópolis-MG, Brazil, where she teaches graduate and postgraduate students. She specializes in biotechnology/molecular biology and has been studying nanomaterials applications (such as delivery and environmental pollution remediation) as well as green synthesis protocols to produce bioactive nanoparticles and the allelopathic potential of secondary metabolites from plants. The research projects developed by Dr. Tonelli have already resulted in 10 patent and two certificates of addition applications, four of them already analyzed and granted. She has published 15 books and authored 20 scientific articles and 57 book chapters with international publishers. She is a member of the editorial board and serves as a reviewer for many prestigious journals. Dr. Tonelli is active in groups of scientists that fight the gender gap in science. Her efforts as a researcher have been recognized with various awards, including For Women in Science Brazil and Under 30 Brazil.

1 Postbiotics

A Promising Concept Beyond Pre- and Probiotics in Human Health

Abrar Hussain
HEJ Research Institute of Chemistry, International Center for Chemical and Biological Sciences, University of Karachi, Karachi, Pakistan

Affaf Shaheen
Federal Urdu University of Arts, Science and Technology, Karachi, Pakistan

Nida Saleem
Shifa Tameer-e-Millat University, Islamabad, Pakistan

Syed Abid Ali
HEJ Research Institute of Chemistry, International Center for Chemical and Biological Sciences, University of Karachi, Karachi, Pakistan

1.1 INTRODUCTION

Probiotics, which are live microorganisms that have health benefits when consumed in sufficient amounts, greatly attracted researchers due to their promising properties. They have a great spectrum of benefits, which makes them an important biotechnological agent [1]. Besides their other properties, probiotics have the potential to produce a number of different metabolites, known as postbiotics. These are bioactive compounds and are produced during the probiotic fermentation process [2]. The term "postbiotics" is increasingly being used in scientific literature and on commercial products, yet it is used inconsistently and without a clear definition. Hence, in 2019, the International Scientific Association for Probiotics and Prebiotics (ISAPP) panel reviewed the definition and scope of postbiotics. According to them, postbiotics are a "preparation of inanimate microorganisms and/or their components that confers a health benefit on the host" [3]. Postbiotics include metabolites, peptides, organic acids, cell wall fragments, and enzymes that provide health benefits to the host. Postbiotics are nonliving components that exert physiological effects without requiring any viability [4].

The postbiotics concept emerged from the broader study of probiotics and their action on gut health. The work of Élie Metchnikoff in the early 20th century about the health benefits of fermented dairy products formed the foundation for probiotic research [5]. The studies gradually concentrated on finding specific bioactive compounds that are responsible for such health benefits, culminating in the recognition of postbiotics as a separate entity in the early 21st century. Advances in microbiology

DOI: 10.1201/9781003607779-1

and genomic techniques made it easier to identify and characterize these metabolites. Postbiotics have gained significant attention in modern science due to their better safety, stability, and efficacy. Unlike probiotics, which require careful storage and handling to maintain viability, postbiotics are nonliving and therefore do not pose risks of infection, which makes them ideal for immunocompromised individuals. Recent genomics and metabolomic technologies have facilitated researchers to identify specific postbiotic compounds and their mechanisms and open new avenues for the use of personalized medicine and functional foods. Postbiotics offer a promising future in microbiome research, contributing to better health results, disease prevention, and innovations in therapy [6, 7].

Postbiotics offer a safer and more stable alternative to probiotics, especially for immunocompromised individuals. They improve the composition of gut microbiota and help in digestion [8]. Postbiotics boost the immune system as they increase the production of antimicrobial peptides and enhance macrophage activity. Postbiotics are also used to alleviate inflammation-related diseases like irritable bowel syndrome (IBS) and colitis [9]. They also aid in weight management and improve insulin sensitivity [10]. In the cosmetic industry, they are used in skin care products due to their anti-inflammatory and antimicrobial properties, and they can promote the skin barrier function and treat dermatological conditions [11]. Postbiotics are used in the pharmaceutical industry as they are incorporated into drugs and supplements to treat gastrointestinal and metabolic disorders [12]. They are used in the food industry as they are added to yogurts, drinks, and snacks to enhance nutritional value [10]. They are also being added to formula milk to mimic the benefits of breastfeeding [13]. Postbiotics are also used to improve gut health and immunity in livestock and poultry [14].

Probiotics are live microorganisms that provide health benefits when administered in adequate amounts, primarily by enhancing gut flora and immunity and maintaining the balance of microbes [15, 16]. Besides other functions, probiotics are also extensively used in various biotechnological processes [17]. Probiotic applications can be enhanced while using probiotics in combination with other encapsulating materials [18]. Prebiotics, in contrast, are nondigestible food ingredients that act as a food source for probiotics, stimulating their growth and activity to help support gut health. Synbiotics are a combination of both probiotics and prebiotics, which work synergistically to enhance digestive and metabolic functions by promoting a balanced gut microbiome. The three (probiotics, prebiotics, and synbiotics) are distinct in that postbiotics are bioactive compounds produced during the fermentation process by probiotics. These include enzymes, peptides, organic acids, and cell wall fragments that provide health benefits without requiring live microorganisms. The stability, safety, and convenience of postbiotics, which can be given to immunocompromised patients, make them a relatively versatile and promising tool in modern health science [19, 20].

Postbiotics have diverse mechanisms, numerous health benefits, and broad applications; therefore they are emerging as an essential component of microbiome-based therapies and functional nutrition. Their stability and safety profile also underscore their relevance in the fight against modern health challenges. Due to their overwhelming applications and properties, it is important to explore the different aspects of postbiotics. Hence, this chapter aims to explore various aspects including properties, types, and applications.

1.2 POSTBIOTICS AND THEIR TYPES

Postbiotics are bioactive substances secreted by probiotic microorganisms during either fermentation or metabolism. These are strain-dependent substances and thus have varying composition in different microorganisms, depending on fermentation substrates and environmental conditions. Recently, there is a notion that viability might not be strictly required for all health benefits linked with microorganisms—namely, emerging para-probiotics and postbiotics concepts. This way, cell lysates or microbial fractions and nonviable microbial cells could provide physiological benefits to the individual by offering extra bioactivity [21]. These nonviable microbial products are bioactive compounds that include organic acids, enzymes, peptides, and polysaccharides, each contributing to distinct health benefits and derived from the metabolic activities of probiotics, which play a

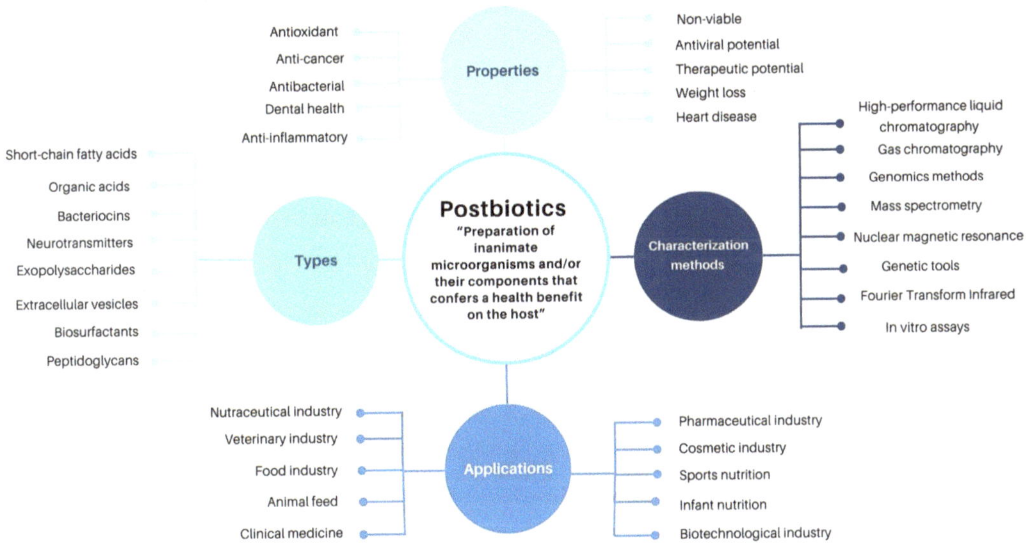

FIGURE 1.1 The different spectra of postbiotics.

crucial role in promoting health by exerting immunomodulatory, antimicrobial, anti-inflammatory, and metabolic regulatory effects. Such bioactive substances are very important for modulating the gut microbiota, enhancing the intestinal barrier function, reducing inflammation, and increasing immune system functioning [2]. The composition of postbiotics can be classified into several categories based on their chemical nature and biological functions, which highlights their versatility and functionality, making them valuable for applications in pharmaceuticals, food, cosmetics, and biotechnology [22]. Advances in research on the microbiome, biotechnological tools, and understanding of postbiotics make them an exciting avenue of effective, safe, and stable alternatives to probiotics and open up opportunities for innovative therapeutic and nutritional applications [23]. The different spectra of postbiotics are summarized in Figure 1.1.

Chemically, they encompass metabolic byproducts such as short-chain fatty acids (SCFAs), lactic acid, acetic acid, proteins, and peptides like antimicrobial peptides and enzymes, exopolysaccharides, lipids and fatty acids, vitamins, minerals, and even gaseous compounds such as hydrogen and methane. These bioactive molecules are crucial to gut health and immunity [10, 24]. Postbiotics can be classified on the basis of several other factors as well. They can also be classified as high- and low-molecular-weight substances. Based on composition, they are classified as carbohydrates (e.g., teichoic acids and galactose-rich polysaccharides), proteins (e.g., p40, p75 molecule, lactocepin), lipids (e.g., butyrate, acetate, propionate, lactate, dimethyl acetyl-derived plasmalogen), vitamins (e.g., B-group vitamins, organic acids like 3-phenyllactic acid and propionic acid), and other complex molecules (e.g., lipoteichoic acids, peptidoglycan-derived muropeptides). Finally, considering their physiological function, they are classified as anti-obesogenic, antioxidant, anti-inflammatory, hypocholesterolemic, antihypertensive, and antiproliferative agents [22].

Postbiotics such as SCFAs (i.e., acetate, propionate, and butyrate) are produced during the fermentation of dietary fibers and play a key role in gut and colon health while controlling the inflammatory processes. Other important postbiotics are antimicrobial peptides, proteins with roles in immune modulation, polysaccharides that maintain gut health, and lipids that affect cell signaling [4, 12]. Overall, postbiotics contribute to a wide range of physiological benefits through interactions with the host's microbiota, immune system, and metabolism (Table 1.1). The following are different types of postbiotics.

TABLE 1.1
The Types of Postbiotics, Source of Production, Function, and Biochemical Composition

Type of Postbiotic	Source of Production	Functions	Biochemical Composition
Short-chain fatty acids (SCFAs)	Probiotic bacteria (e.g., *Lactobacillus*, *Bifidobacterium*)	Regulation of gut pH, enhancement of intestinal barrier, reduction of inflammation, and promotion of gut health	Acetate, propionate, butyrate
Lactic acid	Lactic acid bacteria	Maintenance of acidic pH in the gut, inhibition of pathogenic bacteria	Organic acids
Exopolysaccharides (EPS)	Probiotic bacteria (e.g., *Lactobacillus*)	Formation of biofilms, improvement of gut protection, modulation of immune response	Polysaccharides
Bacteriocins	Probiotic bacteria (e.g., *Lactococcus*, *Enterococcus*)	Inhibition of the growth of pathogenic microbes, regulation of gut microbiota	Proteinaceous peptides (e.g., nisin)
Antimicrobial peptides	Probiotic bacteria and fungi	Inhibition of pathogenic bacteria and fungi, enhancement of immune function	Peptides (e.g., defensins, lantibiotics)
Lipopeptides and lipoproteins	Probiotic bacteria (e.g., *Bacillus*, *Lactobacillus*)	Antimicrobial, immunomodulatory, and anti-inflammatory effects	Lipids, peptides
Bile salt hydrolases (BSH)	Probiotic bacteria (e.g., *Lactobacillus*, *Bifidobacterium*)	Modulation of bile acid metabolism, support for cholesterol and lipid metabolism	Enzyme (hydrolase activity)
Vitamins (e.g., vitamins B, K)	Probiotic bacteria, yeasts	Improvement of metabolic processes, support for immune health, and increased gut microbiome function	Water-soluble vitamins (e.g., B_1, B_2, B_{12}, K)
Enzymes (e.g., proteases, amylases)	Probiotic bacteria, fungi, and yeasts	Help in digestion and breaking down of complex molecules (e.g., proteins, starches)	Proteins (e.g., amylase, protease)
Gaseous compounds (e.g., hydrogen, methane)	Probiotic bacteria, yeasts	Affecting gut fermentation processes, aid in gastrointestinal health	Gases (e.g., hydrogen, methane, hydrogen sulfide)
Exosomes	Probiotic bacteria, fungi	Facilitating transfer of bioactive molecules, modulation of immune response, increasing gut barrier function	Lipids, RNA, proteins

Data from [1, 19, 25–29].

1.2.1 Metabolite-Based Postbiotics

These are the products of metabolic activity carried out by probiotics through fermentation or digestion. They comprise organic acids, SCFAs, and bioactive peptides produced during microbial metabolism [21]. They help to improve gut barrier integrity, inhibit pathogens, and regulate metabolism. Common examples are as follows.

1.2.1.1 Short-Chain Fatty Acids (SCFAs)

SCFAs like butyrate, propionate, and acetate are produced during the fermentation of dietary fiber. They exert anti-inflammatory activities and help to promote gut health and maintain intestinal barrier integrity. They also provide energy to the colonocytes, maintain the pH of the gut, strengthen the intestinal barriers, and exhibit anti-inflammatory actions [1, 19, 25].

1.2.1.2 Lactic Acid

This is produced by lactic acid bacteria during the fermentation process. It contributes to maintaining the acidic pH in the gut, which can suppress pathogenic bacteria growth.

1.2.2 Cell-Free Supernatant (CFS) Postbiotics

These contain soluble bioactive molecules such as enzymes, peptides, and antimicrobial compounds extracted from probiotic cultures. Examples include protein-rich extracellular vesicles and enzymes like lactase that are used in pharmaceuticals and functional foods for their antimicrobial and immunomodulatory effects [25]. These are produced to inhibit the growth of pathogenic organisms and to help in the breakdown of food components. They include the following.

1.2.2.1 Antimicrobial Peptides

These are peptides produced by probiotics that can inhibit the growth of harmful bacteria and viruses, thus providing protection against infections. Peptides also show antimicrobial activity while disturbing the membranes of pathogenic bacteria.

1.2.2.2 Enzymes-Based Postbiotics

Probiotics produce various enzymes during fermentation processes that help in the digestion of lactose in dairy products or fiber in vegetables. They help to promote anti-inflammatory responses and break down biofilms of pathogens as well. They facilitate the breakdown of dietary compounds, help in digestion, and exhibit antimicrobial potential [26].

1.2.3 Cell Wall Fragments–Based Postbiotics

These are complex carbohydrates secreted by certain strains of probiotics and include peptidoglycans, teichoic acids, lipoteichoic acids, and polysaccharides derived from microbial cell walls. They help to activate immune responses and maintain gut integrity. These modulate the immune responses and promote gut barrier integrity [25, 27].

1.2.4 Extracellular Polysaccharide-Based Postbiotics

These include dextran and levan, which help to promote immune modulation and enhance the gut microbial diversity. They can enhance the immune function, support gut barrier integrity, and act as prebiotic substrates for other beneficial bacteria [25].

1.2.5 Peptide- and Protein-Based Postbiotics

These are bioactive peptides, bacteriocins, and protein fragments that exhibit antimicrobial properties, modulate immune responses, and regulate metabolic activity. Bacteriocins contribute to the regulation of gut microbiota by reducing the growth of harmful bacteria. Peptides show antimicrobial activity, disturb the membranes of pathogenic bacteria, and prevent infections.

1.2.6 Inactivated Microbial Cells (Para-Probiotics)–Based Postbiotics

These are nonviable bacterial cells or cell fragments that retain their biological activities and provide immunostimulatory effects. They also improve gut health without the need for live microorganisms such as heat-killed *Lactobacillus* or *Bifidobacterium* species [30].

1.2.7 Vitamin- and Mineral-Producing Postbiotics

Some probiotics can synthesize vitamins (e.g., vitamins B and K) and minerals that can benefit the host, particularly in terms of nutrient absorption and immunity. They also aid in the fortification of foods and dietary supplements to support metabolic and immune health [26].

1.2.8 Gaseous Compound Postbiotics

Different gaseous substances like hydrogen, methane, or hydrogen sulfide are also produced as metabolic products. These are involved in specific health matters, depending on the concentration and the situation. These postbiotics are critical in maintaining the gut environment, enhancing immune function, and maintaining the integrity of the gut barrier by providing a protective mechanism against pathogens to promote health and well-being.

1.3 MECHANISM OF ACTION OF POSTBIOTICS

The beneficial effects of postbiotics are mediated by several mechanisms of action, which are generally conducted by bioactive molecules and metabolites.

1.3.1 Action Via Gut and Intestinal Microbial Balancing

Postbiotics have the potential to maintain the gut microbiota. They can eradicate pathogenic bacteria and reinforce the probiotic balance. Similarly, postbiotics can inhibit the binding of pathogenic bacteria and thus aid the health of the host. Likewise, these substances can also enhance the intestinal integrity and lower leaky gut or intestinal permeability [31].

1.3.2 Immunomodulatory Potential

Postbiotics like lipopolysaccharides (LPS) and peptidoglycans trigger immune cells (e.g., macrophages and dendritic cells), initiating anti-inflammatory cytokines (e.g., IL-10) while suppressing pro-inflammatory cytokines (e.g., TNF-α). The greater production of secretory IgA (sIgA) can enhance the mucosal immunity [31].

1.3.3 Antimicrobial Potential

Postbiotics can also perform their action via the production of antimicrobial substances, which include bacteriocins, organic acids, and others. The production of such compounds can kill the pathogenic bacteria in the close vicinity and thus aid the host's health [31].

1.3.4 Antitumor Effect

Postbiotics also exhibit antitumor potential. The SCFAs were found to reduce gastric cancer and help in the regulation of oncogene suppressions [30].

1.3.5 Autophagy Process

In autophagy, damaged cells are removed from the body. Postbiotics like peptidoglycan are also found to cause autophagy in certain cells. For instance, *Lactobacillus fermentum* is found to trigger autophagy in hepatic cells [30].

1.4 PRODUCTION AND CHARACTERIZATION OF POSTBIOTICS

Postbiotics are produced and characterized from different probiotics via different processes. In the laboratory, the production and characterization of these postbiotics require different assays that help to identify them.

1.4.1 Methods of Production of Postbiotics

Postbiotics are the secreted products of probiotics that are prepared by different mechanisms.

1.4.1.1 Fermentation

The most common method of postbiotic production is microbial fermentation. Probiotics (such as *Lactobacillus*, *Bifidobacterium*, or *Saccharomyces cerevisiae*) ferment nutrients in a controlled environment and produce bioactive compounds. In this process, microorganisms are cultured in a medium (liquid or solid) containing nutrients such as sugars and peptides. The fermentation process is carried out under anaerobic or aerobic conditions, depending on the type of microorganism. The production of postbiotics as part of the fermentation process encompasses SCFAs, lactic acid, enzymes, and bacteriocins [11, 32].

These involve a process of microbial fermentation occurring on solid substrates consisting of cereals, sugarcane, legumes, and other kinds of agricultural waste or even residues with trace amounts of added water. In this process, microorganisms are added, corresponding to probiotic bacteria, directly to the solid substrate. Such fermentation that is controlled over temperature and humidity is solid-state fermentation, which is often applied to generate high-value postbiotics, such as enzymes, bioactive peptides, and exopolysaccharides [28].

1.4.1.2 Bioreactor Culture

Bioreactors are large vessels that allow the scalable production of postbiotics in a controlled environment. During this process, microorganisms are grown in a bioreactor that allows the optimal growth conditions of pH, temperature, oxygen, and nutrients and their levels to be maintained. Postbiotics are recovered from the culture medium or cell mass at the end of the fermentation stage [11, 33].

1.4.1.3 Extraction from Probiotic Cultures

From the culture medium or cells, postbiotics may be extracted after microbial fermentation or growth. Postbiotics can be recovered using solvents, centrifugation, or filtration and also from cell lysis through mechanical means. Common extraction methods include precipitation, ultrafiltration, and lyophilization (freeze-drying) [34].

1.4.1.4 Biotransformation

Some postbiotics are derived from the biotransformation of precursor compounds of probiotics or other microorganisms. During biotransformation, the microorganisms metabolize precursors (e.g., bile salts and sugars) into bioactive compounds like bile salt hydrolases or bioactive peptides. This technique is frequently employed in the synthesis of specialized derivatives such as altered bile acids or lipids [29, 35].

1.4.2 Postbiotics Characterization

The characterization of postbiotics can be performed using different methods and instruments. A few of them are highlighted here.

1.4.2.1 Analytical Approaches

High-performance liquid chromatography (HPLC) serves to separate and quantify different postbiotics like SCFAs, organic acids (lactic acid), and vitamins produced by probiotics. Gas chromatography (GC) is also applied in the analysis of volatile compounds, gases, and bioactive peptides and in determining their chemical structure. Similarly, mass spectrometry (MS) is commonly used in combination with other instruments to identify and quantify complex bioactive compounds such as peptides, exopolysaccharides, and enzymes [36–39]. Likewise, sodium dodecyl sulfate–polyacrylamide gel electrophoresis (SDS-PAGE) is used to separate the postbiotics based on their molecular weight. It helps to characterize antimicrobial peptides or enzymes produced during fermentation [40]. Nuclear magnetic resonance (NMR) spectroscopy determines the molecular structure of postbiotics, including exopolysaccharides, lipopeptides, and bioactive compounds [41]. Fourier transform infrared (FTIR) spectroscopy identifies functional groups in the chemical structure of postbiotics, particularly for complex carbohydrates and peptides [42, 43].

1.4.2.2 Microbiological Approaches

The antimicrobial activity of the postbiotics is determined by agar diffusion or broth microdilution techniques. The disk diffusion test reflects the capability of postbiotics while inhibiting the growth of pathogenic bacteria [44, 45]. The molecular assessment of postbiotics is also elucidated using polymerase chain reaction (PCR). It is used to identify genes involved in the production of specific postbiotics, such as bacteriocins or bioactive peptides [46]. The genetic tools are also used to explore the genetic basis of postbiotics. For instance, the next-generation sequencing (NGS) technologies can provide insights into the genetic pathways involved in the biosynthesis of postbiotics, helping to identify new bioactive compounds and their genetic regulators [47, 48]. In vitro cell cultures are used to determine the biological activity of postbiotics, such as how they affect gut epithelial cells, immune cells, or other tissues. Examples include anti-inflammatory, antioxidant, or gut microbiome modulation effects [49].

1.5 APPLICATIONS OF POSTBIOTICS IN DIFFERENT INDUSTRIES

The stability, safety, long shelf life, and efficacy of postbiotics make them versatile components within modern health and wellness industries, ensuring their continued use in preventive and therapeutic approaches. In the pharmaceutical industry, they are used to develop therapeutic drugs and supplements [4], while in the food and beverage industry they are used in the making of functional foods, fermented products, and beverages to enhance their nutritional value, promote gut health, and act as natural preservatives and extend shelf life [50]. In the cosmetic and personal care industry, postbiotics are incorporated in skincare products to improve the hydration of the skin, repair the skin barrier, and have anti-aging effects [11]. The animal feed industry utilizes postbiotics to promote and support growth and improve gut health and immunity in livestock and poultry, reducing the need for antibiotics [51]. In the nutraceutical industry, they are used in formulating capsules, powders, and tablets to target and promote metabolic health, manage weight, and support the immune system [12]. The biotechnology sector utilizes postbiotics to produce enzymes, bioactive peptides, and drug delivery systems [6]. The veterinary industry uses postbiotics to make supplements to enhance the health of pet animals [49]. Postbiotics are also used for infant nutrition as they mimic breast milk benefits, promoting the development of gut and immunity in infants [13]. In the sports nutrition sector, they are included in protein bars and recovery beverages to promote muscle healing and minimize inflammation [52]. In clinical medicine, postbiotics serve as adjunct therapies for inflammatory bowel disease (IBD), IBS, and antibiotic-associated diarrhea [53].

1.5.1 Role of Postbiotics in Human Health

Numerous studies underscoring the therapeutic effects of postbiotics demonstrate their efficacy over probiotic bacteria. Antioxidant potentials of postbiotics are one of their significant properties related to antioxidant enzymes, metabolites, and metal ions. In human health, postbiotics offer promising results and thus have gotten substantial attention.

1.5.2 Postbiotics in Metabolic Syndrome

Multiple preclinical and clinical studies have assessed the beneficial role that postbiotics play in improving the symptoms of metabolic syndrome [54]. Sodium butyrate was found to improve the serum lipid profile, glucose tolerance, and insulin sensitivity and to reduce adipose tissue inflammation [55]. In a clinical trial, while evaluating the role of postbiotic effects of butyrate in pediatric obesity and metabolic complications, it was revealed that a 6-month supplementation can reduce body mass index, alleviate glucose metabolism, and reduce inflammation [56].

1.5.3 Antidiabetic Potential

Diabetes mellitus is a chronic and irreversible condition that is characterized by dysregulated insulin activity or synthesis. Postbiotics, for instance *Lactobacillus plantarum* L-14 exopolysaccharides, were found to inhibit the immature cell conversion into mature adipocytes, leading to body weight and lipid profile regulation in mice [57]. Likewise, intestinal inflammation amelioration and intestinal barrier strengthening via extracellular regulated protein kinase (ERK) signaling pathway by *Lactobacillus brevis* long-chain polyphosphate is also reported [58].

1.5.4 Antiviral Potential

The postpandemic world has gotten more serious about viral disease prevention and treatment. Postbiotics can exert antiviral effects by preventing viral entry, strengthening the host immune system, and interfering with the viral attachment to host cells, leading to a delay in infection onset. In addition, postbiotics fortify intestinal permeability and aid virus-specific antibody production [59].

1.5.5 Antibacterial Potentials

Organic acids, including acetic acid, lactic acid, tartaric acid, and citric acid, are predominantly bioactive components of postbiotics. These acids perturb the optimum pH levels required for bacterial growth and change membrane integrity [34].

1.6 CHALLENGES AND LIMITATIONS

Despite their promising therapeutic potential, postbiotics face a series of challenges and limitations that make them difficult to use extensively. Standardization and quality control are two of the major problems because the complex and diverse composition of postbiotics makes it challenging to ensure batch-to-batch consistency and accurate measurement. Stability and storage problems become significant issues since many postbiotic compounds are thermally sensitive and undergo degradation when exposed to oxygen, light, or moisture and require advanced preservation techniques like encapsulation or freeze-drying, which can increase production costs. Scalability and economic feasibility are also issues, mainly for large-scale production that does not compromise efficacy. Regulatory and safety concerns further complicate matters since postbiotics fall into a gray area of regulation regarding food, supplements, or pharmaceuticals. Their safety would require consideration of various risks of contaminants, residual toxins, and allergenic components potentially resulting from bacterial fermentation, as well as extensive toxicological assessments and clinical trials to ensure their long-term safety, especially for infants, pregnant women, or immunocompromised individuals. Mechanisms of action remain largely poorly understood for many components of postbiotics; hence, their targeted uses are limited, and this challenges product design for the treatment of specific health conditions. Finally, consumers' awareness and education deficit on the benefits of postbiotics and how these are generated may lower their commercialization.

1.7 FUTURE PERSPECTIVE

The future of postbiotics is bright as research and technological advancements are now unfolding new avenues for their development and application in health, nutrition, and medicine. As the understanding of gut microbiota and host–microbe interactions expands, postbiotics are explored to play a critical role in personalized medicine and precision nutrition, allowing tailored formulations based on individual microbiomes and health needs. Advances in metagenomics, metabolomics, and bioinformatics will lead to further identification and characterization of novel bioactive compounds that will improve functionality and targeted therapeutic applications. Next-generation pharmaceuticals and functional foods will be designed by using postbiotics, such as the development of medications for different diseases. Improvements in nanotechnology and encapsulation methods are going to enhance stability, bioavailability, and controlled delivery of postbiotics with longer shelf life and wider commercial applicability.

1.8 CONCLUSION

Postbiotics are emerging as a revolutionary innovation that will help in microbiome research, promising to restore human health through the benefits of bioactive compounds arising from probiotics. Such products are nonviable but retain significant functional properties due to their stability and safety aspects, thus providing targeted benefits for health. Their mechanisms of action include immunomodulation, anti-inflammation, antimicrobial activity, enhancement of the gut barrier, and metabolic regulation. They hold value as a treatment for many gastrointestinal conditions, metabolic syndromes, allergies, and neurodegenerative conditions. However, many challenges in the realms of standardization, stability, safety assurance, and regulatory frameworks still exist. Future research in the areas of personalized medicine, advanced delivery systems, and clinical validation will unlock their full potential, positioning postbiotics as a key component of functional foods, pharmaceuticals, and preventive healthcare strategies.

REFERENCES

[1] Hussain, A., Koser, N., Aun, S. M., Siddiqui, M. F., Malik, S., & Ali, S. A. (2024). Deciphering the Role of Probiotics in Mental Health: A Systematic Literature Review of Psychobiotics. *Beneficial Microbes, 0*, 1–22. https://doi.org/10.1163/18762891-bja00053

[2] Wegh, C. A. M., Geerlings, S. Y., Knol, J., Roeselers, G., & Belzer, C. (2019). Postbiotics and Their Potential Applications in Early Life Nutrition and Beyond. *International Journal of Molecular Sciences, 20*(19), 4673. https://doi.org/10.3390/ijms20194673

[3] Salminen, S., Collado, M. C., Endo, A., Hill, C., Lebeer, S., Quigley, E. M. M., Sanders, M. E., Shamir, R., Swann, J. R., Szajewska, H., & Vinderola, G. (2021). The International Scientific Association of Probiotics and Prebiotics (ISAPP) Consensus Statement on the Definition and Scope of Postbiotics. *Nature Reviews Gastroenterology and Hepatology, 18*(9), 649–667. https://doi.org/10.1038/s41575-021-00440-6

[4] Vinderola, G., Sanders, M. E., & Salminen, S. (2022). The Concept of Postbiotics. *Foods, 11*(8), 1077. https://doi.org/10.3390/foods11081077

[5] Hussain, A., & Ali, S. A. (2024). Probiotics and Immunity: Exploring the Development of Probiotics Concept. *EC Pharmacology and Toxicology, 12*(10), 01–06. https://ecronicon.net/ecpt/probiotics-and-immunity-exploring-the-development-of-probiotics-concept

[6] Aggarwal, S., Sabharwal, V., Kaushik, P., Joshi, A., Aayushi, A., & Suri, M. (2022). Postbiotics: From Emerging Concept to Application. *Frontiers in Sustainable Food Systems, 6*. https://doi.org/10.3389/fsufs.2022.887642

[7] Zhao, X., Liu, S., Li, S., Jiang, W., Wang, J., Xiao, J., Chen, T., Ma, J., Khan, M. Z., Wang, W., Li, M., Li, S., & Cao, Z. (2024). Unlocking the Power of Postbiotics: A Revolutionary Approach to Nutrition for Humans and Animals. *Cell Metabolism, 36*(4), 725–744. https://doi.org/10.1016/j.cmet.2024.03.004

[8] Mosca, A., Abreu, Y., Abreu, A. T., Gwee, K. A., Ianiro, G., Tack, J., Nguyen, T. V. H., & Hill, C. (2022). The Clinical Evidence for Postbiotics as Microbial Therapeutics. *Gut Microbes, 14*(1). https://doi.org/10.1080/19490976.2022.2117508

[9] Sharma, S., Kumar, S., Sajjad, S., & Sharma, S. (2023). Probiotics in Irritable Bowel Syndrome: A Review Article. *Cureus*, *15*(3), 1–6. https://doi.org/10.7759/cureus.36565

[10] Gurunathan, S., Thangaraj, P., & Kim, J.-H. (2023). Postbiotics: Functional Food Materials and Therapeutic Agents for Cancer, Diabetes, and Inflammatory Diseases. *Foods*, *13*(1), 89. https://doi.org/10.3390/foods13010089

[11] da Silva Vale, A., de Melo Pereira, G. V., de Oliveira, A. C., de Carvalho Neto, D. P., Herrmann, L. W., Karp, S. G., Soccol, V. T., & Soccol, C. R. (2023). Production, Formulation, and Application of Postbiotics in the Treatment of Skin Conditions. *Fermentation*, *9*(3), 264. https://doi.org/10.3390/fermentation9030264

[12] Thorakkattu, P., Khanashyam, A. C., Shah, K., Babu, K. S., Mundanat, A. S., Deliephan, A., Deokar, G. S., Santivarangkna, C., & Nirmal, N. P. (2022). Postbiotics: Current Trends in Food and Pharmaceutical Industry. *Foods*, *11*(19), 3094. https://doi.org/10.3390/foods11193094

[13] Lemoine, A., Tounian, P., Adel-Patient, K., & Thomas, M. (2023). Pre-, Pro-, Syn-, and Postbiotics in Infant Formulas: What Are the Immune Benefits for Infants? *Nutrients*, *15*(5), 1–25. https://doi.org/10.3390/nu15051231

[14] Saeed, M., Afzal, Z., Afzal, F., Khan, R. U., Elnesr, S. S., Alagawany, M., & Chen, H. (2023). Use of Postbiotic as Growth Promoter in Poultry Industry: A Review of Current Knowledge and Future Prospects. *Food Science of Animal Resources*, *43*(6), 1111–1127. https://doi.org/10.5851/kosfa.2023.e52

[15] Hussain, A., & Abid Ali, S. (2024). Exploring *Enterococcus* Species for their Next-Generation Probiotics Potential. In *Probiotics, Prebiotics, and Postbiotics in Human Health and Sustainable Food Systems [Working Title]* (p. 13). IntechOpen. https://doi.org/10.5772/intechopen.1007306

[16] Hussain, A., Qureshi, S. R., & Parveen, S. (2024). The Potential Use of Probiotics as Medicine. *Gut Heath, Microbiota and Animal Diseases*, 30–43. https://doi.org/10.47278/book.CAM/2024.058

[17] Hussain, A., Shakeel, M., Khan, M. T., & Ali, S. A. (2024). Probiotic Potential and Biotechnological Characterization of the Selected Enterococcal Strains. *Food Bioscience*, *63*(December 2024), 105792. https://doi.org/10.1016/j.fbio.2024.105792

[18] Hussain, A., Parveen, S., Riaz, M., Zia, A., & Abid Ali, S. (2024). Probiotics and Vegetable Oil Association. A Review. *IOP Conference Series: Earth and Environmental Science*, *1379*(1), 012001. https://doi.org/10.1088/1755-1315/1379/1/012001

[19] Ji, J., Jin, W., Liu, S. J., Jiao, Z., & Li, X. (2023). Probiotics, Prebiotics, and Postbiotics in Health and Disease. *MedComm*, *4*(6), 1–23. https://doi.org/10.1002/mco2.420

[20] Martyniak, A., Zakrzewska, Z., Schab, M., Zawartka, A., Wędrychowicz, A., Skoczeń, S., & Tomasik, P. J. (2023). Prevention and Health Benefits of Prebiotics, Probiotics and Postbiotics in Acute Lymphoblastic Leukemia. *Microorganisms*, *11*(7), 1775. https://doi.org/10.3390/microorganisms11071775

[21] Pimentel, T. C., Cruz, A. G., Pereira, E., da Costa, W. K.A., da Silva Rocha, R., de Souza Pedrosa, G.T., Rocha, C., Dos, S., Alves, J. M., Alvarenga, V. O., Sant'Ana, A. S., & Magnani, M. (2023). Postbiotics: An Overview of Concepts, Inactivation Technologies, Health Effects, and Driver Trends. *Trends in Food Science & Technology*, *138*(May), 199–214. https://doi.org/10.1016/j.tifs.2023.06.009

[22] Aguilar-Toalá, J. E., Garcia-Varela, R., Garcia, H. S., Mata-Haro, V., González-Córdova, A. F., Vallejo-Cordoba, B., & Hernández-Mendoza, A. (2018). Postbiotics: An Evolving Term Within the Functional Foods Field. *Trends in Food Science & Technology*, *75*(February), 105–114. https://doi.org/10.1016/j.tifs.2018.03.009

[23] Kumar, A., Green, K. M., & Rawat, M. (2024). A Comprehensive Overview of Postbiotics with a Special Focus on Discovery Techniques and Clinical Applications. *Foods*, *13*(18), 2937. https://doi.org/10.3390/foods13182937

[24] Santana, G. B., Quelemes, P. V., da Silva Neta, E. R., de Lima, S. G., & Vale, G. C. (2024). Chemical Characterization and Effect of a *Lactobacilli*-Postbiotic on *Streptococcus mutans* Biofilm In Vitro. *Microorganisms*, *12*(5), 843. https://doi.org/10.3390/microorganisms12050843

[25] Islam, F., Azmat, F., Imran, A., Zippi, M., Hong, W., Tariq, F., Shehzadi, U., Fatima, A., Safdar, M., Ahmed, H., Ijaz, M., Zeeshan, F., Ali, R., Asif Shah, M., & Suleria, H. A. R. (2024). Role of Postbiotics in Food and Health: A Comprehensive Review. *CYTA - Journal of Food*, *22*(1). https://doi.org/10.1080/19476337.2024.2386412

[26] Bourebaba, Y., Marycz, K., Mularczyk, M., & Bourebaba, L. (2022). Postbiotics as Potential New Therapeutic Agents for Metabolic Disorders Management. *Biomedicine and Pharmacotherapy*, *153*(March), 113138. https://doi.org/10.1016/j.biopha.2022.113138

[27] Liang, B., & Xing, D. (2023). The Current and Future Perspectives of Postbiotics. *Probiotics and Antimicrobial Proteins*, *15*(6), 1626–1643. https://doi.org/10.1007/s12602-023-10045-x
[28] Tong, Y., Guo, H., Abbas, Z., Zhang, J., Wang, J., Cheng, Q., Peng, S., Yang, T., Bai, T., Zhou, Y., Li, J., Wei, X., Si, D., & Zhang, R. (2023). Optimizing Postbiotic Production Through Solid-State Fermentation with *Bacillus amyloliquefaciens* J and *Lactiplantibacillus plantarum* SN4 Enhances Antibacterial, Antioxidant, and Anti-inflammatory Activities. *Frontiers in Microbiology*, *14*. https://doi.org/10.3389/fmicb.2023.1229952
[29] Pan, Z., Mao, B., Zhang, Q., Tang, X., Yang, B., Zhao, J., Cui, S., & Zhang, H. (2022). Postbiotics Prepared Using *Lactobacillus paracasei* CCFM1224 Prevent Nonalcoholic Fatty Liver Disease by Modulating the Gut Microbiota and Liver Metabolism. *International Journal of Molecular Sciences*, *23*(21), 13522. https://doi.org/10.3390/ijms232113522
[30] Zółkiewicz, J., Marzec, A., Ruszczyński, M., & Feleszko, W. (2020). Postbiotics—A Step Beyond Pre- and Probiotics. *Nutrients*, *12*(8), 2189. https://doi.org/10.3390/nu12082189
[31] Ma, L., Tu, H., & Chen, T. (2023). Postbiotics in Human Health: A Narrative Review. *Nutrients*, *15*(2), 1–17. https://doi.org/10.3390/nu15020291
[32] Danova, S., Yankov, D., Dobreva, L., Dobreva, A., Armenova, N., Apostolov, A., & Mileva, M. (2023). Postbiotics Production of Candidate-Probiotic *Lactiplantibacillus plantarum* AC131 with Renewable Bio Resources. *Life*, *13*(10), 2006. https://doi.org/10.3390/life13102006
[33] Dunand, E., Burns, P., Binetti, A., Bergamini, C., Peralta, G. H., Forzani, L., Reinheimer, J., & Vinderola, G. (2019). Postbiotics Produced at Laboratory and Industrial Level as Potential Functional Food Ingredients with the Capacity to Protect Mice Against Salmonella Infection. *Journal of Applied Microbiology*, *127*(1), 219–229. https://doi.org/10.1111/jam.14276
[34] Rad, A. H., Hosseini, S., & Pourjafar, H. (2021). Postbiotics as Dynamic Biological Molecules for Antimicrobial Activity: A Mini-Review. *Biointerface Research in Applied Chemistry*, *12*(5), 6543–6556. https://doi.org/10.33263/BRIAC125.65436556
[35] Peluzio, M., Do, C. G., Martinez, J. A., & Milagro, F. I. (2021). Postbiotics: Metabolites and Mechanisms Involved in Microbiota–Host Interactions. *Trends in Food Science & Technology*, *108*(October 2020), 11–26. https://doi.org/10.1016/j.tifs.2020.12.004
[36] Asif, A., Afzaal, M., Shahid, H., Saeed, F., Ahmed, A., Shah, Y. A., Ejaz, A., Ghani, S., Ateeq, H., & Khan, M. R. (2023). Probing the Functional and Therapeutic Properties of Postbiotics in Relation to Their Industrial Application. *Food Science & Nutrition*, *11*(8), 4472–4484. https://doi.org/10.1002/fsn3.3465
[37] Benkeblia, N. (2023). Gas Chromatography–Mass Spectrometry and Liquid Chromatography–Mass Spectrometry Metabolomics Platforms: Tools for Plant Oligosaccharides Analysis. *Carbohydrate Polymer Technologies and Applications*, *5*(February), 100304. https://doi.org/10.1016/j.carpta.2023.100304
[38] Gould, O., Nguyen, N., & Honeychurch, K. C. (2023). New Applications of Gas Chromatography and Gas Chromatography–Mass Spectrometry for Novel Sample Matrices in the Forensic Sciences: A Literature Review. *Chemosensors*, *11*(10), 527. https://doi.org/10.3390/chemosensors11100527
[39] Moradi, M., Molaei, R., & Guimarães, J. T. (2021). A Review on Preparation and Chemical Analysis of Postbiotics from Lactic Acid Bacteria. *Enzyme and Microbial Technology*, *143*(August 2020), 109722. https://doi.org/10.1016/j.enzmictec.2020.109722
[40] Freitas, A. Dos S., Barroso, F. A. L., Campos, G. M., Américo, M. F., Viegas, R. C. Dos S., Gomes, G. C., Vital, K. D., Fernandes, S. O. A., Carvalho, R. D. De O., Jardin, J., Miranda, A. P. G. Dos S., Ferreira, E., Martins, F. S., Laguna, J. G., Jan, G., Azevedo, V., & De Jesus, L. C. L. (2024). Exploring the Anti-inflammatory Effects of Postbiotic Proteins from *Lactobacillus delbrueckii* CIDCA 133 on Inflammatory Bowel Disease Model. *International Journal of Biological Macromolecules*, *277*(July), 134216. https://doi.org/10.1016/j.ijbiomac.2024.134216
[41] He, W., & Bertram, H. C. (2022). NMR-Based Metabolomics to Decipher the Molecular Mechanisms in the Action of Gut-Modulating Foods. *Foods*, *11*(17), 2707. https://doi.org/10.3390/foods11172707
[42] Hussain, A., Akram, S., Ahmad, D., Rehman, M., Ahmed, A., & Ali, S. A. (2023). Molecular Assessment and Validation of the Selected Enterococcal Strains as Probiotics. *Probiotics and Antimicrobial Proteins*, *0123456789*, 1–15. https://doi.org/10.1007/s12602-023-10163-6
[43] Papadopoulou, O. S., Argyri, A. A., Kounani, V., Tassou, C. C., & Chorianopoulos, N. (2021). Use of Fourier Transform Infrared Spectroscopy for Monitoring the Shelf Life and Safety of Yogurts Supplemented with a *Lactobacillus plantarum* Strain with Probiotic Potential. *Frontiers in Microbiology*, *12*(June), 1–16. https://doi.org/10.3389/fmicb.2021.678356

[44] Noori, S. M. A., Behfar, A., Saadat, A., Ameri, A., Atashi Yazdi, S. S., & Siahpoosh, A. (2022). Antimicrobial and Antioxidant Properties of Natural Postbiotics Derived from Five Lactic Acid Bacteria. *Jundishapur Journal of Natural Pharmaceutical Products*, *18*(1). https://doi.org/10.5812/jjnpp-130785

[45] Balouiri, M., Sadiki, M., & Ibnsouda, S. K. (2016). Methods for In Vitro Evaluating Antimicrobial Activity: A Review. *Journal of Pharmaceutical Analysis*, *6*(2), 71–79. https://doi.org/10.1016/j.jpha.2015.11.005

[46] Boyte, M.-E., Benkowski, A., Pane, M., & Shehata, H. R. (2023). Probiotic and Postbiotic Analytical Methods: A Perspective of Available Enumeration Techniques. *Frontiers in Microbiology*, *14*(December), 1–15. https://doi.org/10.3389/fmicb.2023.1304621

[47] Morán, J., & Kilasoniya, A. (2024). Integration of Postbiotics in Food Products Through Attenuated Probiotics: A Case Study with Lactic Acid Bacteria in Bread. *Foods*, *13*(13), 2042. https://doi.org/10.3390/foods13132042

[48] Jagadeesan, B., Gerner-Smidt, P., Allard, M. W., Leuillet, S., Winkler, A., Xiao, Y., Chaffron, S., Van Der Vossen, J., Tang, S., Katase, M., McClure, P., Kimura, B., Ching Chai, L., Chapman, J., & Grant, K. (2019). The Use of Next Generation Sequencing for Improving Food Safety: Translation into Practice. *Food Microbiology*, *79*(June 2018), 96–115. https://doi.org/10.1016/j.fm.2018.11.005

[49] Hosseini, S. H., Farhangfar, A., Moradi, M., & Dalir-Naghadeh, B. (2024). Beyond Probiotics: Exploring the Potential of Postbiotics and Parabiotics in Veterinary Medicine. *Research in Veterinary Science*, *167*(December 2023), 105133. https://doi.org/10.1016/j.rvsc.2023.105133

[50] Da, M., Sun, J., Ma, C., Li, D., Dong, L., Wang, L. S., & Chen, F. (2024). Postbiotics: Enhancing Human Health with a Novel Concept. *EFood*, *5*(5), 1–14. https://doi.org/10.1002/efd2.180

[51] Urban, J., Kareem, K. Y., Atanasov, A. G., Matuszewski, A., Bień, D., Ciborowska, P., Rygało-Galewska, A., & Michalczuk, M. (2024). Postbiotics, a Natural Feed Additive for Growth Performance, Gut Microbiota and Quality of Poultry Products. *Current Research in Biotechnology*, *8*(June), 100247. https://doi.org/10.1016/j.crbiot.2024.100247

[52] Kerksick, C. M., Moon, J. M., & Jäger, R. (2024). It's Dead! Can Postbiotics Really Help Performance and Recovery? A Systematic Review. *Nutrients*, *16*(5), 720. https://doi.org/10.3390/nu16050720

[53] Babu, A. S., Guruprasath, N., Adeyeye, S. A. O., Sankarganesh, P., Kumar, A. G., & Sivapriya, T. (2023). A Critical Analysis of Postbiotics: Exploring Their Potential Impact on the Health and Food Industries. *Journal of Pure and Applied Microbiology*, *17*(4), 2041–2059. https://doi.org/10.22207/JPAM.17.4.18

[54] Yoshitake, R., Hirose, Y., Murosaki, S., & Matsuzaki, G. (2021). Heat-Killed *Lactobacillus plantarum* L-137 Attenuates Obesity and Associated Metabolic Abnormalities in C57BL/6 J Mice on a High-Fat Diet. *Bioscience of Microbiota, Food and Health*, *40*(2), 84–91. https://doi.org/10.12938/bmfh.2020-040

[55] Chiang, C.-J., Tsai, B. C.-K., Lu, T.-L., Chao, Y.-P., Day, C. H., Ho, T.-J., Wang, P.-N., Lin, S.-C., Padma, V. V., Kuo, W.-W., & Huang, C.-Y. (2021). Diabetes-Induced Cardiomyopathy Is Ameliorated by Heat-Killed *Lactobacillus reuteri* GMNL-263 in Diabetic Rats via the Repression of the Toll-like Receptor 4 Pathway. *European Journal of Nutrition*, *60*(6), 3211–3223. https://doi.org/10.1007/s00394-020-02474-z

[56] Coppola, S., Nocerino, R., Paparo, L., Bedogni, G., Calignano, A., Di Scala, C., De Giovanni Di Santa Severina, A. F., De Filippis, F., Ercolini, D., & Berni Canani, R. (2022). Therapeutic Effects of Butyrate on Pediatric Obesity. *JAMA Network Open*, *5*(12), e2244912. https://doi.org/10.1001/jamanetworkopen.2022.44912

[57] Li, H.-Y., Zhou, D.-D., Gan, R.-Y., Huang, S.-Y., Zhao, C.-N., Shang, A., Xu, X.-Y., & Li, H.-B. (2021). Effects and Mechanisms of Probiotics, Prebiotics, Synbiotics, and Postbiotics on Metabolic Diseases Targeting Gut Microbiota: A Narrative Review. *Nutrients*, *13*(9), 3211. https://doi.org/10.3390/nu13093211

[58] Isozaki, S., Konishi, H., Fujiya, M., Tanaka, H., Murakami, Y., Kashima, S., Ando, K., Ueno, N., Moriichi, K., & Okumura, T. (2021). Probiotic-Derived Polyphosphate Accelerates Intestinal Epithelia Wound Healing Through Inducing Platelet-Derived Mediators. *Mediators of Inflammation*, *2021*, 1–14. https://doi.org/10.1155/2021/5582943

[59] Sundararaman, A., Ray, M., Ravindra, P. V., & Halami, P. M. (2020). Role of Probiotics to Combat Viral Infections with Emphasis on COVID-19. *Applied Microbiology and Biotechnology*, *104*(19), 8089–8104. https://doi.org/10.1007/s00253-020-10832-4

2 Postbiotics
Beneficial Impact on Cardiovascular Disorders

Gourab Dey, Srabani Podder and Shankhadip Nandi
Eminent College of Pharmaceutical Technology, Kolkata, West Bengal, India

Mohammad Adnan
Integral University, Lucknow, India

Imran Ataurrahman Sheikh
Anjuman I Islam's Kalsekar Technical Campus School of Pharmacy, Navi Mumbai, Maharashtra India

Sumel Ashique and Mohhammad Ramzan
Lovely Professional University, Phagwara, Punjab, India
Biplab Debnath: Bharat Technology, Uluberia-711316, West Bengal, India

2.1 INTRODUCTION

Cardiovascular diseases (CVDs) encompass a range of conditions affecting the heart and blood vessels. These include coronary heart disease (CHD), also known as coronary artery disease or ischemic heart disease, as well as cerebrovascular disease, rheumatic heart disease, and peripheral vascular disease, among others. These conditions disrupt the flow of blood to vital organs, such as the heart (leading to ischemia and potential myocardial injury), the brain (potentially causing an ischemic stroke), and the extremities [1–5]. CVDs are a primary cause of illness and death in developing nations. Unhealthy lifestyle choices, such as poor diet, alcohol consumption, smoking, and lack of physical activity, significantly increase the risk of developing and experiencing worsening CVDs [6, 7]. In the past two decades, more attention has been given to interaction with gut microbiota, which have became one of the most inventive fields regarding the abovementioned diseases. The gut microbiota, a community of microorganisms residing in our intestinal tract, act as a vast endocrine organ [8]. This ever-evolving ensemble profoundly impacts our health by generating bioactive metabolites that influence various physiological processes. Alterations in the composition and functionality of the microbial population, termed intestinal dysbiosis, are linked to several diseased conditions, including acute or chronic impairments of the patient's cardiovascular system. This intricate ecosystem of microorganisms residing in our digestive tract plays a crucial role in host metabolism, immunity, and overall health. Imbalances in this microbiome have been implicated in various CVDs, including atherosclerosis, hypertension, and heart failure [9–11]. The ecosystem of microorganisms residing within our gut is an integral part of modulation of host metabolism and immune system.

 DOI: 10.1201/9781003607779-2

The advantageous effects of probiotic microbes on the host are maintained by postbiotics, which are composed of dead microbial products or metabolites. In contrast to administration of probiotics as living microorganisms, postbiotics provide a stable approach toward prometabolic modulation. These include short-chain fatty acids (SCFAs), extracellular polysaccharides, microbial cell components, vitamins, and other products [12, 13]. Two postbiotic derivatives—Amuc_1100, derived from *Akkermansia muciniphila*, and muramyl dipeptide (MDP)—have shown promising results to modulate cardiometabolism through activation of the innate immune system. Other microbial antigens like certain lipopolysaccharides (LPSs) and trimethylamine N-oxide (TMAO) are pathognomonic regarding immunometabolic connectivity, exacerbating risks of CVDs [14]. Besides all these chemical entities, potential probiotic-related immune regulation can be achieved through many postbiotic candidates that recently gained interest in particular exopolysaccharides produced by *Lactobacillus reuteri* [15–17]. Postbiotics possess several advantages over traditional therapies. They exhibit better shelf-life stability compared to their living counterparts because they do not divide at room temperature or lower temperatures for long periods. One of the key benefits of utilizing these well-defined molecules is the ability for precise characterization. This eliminates the uncertainty that comes with using live microbes or undefined mixtures, where the composition and activity can vary significantly. Furthermore, postbiotics can be safely incorporated into food products as FDA-defined ingredients. Unlike live microbes, which may pose health risks for certain populations, these purified compounds offer a safe alternative, particularly for immunocompromised individuals. By using these defined molecules, the potential health concerns associated with hosting live microorganisms into the body can be avoided. This provides a more controlled and predictable approach to influencing the gut microbiota and promoting health [18–23]. Generally, the probiotics that contain living organisms can sometimes be damaged by normal processing methods or extreme heat conditions. This makes them more stable and versatile for use in various preparations as ingredients in food and pharmaceutical products, as they are able to withstand a wider range of heating, cooling, and freezing processes without losing their physiological functions [24]. Furthermore, postbiotics do not pose risk of live bacteria translocation, which is a significant concern with probiotics, particularly in immunocompromised individuals such as those who are critically ill. Translocation occurs when live bacteria manage to cross the gut barrier and enter into the blood circulation, ultimately causing infections. This risk is minimal with the usage of postbiotics as they are nonliving and cannot colonize or translocate. Individuals who show sensitivity or are allergic toward use of live bacteria products (probiotic formulations) may comfortably use postbiotic products. In fact, many researchers consider them an ideal alternative for high-risk patients having compromised immunity or comorbidities where exposure with live ecobionts would not be safe or indicated for therapy or prevention application [20]. Postbiotics can be biochemically defined entities as they possess clear chemical structure; for example, lipoteichoic acid is a kind of teichoic acid, and peptidoglycans present in bacterial cell wall can further be classified into L-lysine type, L-ornithine type, and so forth. Due to this aspect, composition and dosage of postbiotics can be precisely characterized and standardized, which will lead to more predictable and consistent effects [23]. Functionally, postbiotics also offer a similar range of health benefits like those from probiotics such as improved gut health, immune modulation, and enhanced nutrient absorption by affecting gut microbiota composition, stimulating growth of certain beneficial bacteria, and inhibiting harmful or pathogenic bacteria [21].

2.2 MECHANISM OF ACTION OF POSTBIOTICS

Postbiotics exert their cardio protective effects through a complex interplay of several mechanisms that target multiple aspects of cardiovascular health. One of the primary mechanisms through which they mediate this protection is by modulating gut microbiota composition [25]. Dysbiosis, an alteration in the gut microbial ecosystem, has been implicated in the development and propagation of CVDs. Postbiotics can contribute to restoring healthy dysbiotic alterations by facilitating the growth and proliferation of beneficial bacteria (e.g., those producing SCFAs like butyrate) while inhibiting that of potentially harmful bacterial species associated with inflammation and metabolic derangement [26]. Short-chain fatty acids, primarily butyrate, play an important role in strengthening intestinal barrier integrity, reducing inflammation, and lowering blood pressure via their dual activity as end products of microbial fermentation in the intestine as well as signaling molecules addressing various tissues at distal anatomical sites beyond the gastrointestinal tract such as adipose tissue, muscle, liver, and heart. By enhancing intestinal barrier function, postbiotics can prevent bacterial-associated molecular patterns (e.g., LPS) from being translocated into systemic circulation, a phenomenon shown to trigger low-grade immune activation, promoting systemic inflammation and oxidative stress commonly found with patients suffering from CVDs if left uncontrolled [27, 28]. Furthermore, postbiotics can also exert effects on the cardiovascular system through their direct interaction with host cells and metabolic processes. Some postbiotics have been shown to improve lipid metabolism by reducing cholesterol absorption in the gut and promoting cholesterol excretion. This reduction in cholesterol levels is important because LDL cholesterol (also known as "bad" cholesterol) is a major risk factor for CVDs [29]. Other postbiotics have been found to impact blood pressure regulation by interacting with the renin–angiotensin system [30]. Postbiotics have the ability to lower blood pressure, thereby alleviating the workload of the heart and vessels. In addition, some postbiotics possess antioxidant characteristics; oxidative stress induces cell damage that contributes to CVD development by creating an imbalance between reactive oxygen species production and elimination in vivo [31]. By removing harmful free radicals, these antioxidants provide protection from oxidative injury and reduce risk of CVDs. Blood clotting is another process regulated by certain types of postbiotics. Although clotting is necessary when bleeding occurs, it poses risks associated with CVDs if dysregulated, leading to arterial blockage-induced myocardial infarction or stroke [32]. Postbiotics mediate cardio protection through a multitasking mechanism network, exerting influences on different cardiovascular pathologies. They shape gut microbiota, regulate immune response, interact with the host tissue and metabolism directly or indirectly, alleviate oxidative stress and inflammation, as well as modulate blood coagulation. Hence, especially from the perspectives of multiple targets, postbiotics are a key potential therapeutic approach on obesity-related CVDs (the primary factor reflects microenvironment change in vivo) [33, 34] (Figure 2.1).

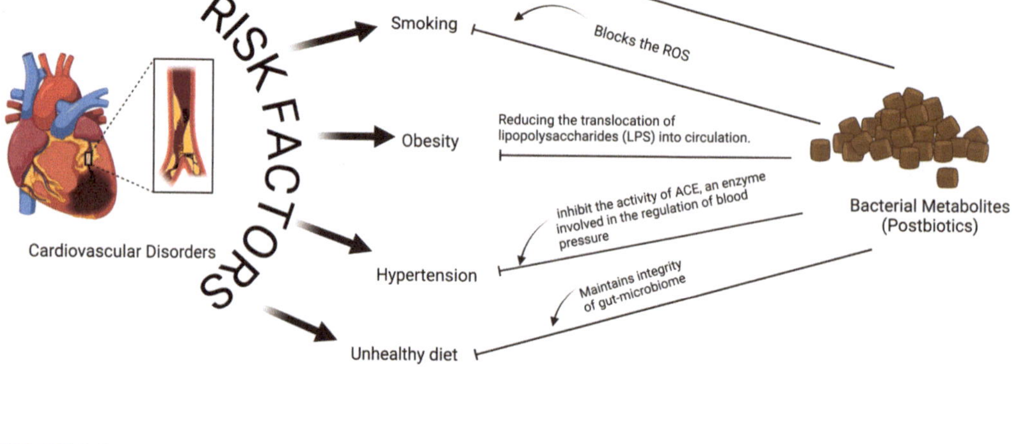

FIGURE 2.1 Mechanism of postbiotics in mitigating CVDs.

2.3 ROLE OF POSTBIOTICS IN MANAGEMENT OF VARIOUS CVDs

2.3.1 In Hypertension

Postbiotics, the nonliving bioactive compounds derived from probiotics during fermentation, have emerged as potential therapeutic agents for various health conditions, including hypertension [35]. These substances exert their effects by modulating the gut microbiome, reducing oxidative stress, and curbing inflammation [36]. By encouraging the growth of beneficial bacteria and suppressing harmful strains, postbiotics improve blood pressure regulation through the various pathways [37]. SCFAs such as acetate, propionate, and butyrate, derived from dietary fiber fermentation, promote vasodilation via G-protein-coupled receptors (e.g., GPR41, GPR43) [38]. They also reduce systemic inflammation by lowering pro-inflammatory cytokines and boosting anti-inflammatory markers [39]. Postbiotics enhance gut barrier function, preventing harmful substances like LPS from entering the bloodstream [40]. This reduces endotoxemia, systemic inflammation, and oxidative stress, all contributors to hypertension [36]. Postbiotics offer a multifaceted defense against oxidative stress, a key contributor to hypertension and blood vessel damage. They act by directly scavenging excess reactive oxygen species (ROS), preserving vasodilatory nitric oxide (NO) levels, and boosting the activity of the body's own antioxidant enzymes like superoxide dismutase (SOD) and catalase. This combined approach helps protect endothelial cells and maintain vascular health [41–44]. Postbiotics combat chronic inflammation, a significant factor in hypertension, through a multipronged approach. They modulate cytokine levels, decreasing harmful ones like TNF-α and IL-6 while increasing beneficial ones like IL-10, thereby protecting blood vessels from damage. Additionally, postbiotics suppress systemic inflammation by inhibiting NF-κB activation, a key regulator of pro-inflammatory gene expression, thus preserving vascular health. This protective effect extends to the endothelium, where reduced inflammation promotes vasodilation and improves blood flow, contributing to lower blood pressure [40, 45, 46]. Postbiotics offer a promising approach to regulating the renin–angiotensin–aldosterone system (RAAS), a key player in hypertension. Bioactive peptides found in postbiotics can inhibit angiotensin-converting enzyme (ACE), which in turn reduces the production of the potent vasoconstrictor angiotensin II, lowering blood pressure. The resulting decrease in angiotensin II levels promotes vasodilation by increasing nitric oxide bioavailability. Moreover, postbiotics help mitigate the overactivation of the RAAS by reducing associated inflammation and oxidative stress, contributing to long-term vascular health [47–49]. Postbiotics enhance endothelial function, which is vital for regulating vascular tone and blood pressure, through several mechanisms. They increase nitric oxide (NO) bioavailability by preventing its degradation by reactive oxygen species (ROS) and stimulating NO synthesis, thereby promoting vasodilation. Additionally, postbiotics reduce vascular stiffness by mitigating inflammation and oxidative stress, restoring the elasticity of blood vessels. This, in turn, strengthens endothelial barriers and improves communication between the gut and the cardiovascular system, enhancing overall vascular health [50–52]. Postbiotics contain several bioactive components that contribute to their antihypertensive effects. SCFAs promote vasodilation, reduce inflammation, and indirectly support blood pressure control by strengthening the gut barrier. Peptides derived from *Lactobacillus bacteria* act as natural ACE inhibitors, regulate vascular tone, and reduce oxidative stress, thus lowering blood pressure. Exopolysaccharides (EPSs) modulate immune responses, enhance gut health, and improve vascular function through the gut–heart axis [53, 54].

2.3.2 In Atherosclerosis

Postbiotics contribute to lipid metabolism and may diminish the likelihood of cardiovascular events. Kefir grains mostly comprise lactic acid bacteria (*Lactobacillus*, *Lactococcus*, *Leuconostoc*, and *Streptococcus*) and yeasts (*Kluyveromyces*, *Candida*, *Saccharomyces*, and *Pichia*) encapsulated in a polysaccharide termed kefiran, a postbiotic substance [55]. It may mitigate CVDs by postponing atherosclerosis, regulating blood glucose levels, and averting hypertension [19]. Ghizi et al. revealed that kefir lowers blood pressure via enhancing baroreflex and endothelial function [56, 57] and contractility, also reducing sympathetic activity [58] and inhibiting the activity of the ACE in spontaneously hypertensive rats [59]. Amorim et al. recently found 35 peptides exhibiting ACE inhibitory action in kefiran, underscoring its potential to lower blood pressure [60]. SCFAs, the postbiotics generated by the fermentation of dietary fiber by intestinal microbiota, have demonstrated potential in mitigating atherosclerosis. Butyrate, an SCFA, can reduce inflammation and oxidative stress, both of which are implicated in atherosclerosis. Regular use of the postbiotic may yield therapeutic and anti-inflammatory benefits, enhance immunological function, reduce cholesterol levels, and provide antioxidant protection against diabetes onset [55]. Certain postbiotics, such as gamma-aminobutyric acid (GABA) synthesized by specific bacteria, can enhance insulin secretion in the pancreas. This can enhance glucose absorption and use by cells. The antiatherogenic characteristics of kefiran may stem from its ability to diminish inflammation, inhibit cholesterol formation in macrophages, and lower lipid content [61]. Few experimental studies evidenced that postbiotics may lower the risk of cardiovascular events and have a major impact in lipid metabolism. For instance, the SCFA propionate has statin-like actions by preventing the condensation of cholesterol precursors [62]. Additionally, *Lactobacillus* BLs were observed to diminish LDL cholesterol and triglyceride attributes while elevating healthy HDL cholesterol levels in an obese murine model. The advantageous impact of fragmented *Lactobacillus* bacteria on the lipid profile resulted from the stimulation of the peroxisome proliferator-activated receptor (PPAR)α, which serves as the therapeutic target for the fibrate class of cholesterol-lowering medications [63].

2.3.3 Heart Failure

Postbiotics, especially SCFAs such as butyrate, acetate, and propionate, may have therapeutic potential in heart failure by targeting systemic inflammation, oxidative stress, and gut dysbiosis. SCFAs decrease the fabrication of proinflammatory cytokines and increase anti-inflammatory cytokine levels [64]; they possess antioxidative properties limiting oxidative damage to cardiac cells [64]. Gut dysbiosis leads to an enhanced inflammatory state with reduced nutrient absorption [65]. Postbiotics restore eubiosis, thereby decreasing blood pressure, improving lipid and glucose metabolism and enhancing glucose homeostasis. Furthermore, SCFAs positively influence endothelial function through increased nitric oxide production as well as decreased arterial stiffness resulting in improved vascular health [66]. Postbiotics modulate RAAS activity, pointing toward a RAAS-inhibiting effect, which could lead to diminished fluid retention and reduced cardiac work load [67]. The interaction between postbiotics and the enteric nervous system may also have an impact on central neurohormonal systems leading to reduction of sympathetic tone. Through tightening, the gut-barrier defense against bacterial translocation is enhanced, further reducing systemic inflammation. In addition to these effects, postbiotics might enhance metabolic health via direct regulation in lipidome and metabolome or indirectly by increasing fiber metabolism. Also, GI symptom relief might contribute since particularly fatigue seems correlated with α-diversity. However, clinical studies exploring the cardiovascular benefit are sparse. Thus, at present, clear recommendations relating to postbiotic supplementation cannot be given [68].

2.3.4 Coronary Artery Disease

Postbiotics can fight against coronary artery disease (CAD) by acting on several fronts. They can improve lipid profiles by reducing total and LDL cholesterol, key culprits in plaque buildup. Furthermore, postbiotics combat chronic inflammation, a major driver of CAD progression, by modulating the immune system and reducing inflammation in the arteries [69]. A healthy gut is essential for cardiovascular health, and postbiotics promote beneficial gut bacteria and improve gut barrier function, which in turn reduces systemic inflammation. Some postbiotics may even help regulate blood pressure by influencing the renin–angiotensin system. Finally, they can improve endothelial function, the lining of blood vessels, promoting vasodilation, and reducing the risk of blood clots. While research on postbiotics in CAD management is still in its early stages, several studies have shown promising results. For example, research suggests that SCFAs may improve lipid profiles and reduce inflammation in individuals with CAD [70].

2.3.5 Cardiomyopathy

Postbiotics show promise in supporting cardiomyopathy management in several ways. They can modulate inflammation in the heart muscle, particularly in inflammatory types like myocarditis, by regulating the immune response and reducing harmful molecules. Additionally, postbiotics promote a healthy gut microbiome, which can lessen systemic inflammation that may worsen cardiomyopathy. Some studies suggest postbiotics may even influence energy metabolism in heart muscle cells and protect them from oxidative damage. However, more research is needed to confirm these effects and identify the most beneficial postbiotics. As research advances, postbiotics may become a valuable tool in managing this complex condition [71].

2.4 CLINICAL TRIALS

Regulating cardiovascular health using postbiotics is a growing area of research interest. Research is still at an early stage; however, some clinical trials and studies have indicated promising therapeutic applications. A clinical trial study examined the effect of pasteurized *Akkermansia muciniphila*, a postbiotic on insulin resistant overweight or obese individuals, indicating an improved insulin sensitivity and a reduction in total cholesterol with implications for management of cardiometabolic

TABLE 2.1
Postbiotics and Their Application in Different CVDs

Cardiovascular Disease	Postbiotic Effect	Possible Mechanisms	Example Postbiotics	Reference
Atherosclerosis	- Reduced plaque formation - Improved lipid profile - Decreased inflammation	- Inhibition of cholesterol synthesis - Increased cholesterol excretion - Modulation of immune response	- SCFAs - Cell wall components of *Lactobacillus* and *Bifidobacterium* strains	[19, 55]
Hypertension	- Lowered blood pressure	- Improved vascular function - Reduced renin–angiotensin system activity	- SCFAs (e.g., butyrate) - Peptides from fermented milk	[36, 39]
Heart Failure	- Improved lipid profile - Reduced inflammation	- Modulation of gut microbiota - Decreased gut permeability - Enhanced antioxidant defences	- SCFAs Lipoteichoic acid from *Lactobacillus plantarum*	[66, 67]
Stroke	- Reduced neurological damage - Improved recovery	- Anti-inflammatory effects - Modulation of gut-brain axis	- SCFAs - Exopolysaccharides from *Lactobacillus* strains	[68]
Arrhythmias	- Stabilized heart rhythm	- Improved autonomic nervous system balance - Reduced inflammation	- SCFAs - Cell-free supernatants of *Lactobacillus rhamnosus*	[72]

risk factors [73]. Postbiotics have been considered for their endothelial protective and anti-inflammatory effects among renal patients due to its high prevalence as a risk factor associated with CVDs. Although clinical trials specifically regarding postbiotics for CKD-related cardiovascular complication smay be scarce, literature has indicated their potential through mitigating inflammation as well as oxidative stress, which plays a pivotal role in cardiovascular damage among CKD [74]. More specific clinical trials are required to validate these preliminary findings and to determine optimum dosages, long-term effects, as well as possible drug interactions of postbiotics.

2.5 CHALLENGES AND FUTURE ASPECTS

Postbiotics show promise in the management of CVDs by modulating gut microbiota and improving metabolic health [12]. However, certain challenges need to be addressed before realizing their promising applications.

2.5.1 Standardization and Characterization

Lack of clear definition and classification: The definition of postbiotics is still evolving, which raises confusion in research and product development [20].

Variability in composition and efficacy: The composition of postbiotics depends on the microbial strain, substrate, and production process, with a consequent impact on their therapeutic properties. This fact makes it difficult to compare results from different studies [22].

Standardized protocols for production and quality control: Standardization of production processes and quality control is essential in ensuring the safety and therapeutic effects of postbiotic-based products [75].

2.5.2 Mechanism of Action

While few mechanisms are recognized, such as anti-inflammatory effects or gut barrier integrity, a comprehensive understanding of how postbiotics interact with the host and affect CVDs is lacking. Further research needs to find and characterize the precise bioactive components responsible for the observed effects [76].

2.5.3 Clinical Evidence

The majority of studies on postbiotics and CVDs are either preclinical or have small sample size. There is a need for large-scale, well-designed human clinical trials to establish their efficacy and safety in human subjects. More research is required to assess the long-term effects as well as safety concerns of postbiotic interventions, particularly in vulnerable populations [77].

2.5.4 Regulatory and Market Challenges

The absence of clear regulatory guidelines for postbiotic products in many regions poses a significant challenge. This lack of clarity creates uncertainty for both producers and consumers. To foster market acceptance and widespread use, it is crucial to increase public understanding and awareness of postbiotics [78].

2.5.5 Delivery and Formulation

Ensuring the stability and bioavailability of postbiotics within the gut requires the development of effective delivery mechanisms. Furthermore, transforming postbiotics into consumer-friendly products that are both palatable and convenient, while preserving their efficacy, presents a significant formulation challenge [79].

2.6 CONCLUSION

Postbiotics are capable of developing new approaches to cardiovascular health by positively influencing metabolism, inflammation, and gut health. They have potential in the management of cholesterol and blood pressure, as well as strengthening the gut barrier over and above what the traditional treatment may achieve. Though promising, preclinical and clinical studies are required to understand their mechanisms, optimal dosages, and long-term safety. While current research is compelling, further investigation is needed to understand their mechanisms, optimal dosages, and long-term effects. Standardized guidelines are crucial to ensure their safe and effective use. As our knowledge of the gut microbiome and its impact on cardiovascular health grows, postbiotics are likely to play an increasingly important role in preventing and managing CVDs, offering new therapeutic possibilities through targeted microbiome modulation.

REFERENCES

1. Thiriet M. Cardiovascular disease: an introduction. *Vasculopathies*. 2019;8:1–90.
2. Afjeh-Dana E, Naserzadeh P, Moradi E, Hosseini N, Seifalian AM, Ashtari B. Stem cell differentiation into cardiomyocytes: current methods and emerging approaches. *Stem Cell Rev Rep*. 2022;18:2566–92.
3. Flora GD, Nayak MK. A brief review of cardiovascular diseases, associated risk factors and current treatment regimes. *Curr Pharm Des*. 2019;25:4063–84.
4. Gadó K, Szabo A, Markovics D, Virág A. Most common cardiovascular diseases of the elderly—a review article. *Dev Health Sci*. 2022;4:27–32.
5. Netala VR, Teertam SK, Li H, Zhang Z. A comprehensive review of cardiovascular disease management: cardiac biomarkers, imaging modalities, pharmacotherapy, surgical interventions, and herbal remedies. *Cells*. 2024;13(17):1471.
6. Schultz WM, Kelli HM, Lisko JC, Varghese T, Shen J, Sandesara P, et al. Socioeconomic status and cardiovascular outcomes: challenges and interventions. *Circulation*. 2018;137:2166–78.
7. Ghodeshwar GK, Dube A, Khobragade D. Impact of lifestyle modifications on cardiovascular health: a narrative review. *Cureus*. 2023;15:e42616.
8. Hou K, Wu ZX, Chen XY, et al. Microbiota in health and diseases. *Sig Transduct Target Ther*. 2022 Apr 23;7(1):135.
9. Tang WHW, Kitai T, Hazen SL. Gut microbiota in cardiovascular health and disease. *Circ Res*. 2017 Mar 3;120(7):1183–96.
10. Ahmadmehrabi S, Tang WHW. Gut microbiome and its role in cardiovascular diseases. *Curr Opin Cardiol*. 2017 Nov;32(6):761–6.
11. Trøseid M, Ueland T, Hov JR. The gut microbiome in coronary artery disease and heart failure. *Curr Opin Clin Nutr Metab Care*. 2020 Mar;23(2):103–14.
12. Heniedy AM, Mahdy DM, Abo Elenien WI, Mourad S, El-Kadi RA. Postbiotics as a health-promoting technique: A review article on scientific and commercial interest. *Process Biochem*. 2024;144:6–19.
13. Rafique N, Jan SY, Dar AH, Dash KK, Sarkar A, Shams R, et al. Promising bioactivities of postbiotics: a comprehensive review. *J Agric Food Res*. 2023;14:100708.
14. Niu H, Zhou M, Zogona D, Xing Z, Wu T, Chen R, et al. Akkermansia muciniphila: a potential candidate for ameliorating metabolic diseases. *Front Immunol*. 2024;15:1370658.
15. Fanning S, Hall LJ, Cronin M, Zomer A, MacSharry J, Goulding D, et al. Bifidobacterial surface-exopolysaccharide facilitates commensal-host interaction through immune modulation and pathogen protection. *Proc Natl Acad Sci U S A*. 2012 Feb 7;109(6):2108–13.
16. Thoda C, Touraki M. Immunomodulatory properties of probiotics and their derived bioactive compounds. *Appl Sci*. 2023;13(8):4726.
17. Lebeer S, Claes I, Tytgat HL, Verhoeven TL, Marien E, von Ossowski I, et al. Functional analysis of *Lactobacillus rhamnosus* GG exopolysaccharide in innate immunity. *PLoS One*. 2012;6(3):e17251.

18. Ikram A, Safdar SZ, Arshad MT, Rasheed A, Gnedeka KT. An overview of postbiotics: unveiling their distinct role in gut health. *Food Agric Immunol.* 2024;35(1):1–18.
19. Zółkiewicz J, Marzec A, Ruszczyński M, Feleszko W. Postbiotics—A step beyond pre- and probiotics. *Nutrients.* 2020 Aug 19;12(8):2189.
20. Salminen S, Collado MC, Endo A, Hill C, Lebeer S, Quigley EM, et al. The International Scientific Association for Probiotics and Prebiotics (ISAPP) consensus statement on the definition and scope of postbiotics. *Nat Rev Gastroenterol Hepatol.* 2021 Sep;18(9):649–67.
21. Islam F, Azmat F, Imran A, Zippi M, Hong W, Tariq F, et al. Role of postbiotics in food and health: a comprehensive review. *CyTA - J Food.* 2024;22(1):1–20.
22. Aguilar-Toalá JE, Garcia-Varela R, Garcia HS, Mata-Haro V, González-Córdova AF, Vallejo-Cordoba B, et al. Postbiotics: An evolving term within the functional foods field. *Trends Food Sci Technol.* 2018;75:105–14.
23. Wei L, Wang B, Bai J, Zhang Y, Liu C, Suo H, et al. Postbiotics are a candidate for new functional foods. *Food Chem X.* 2024;23:101650.
24. Tripathi MK, Giri SK. Probiotic functional foods: survival of probiotics during processing and storage. *J Funct Foods.* 2014;9:225–41.
25. Chen X, Liu X, Wang B, Zhao J, Wang L. Gut microbiota and microbiota-derived metabolites in cardiovascular diseases. *Chin Med J (Engl).* 2023 Oct 5;136(19):2269–84.
26. Ghosh TS, Valdes AM. Evidence for clinical interventions targeting the gut microbiome in cardiometabolic disease. *BMJ.* 2023 Oct 9;383:e075180.
27. Peng L, He Z, Chen W, Holst JJ, Jeppesen PB, Zhao L, et al. Butyrate enhances the intestinal barrier by facilitating tight junction assembly via activation of AMP-activated protein kinase in Caco-2 cell monolayers. *J Nutr.* 2009 Sep;139(9):1619–25.
28. Clemente JC, Ursell LK, Parfrey LW, Knight R. The impact of the gut microbiota on human health: an integrative view. *Cell.* 2012 Jun 15;148(6):1258–70.
29. Liu Y, Zhu D, Liu J, Sun X, Gao F, Duan H, et al. *Pediococcus pentosaceus* PR-1 modulates high-fat-diet-induced alterations in gut microbiota, inflammation, and lipid metabolism in zebrafish. *Front Nutr.* 2023 Feb 1;10:1087703.
30. Jaworska K, Koper M, Ufnal M. Gut microbiota and renin-angiotensin system: a complex interplay at local and systemic levels. *Am J Physiol Gastrointest Liver Physiol.* 2021 Oct 1;321(4):G355–G366.
31. Lobo V, Patil A, Phatak A, Chandra N. Free radicals, antioxidants and functional foods: impact on human health. *Pharmacogn Rev.* 2010 Jul;4(8):118–26.
32. Liang B, Xing D. The current and future perspectives of postbiotics. *Probiotics & Antimicro. Prot.* 2023;15:1626–43.
33. Lou X, Xue J, Shao R, Mo C, Wang F, Chen G. Postbiotics as potential new therapeutic agents for sepsis. *Burns & Trauma.* 2023;11.
34. Hijová E. Postbiotics as metabolites and their biotherapeutic potential. *Int J Mol Sci.* 2024 May 16;25(10):5441.
35. Bourebaba Y, Marycz K, Mularczyk M, Bourebaba L. Postbiotics as potential new therapeutic agents for metabolic disorders management. *Biomed Pharmacol.* 2022 Sep;1(153):113138.
36. Peluzio MD, Martinez JA, Milagro FI. Postbiotics: metabolites and mechanisms involved in microbiota–host interactions. *Trends Food Sci Technol.* 2021 Feb 1;108:11–26.
37. Lymperopoulos A, Suster MS, Borges JI. Short-chain fatty acid receptors and cardiovascular function. *Int J Mol Sci.* 2022 Mar 18;23(6):3303.
38. Eslick S, Thompson C, Berthon B, Wood L. Short-chain fatty acids as anti-inflammatory agents in overweight and obesity: a systematic review and meta-analysis. *Nutr Rev.* 2022 Apr 1;80(4):838–56.
39. Flori L, Benedetti G, Martelli A, Calderone V. Microbiota alterations associated with vascular diseases: postbiotics as a next-generation magic bullet for gut-vascular axis. *Pharmacol Res.* 2024 Aug;3:107334.
40. Puddu P, Puddu GM, Cravero E, Rosati M, Muscari A. The molecular sources of reactive oxygen species in hypertension. *Blood Press.* 2008 Jan 1;17(2):70–7.
41. Guerrero-Encinas I, González-González JN, Santiago-López L, Muhlia-Almazán A, Garcia HS, Mazorra-Manzano MA, Vallejo-Cordoba B, González-Córdova AF, Hernández-Mendoza A. Protective effect of *Lacticaseibacillus casei* CRL 431 postbiotics on mitochondrial function and oxidative status in rats with aflatoxin B1–induced oxidative stress. *Probiotics Antimicrob Proteins.* 2021 Aug;1:1–1.
42. Aggarwal S, Sabharwal V, Kaushik P, Joshi A, Aayushi A, Suri M. Postbiotics: from emerging concept to application. *Front Sustain Food Syst.* 2022 Sep;21(6):887642.

43. Rezaie N, Bagheri-Amiri F, Aghamohammad S, Khatami S, Talebi M, Sohrabi A, Pourshafie MR, Rohani M. The therapeutic effect of antioxidant postbiotic cocktail in colitis mice: A promising approach to alleviate oxidative stress in two high-fat and normal-diet feeding mice. *J Agric Food Res.* 2024 Nov;26:101547.
44. Jastrząb R, Graczyk D, Siedlecki P. Molecular and cellular mechanisms influenced by postbiotics. *Int J Mol Sci.* 2021 Dec 15;22(24):13475.
45. Mousavi Ghahfarrokhi SS, Mohamadzadeh M, Samadi N, Fazeli MR, Khaki S, Khameneh B, Khameneh Bagheri R. Management of cardiovascular diseases by short-chain fatty acid postbiotics. *Curr Nutr Rep.* 2024 Apr;24:1–20.
46. Tsafack PB, Li C, Tsopmo A. Food peptides, gut microbiota modulation, and antihypertensive effects. *Molecules.* 2022 Dec 12;27(24):8806.
47. Giménez VM, Mikusic NL, Lee HJ, Menéndez SG, Choi MR, Manucha W. Physiopathological mechanisms involved in the development of hypertension associated with gut dysbiosis and the effect of nutritional/pharmacological interventions. *Biochem Pharmacol.* 2022 Oct;1;204:115213.
48. Danilenko V, Devyatkin A, Marsova M, Shibilova M, Ilyasov R, Shmyrev V. Common inflammatory mechanisms in COVID-19 and Parkinson's diseases: the role of microbiome, pharmabiotics and postbiotics in their prevention. *J Inflamm Res.* 2021 Nov;30:6349–81.
49. Ali MA, Kamal MM, Rahman MH, Siddiqui MN, Haque MA, Saha KK, Rahman MA. Functional dairy products as a source of bioactive peptides and probiotics: current trends and future perspectives. *J Food Sci Technol.* 2022 Apr;59(4):1263–79.
50. Wang P, Wang S, Wang D, Li Y, Yip RC, Chen H. Postbiotics-peptidoglycan, lipoteichoic acid, exopolysaccharides, surface layer protein and pili proteins—structure, activity in wounds and their delivery systems. *Int J Biol Macromol.* 2024 Jun;15:133195.
51. Romero M, Duarte J. Probiotics and prebiotics in cardiovascular diseases. *Nutrients.* 2023 Aug 23;15(17):3686.
52. Zhang L, Ko CY, Zeng YM. Immunoregulatory effect of short-chain fatty acids from gut microbiota on obstructive sleep apnea-associated hypertension. *Nat Sci Sleep.* 2022 Mar;10:393–405.
53. Sharma H, Kaushik M, Venishaa S, Pathak R, Farid A, Bhowmick M. Correlation and successive role of synbiotics to manage blood pressure. In: *Synbiotics in Metabolic Disorders* (pp. 103–120). CRC Press; 2022. p. 103–20.
54. Salari A, Ghodrat S, Gheflati A, Jarahi L, Hashemi M, Afshari A. Effect of kefir beverage consumption on glycemic control: a systematic review and meta-analysis of randomized controlled clinical trials. *Complement Ther Clin Pract.* 2021;44:100454.
55. Klippel BF, Duemke LB, Leal MA, Friques AGF, Dantas EM, Dalvi RF, et al. Effects of kefir on the cardiac autonomic tones and baroreflex sensitivity in spontaneously hypertensive rats. *Front Physiol.* 2016;7(Jun):211.
56. Friques AGF, Arpini CM, Kalil IC, Gava AL, Leal MA, Porto ML, et al. Chronic administration of the probiotic kefir improves the endothelial function in spontaneously hypertensive rats. *J Transl Med.* 2015;13(1):1–6.
57. Silva-Cutini MA, Almeida SA, Nascimento AM, Abreu GR, Bissoli NS, Lenz D, et al. Long-term treatment with kefir probiotics ameliorates cardiac function in spontaneously hypertensive rats. *J Nutr Biochem.* 2019;66:79–85.
58. Ghizi Ac Da S, De Almeida Silva M, Moraes Fs De A, Da Silva CL, Endringer DC, Scherer R, et al. Kefir improves blood parameters and reduces cardiovascular risks in patients with metabolic syndrome. *PharmaNutrition.* 2021;16:100266.
59. Amorim FG, Coitinho LB, Dias AT, Friques AGF, Monteiro BL, Rezende LCD De, et al. Identification of new bioactive peptides from kefir milk through proteopeptidomics: bioprospection of antihypertensive molecules. *Food Chem.* 2019;282:126149.
60. Uchida M, Ishii I, Inoue C, Akisato Y, Watanabe K, Hosoyama S, et al. Kefiran reduces atherosclerosis in rabbits fed a high cholesterol diet. *J Atheroscler Thromb.* 2010;17(9):899–907.
61. Bush RS, Milligan LP. Study of the mechanism of inhibition of ketogenesis by propionate in bovine liver. *Can J Anim Sci.* 1971;51(1):67–74.
62. Aimo A, Castiglione V, Borrelli C, Saccaro LF, Franzini M, Masi S, et al. Oxidative stress and inflammation in the evolution of heart failure: from pathophysiology to therapeutic strategies. *Eur J Prev Cardiol.* 2020 Mar;27(5):494–510.
63. Mallet RT, Sun J. Antioxidant properties of myocardial fuels. *Mol Cell Biochem.* 2003 Nov;253:103–11.

64. Tang WW, Li DY, Hazen SL. Dietary metabolism, the gut microbiome, and heart failure. *Nat Rev Cardiol.* 2019 Mar;16(3):137–54.
65. Natarajan N, Hori D, Flavahan S, Steppan J, Flavahan NA, Berkowitz DE, Pluznick JL. Microbial short-chain fatty acid metabolites lower blood pressure via endothelial G protein-coupled receptor 41. *Physiol Genomics.* 2016 Nov 1;48(11):826–34.
66. Vallianou NG, Kounatidis D, Panagopoulos F, Evangelopoulos A, Stamatopoulos V, Papagiorgos A, Geladari E, Dalamaga M. Gut microbiota and its role in the brain-gut-kidney axis in hypertension. *Curr Hypertens Rep.* 2023 Nov;25(11):367–76.
67. Vera-Santander VE, Hernández-Figueroa RH, Jiménez-Munguía MT, Mani-López E, López-Malo A. Health benefits of consuming foods with bacterial probiotics, postbiotics, and their metabolites: a review. *Molecules.* 2023 Jan 27;28(3):1230.
68. Wu H, Chiou J. Potential benefits of probiotics and prebiotics for coronary heart disease and stroke. *Nutrients.* 2021 Aug 21;13(8):2878.
69. Trøseid M, Andersen GØ, Broch K, Hov JR. The gut microbiome in coronary artery disease and heart failure: current knowledge and future directions. *Nat Rev Cardiol.* 2020 Mar;17(3):183–200.
70. Zhao X, Liu S, Li S, Jiang W, Wang J, Xiao J, et al. Unlocking the power of postbiotics: A revolutionary approach to nutrition for humans and animals. *Anim Nutr.* 2023 Jun;9(2):281–93.
71. Depommier C, Everard A, Druart C, Plovier H, Van Hul M, Vieira-Silva S, et al. Pasteurized *Akkermansia muciniphila* increases whole-body energy expenditure and fecal energy excretion in overweight/obese volunteers. *Gut.* 2021 Nov;70(11):2060–2069.
72. Fan H, Liu X, Ren Z, Fei X, Luo J, Yang X, Xue Y, Zhang F, Liang B. Gut microbiota and cardiac arrhythmia. *Front Cell Infect Microbiol.* 2023 Apr 25;13:1147687.
73. Rysz J, Franczyk B, Cieślik B, Gluba-Brzózka A. Postbiotics in cardiovascular disease: A narrative review of the literature. *Nutrients.* 2022 Jul 12;14(14):2823.
74. Wegh CAM, Geerlings SY, Knol J, Roeselers G, Belzer C. Postbiotics and their potential applications in early life nutrition and beyond. *Int J Mol Sci.* 2019 Oct 3;20(19):4673.
75. Cuevas-González PF, Liceaga AM, Aguilar-Toalá JE. Postbiotics and their potential for cardiovascular health: a narrative review. *Nutrients.* 2020 Sep 2;12(9):2808.
76. Christiansen T, Christensen PD, Licht TR. The effect of *Bifidobacterium animalis* subsp. lactis 420 on faecal microbiota composition and metabolic profiles in healthy adults: a randomised, double-blind, placebo-controlled, crossover study. *Br J Nutr.* 2019 Nov;122(6):641–50.
77. Lohner S, Kühnle A, Dierks ML. Postbiotics—a promising avenue for human health? *Trends Food Sci Technol.* 2022 Jun;120:116–28.
78. Siciliano RA, Aguilar-Toalá JE. Postbiotics in food: opportunities, challenges, and perspectives. *Curr Opin Food Sci.* 2023 Apr;49:100949.
79. de Almada CN, Martinez RCR. Beneficial effects of probiotics: a review. *J Clin Gastroenterol.* 2016;50(5):391–99.

3 Postbiotics in Cancer Treatment

Ciniraj Raveendran
Government Medical College Thiruvananthapuram, Thiruvananthapuram, Kerala, India

Riya Dave
Gujarat University, Ahmedabad, Gujarat, India

Vijay Singh
Ajay Sanghal Institute of Medical Science and Research and Ayushman Hospital, Village Rajjak Nagar, Shamli Karnal Road, Shamli, Uttar Pradesh, India
Praduman Singh SPS Pharmacy College, Basti, Uttar Pradesh, India

Amit Kumar Agrahari, Ashutosh Chand Kaushal, Nilesh Kumar and Sumel Ashique
Lovely Professional University, Phagwara, Punjab, India

3.1 INTRODUCTION

The International Scientific Association of Probiotics and Prebiotics (ISAPP) defines postbiotics as the "preparation of inanimate microorganisms and/or their components that confers a health benefit on the host" [1]. Unlike probiotics, which are live bacteria, postbiotics are nonviable bacterial products produced by the bacteria inside the host. These bioactive bacterial products comprise cell wall components, enzymes, peptides, short-chain fatty acids (SCFAs), and other microbial molecules [2]. The applicability of postbiotics is beyond what is conferred by probiotics because of their various mechanism of action [3]. The anti-inflammatory properties could help reduce the influence of the body's inflammatory responses in various chronic inflammatory conditions and metabolic disorders [4–6]. The immunomodulatory effects are mediated through innate immune responses and are well utilized to reduce proinflammatory responses [7]. Even though the antimicrobial effect is not approved for clinical use, postbiotics could be used as a supplement to antibacterials used in infective conditions [8]. The antiproliferative effect of bioactive postbiotics could potentially impact the treatment of solid malignancies such as gastric, breast, colorectal, and pancreatic cancer [9–12].

3.2 TYPES OF POSTBIOTICS

3.2.1 Short-Chain Fatty Acids (SCFAs)

The microbiota converts the ingested starch and dietary fibers into SCFAs in the large intestine [13]. These SCFAs include acetate, butyrate, and propionate. Through the gut–brain axis, they affect numerous functions in the central nervous system, which modulates the function of the brain [14].

DOI: 10.1201/9781003607779-3

These SCFAs, due to their modification of immune function, have a crucial role in the prevention and progression of several cancers [15].

3.2.2 Enzymes

Several enzymes produced by the microbiota, including lactase, amylase, hydrogen peroxide, proteases, and lipase, have immunomodulatory effects and help prevent and control the dissemination of a variety of cancers, including colorectal, pancreatic, and liver cancers [16].

3.2.3 Bacteriocins

Lactic acid bacilli present in the gut produce bacteriocins with antimicrobial effects. These postbiotic bacteriocins include nicin, enterocin, plantaracin, bovicin, colicin, and microcin [17].

3.2.4 Exopolysaccharides

Produced by lactic acid bacilli, exopolysaccharide is a natural polysaccharide found in the intestine that has several health benefits, including the antiproliferative effect, which helps prevent and spread several malignancies [18].

3.2.5 Cell Wall Components

The cell wall components include peptidoglycan, teichoic acid, and extracellular vesicles [19].

3.2.6 Microbial Cell-Free Supernatants

In vitro studies have shown that cell-free supernatants isolated from gut microbes exhibited anticancer effects mainly via apoptosis [20].

3.2.7 Neurotransmitters

Gamma-aminobutyric acid (GABA)–producing lactobacilli can inhibit metastasis and inhibit proliferation of drug-resistant (5FU) colorectal cancer [21].

3.3 MECHANISM OF ACTION OF POSTBIOTICS IN CANCER

3.3.1 Inhibition of Cell Cycle Progression in Cancer Cells

Cyclin-dependent kinase (CDK) inhibitors are currently in use for the treatment of breast cancer [22]. Postbiotics effectively arrest the cell cycle by upregulating CDK inhibitors, preventing cancer cells from progressing through the various stages of cell division [23]. Postbiotics, especially nisin, bacterial cytoplasmic extract, and cell wall, further inhibit the cell cycle by downregulating CDKs and cyclins [24]. Postbiotics can induce cell cycle arrest at the early stages of the cell cycle G0/G1 or G2/M phase [19]. Postbiotics also promote the activity of tumor suppressor gene P53, thereby preventing aberrant cell growth. The administration of postbiotics consisting of *B. bifidum* could decrease the incidence and spread of breast cancer, increase the expression of P53, and decrease Ki67 activity [25].

3.4 INDUCTION OF AUTOPHAGY IN CANCER CELLS

Postbiotics can induce autophagy in cancer cells through several mechanisms, most notably via the activation of 5' AMP-activated protein kinase (AMPK) that regulates cellular energy homeostasis by activating glucose and fatty acid uptake and oxidation and inhibiting the mammalian target of the rapamycin (mTOR) signaling pathway [26, 27]. Also, postbiotics can upregulate the expression of autophagy-related genes (ATGs) required to form and mature autophagosomes such as Beclin-1, LC3, and ATG5 [23]. Induction of endoplasmic reticulum (ER) stress in cancer cells by postbiotics triggers the unfolded protein response (UPR), which is linked to autophagy induction. Other ways autophagy is enhanced include producing reactive oxygen species, enhancing the immune defense system, and altering intracellular calcium, which regulates the signaling pathways. Together, these mechanisms contribute to the induction of autophagy in cancer cells, leading to the death of cancer cells, enhanced sensitivity to chemotherapy drugs, and slowing tumor growth and metastasis.

3.5 MODULATION OF IMMUNE RESPONSE

The abundance of NK cells in tumor cells indicates an active immune response and a good prognosis [28]. Postbiotics can enhance NK cell activation through the gut microbiome axis [29]. Short-chain fatty acids (SFAs) can boost the NK cell's antitumor activity and increase cellular killing by targeting cancer cells [30]. Postbiotics upregulate human leucocyte antigen (HLA) and enhance the recognition of cancer cells by T cells, making them susceptible to immune-mediated death [31]. During carcinogenesis, an increased cellular inflammation is mediated by a set of proteins called cytokines, creating a hostile environment for tumor cell proliferation and metastasis [32]. Postbiotics could potentiate the action of proinflammatory cytokines like interleukin-10 (IL-10) and transforming growth factor-beta (TGF-β) while simultaneously suppressing proinflammatory cytokines like tumor necrosis factor-alpha (TNF-α) and interleukin-6 (IL-6) [33–36]. Another class of molecules by which postbiotics act is pattern recognition receptors (PRRs) [37]. PRRs that can recognize postbiotic molecules include Toll-like receptors (TLRs) and NOD-like receptors (NLRs) [38]. Postbiotics can also influence epigenetic regulation in immune cells. Epigenetic regulation controls cellular processes like cell proliferation, apoptosis, invasion, and senescence. Studies have shown that postbiotics can modulate DNA methylation and histone modifications in immune cells to enhance their ability to recognize and eliminate cancer cells [39].

3.6 INDUCTION OF APOPTOSIS IN CANCER CELLS

Postbiotics could exert anticancer effects by facilitating apoptosis in cells [40]. Postbiotic metabolites could downregulate the antiapoptotic genes, shifting the balance from antiapoptotic to proapoptotic [40]. The proapoptotic proteins, Bax and Bad, are upregulated by the postbiotic metabolites. When stimuli for uncontrolled cell proliferation are recognized, the proapoptotic pathway takes the lead. Postbiotics also stimulate the mitochondria and activate the release of cytochrome c [41]. Once released into the cytoplasm, cytochrome c interacts with other proteins and forms apoptosome, which activates the caspase cascade, leading to the execution of the final step in the apoptotic pathway [42]. Postbiotics also influence the extrinsic apoptosis pathway by binding to the death domain on the cell surface to activate the intracellular effector domain, triggering a signal for apoptosis and death cascade [43]. Some postbiotics generate reactive oxygen species (ROS) in the cell, increasing the oxidative stress and triggering apoptosis [44]. ROS concentrations are high inside the cancer cells, and such cells are more prone to mediated apoptosis, a potentially effective strategy for suppressing cancer cells [45]. Postbiotics also act by activating the P53 gene, the guardian of the genome. In animal experiments of triple-negative breast cancer treated with postbiotics, a higher concentration of P53 with a decrease in Ki-67 activity is observed due to its inhibitory effects on cell proliferation [25]. Postbiotics also facilitate the NF-κb and Nrf2 pathways in cellular immunity, inflammation, and carcinogenesis [46]. Postbiotics also act on the PI3K pathway, which regulates cellular growth and survival [47].

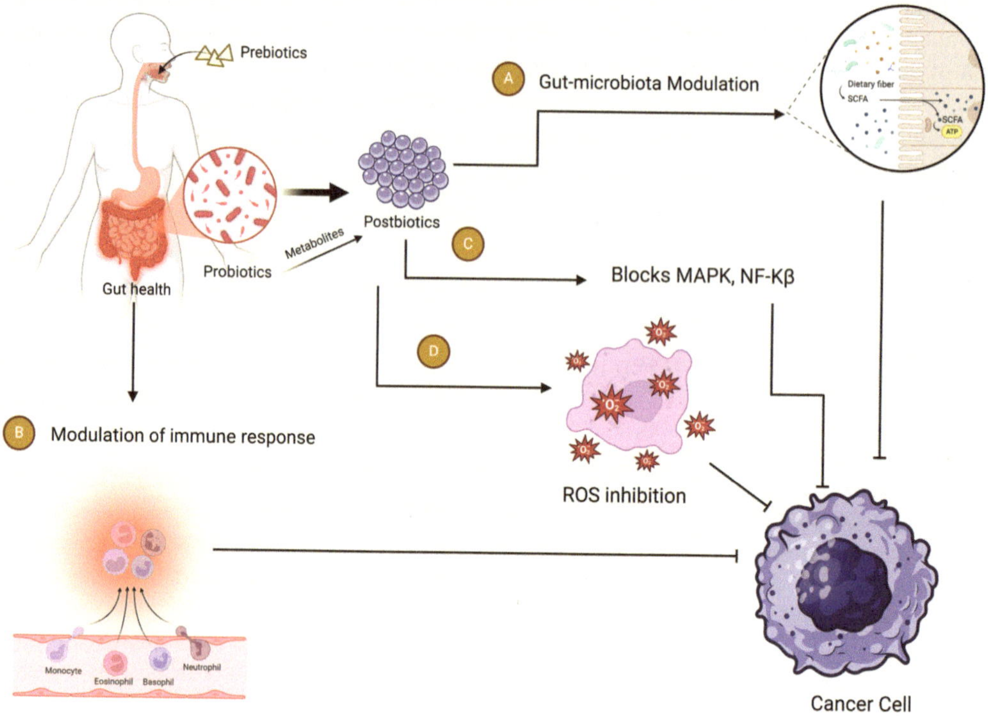

FIGURE 3.1 Role of postbiotics in cancer mitigation.

3.6.1 Modulation of Gut Microbiota

The gut microbiota play a vital role in maintaining the immune system and preventing cancer. In disease states, including cancer, there is an increase in harmful proinflammatory bacteria such as *E. coli* and *Clostridium difficile* and a decrease in beneficial bacteria like lactobacilli and bifidobacterium [48]. Specific subspecies of *Streptococcus* bacteria are implicated in the development of colorectal cancer [49]. Maintaining the gut microbiome axis is imperative to maintain its composition and bioavailability and to receive optimum benefits [50]. Postbiotic compounds could maintain intestinal mucosa integrity by maintaining a tight gut barrier junction with the help of tight junction proteins such as occludin and claudin, thereby decreasing intestinal permeability [51, 52]. After fermentation of dietary food by gut microbiota, the SFAs exert anti-inflammatory and immunosuppressive action on colonocytes to maintain cellular function and prevent carcinogenesis [53, 54] (Figure 3.1).

3.7 APPLICATIONS OF POSTBIOTICS IN DIFFERENT TYPES OF CANCER

3.7.1 Lung Cancer

Better outcomes were linked with probiotics given to lung cancer patients treated with immune checkpoint inhibitors (ICIs) regarding progression-free survival and better overall survival [55, 56]. Postbiotics contribute systemically to better health in cancer patients [57]. Postbiotics may modulate communications between tumor cells, their immune counterparts, and the various microbiota components, thus constituting a supporting therapy strategy in lung cancer management [58–60]. Postbiotics like JK5G, CBS, and compound probiotic preparations have shown some efficacy in enhancing chemotherapy treatment in lung cancer patients. Expected benefits from the influence of postbiotics may be directed toward immune response, anti-inflammation, antimicrobials, and other possible routes of enhancement of the chemotherapeutic efficacy [3, 61, 62].

3.7.2 Breast Cancer

Some postbiotics hold promise for therapeutic use in managing breast cancer to improve the outcomes of therapy [63]. They are possible natural drugs that could be used to support and augment synthetic chemotherapeutic drugs for better cancer treatment outcomes [64]. They could regulate estrogen metabolism through action on its circulation-related enzymes, potentially lowering estrogen levels that could be available to cancer cells. The antitumor activity is relevant in the setting of hormone receptor-positive breast cancers [65, 66]. Treatments with these microorganisms show decreased tumor incidence, volume, proliferation, and survival [67]. Postbiotics modulate crucial molecular pathways and increase BAX/BCL2 ratio [68].

3.7.3 Cervical Cancer

Postbiotics are potential agents for cancer treatment by a targeted approach to induce apoptosis in the cells [69]. Cervical cancer is closely associated with infections by high-risk human papillomavirus (HPV), especially by HPV type 16. Some postbiotics, like certain strains of bifidobacterium, have shown antiviral activity by suppressing HPV oncogenes responsible for cancer development [57, 70]. The metabolites generated by the *Lactobacillus* species have shown antiproliferative and antimetastatic effects on cervical cancer cells [71]. Postbiotic metabolites play a role in cellular growth, differentiation, and the sustenance of tissues [57]. Butyric acid, another important product of the postbiotics, has been shown to suppress the growth of cervical carcinoma [57].

3.7.4 Colorectal Cancer

Postbiotics represent an attractive area of interest for use as both preventive strategies and adjuvant therapies in colorectal cancer (CRC) patients [72]. These bioactive compounds can prevent the proliferation of cancer cells, inhibit migration, and promote apoptosis, thus limiting tumor progression [73]. Postbiotics are emerging as important therapeutic agents in managing CRC because of their anti-inflammatory, antiproliferative, and anticancer properties [58]. They have potential in the context of safe and practical approaches to preventing CRC and as supportive therapies [58]. SCFAs, including butyrate, play an important role in inhibiting the growth of cancer cells and inducing apoptosis in colorectal cancer. The combined action of SCFAs and bioactive peptides has opened great therapeutic potential by targeting multiple mechanisms in inhibiting tumor progression in CRC [74, 75].

3.8 POSTBIOTICS FOR CANCER CARE: PATENTS AND CLINICAL TRIALS

Modern approaches to using advanced analytical techniques ensure the identification and characterization of new postbiotic molecules [76, 77]. These bioactive compounds promise new supportive therapies for various cancers, contributing to prevention and treatment strategies with their biological activities [73, 78]. Various bioactive compounds are patented as postbiotics for cancer care [57–59, 79, 80]. Natural products present promising strategies for addressing cancer-related conditions [81, 82]. The dual-action peptides are engineered to block cancer cells [83–85]. The patents on postbiotics focus on the advanced techniques applied in their preparation and stabilization. Metabolites of postbiotics are investigated as possible antimicrobial and antioxidant agents [86, 87]. Self-emulsifying drug delivery systems (SEDDS) are another innovation and have been developed to realize improved therapeutic efficacy [88–90]. Furthermore, liposomal drug delivery systems are patented to improve treatment efficacy [91–95]. Synergistic effects can be generated with conventional chemotherapy drugs [96]. Postbiotics offer a new, effective way to increase the success of conventional treatments [96]. See Table 3.1 [96–104].

TABLE 3.1
Postbiotics for Cancer Care—Preclinical and Clinical Studies

Study	Postbiotic Type	Study Design	Cancer Type Targeted	Mechanism of Action	Key Findings
Riaz et al. (2019) [96]	Cell-free supernatant from *Lactobacillus rhamnosus* strains (SHA111, SHA112, SHA113)	Laboratory-based study	Cervical cancer	Antioxidant activity, induction of apoptosis	Intense anticancer activity was observed through cytotoxicity and apoptosis induction in HeLa cells. Gene expression analysis confirmed the apoptotic mechanism.
Lee et al. (2019) [97]	Cell-free supernatant (CFS) from *Lactobacillus fermentum*	In-vitro study	Colorectal cancer	Increased transcription of apoptotic genes	Reduced cell viability Apoptosis-related gene expression and caspase 3 activation were pronouncedly upregulated.
Konishi et al. (2016) [98]	Ferrichrome (a molecule produced by *Lactobacillus casei* ATCC334)	In-vitro study	Colon cancer	Induction of apoptosis in colon cancer cells via activation of the c-jun N-terminal kinase (JNK) signaling pathway	Ferrichrome demonstrated a stronger tumor-suppressive effect on colon cancer cells compared to cisplatin and 5-fluorouracil.
Bahmani et al. (2019) [99]	Cell-free supernatant (CFS) derived from *Bifidobacterium bifidum*	Experimental laboratory-based study	Colon cancer	Inhibition of colon cancer cell growth through cytotoxic effects of CFS	Two isolates (from infant feces and probiotic capsules) showed the highest cancer cell growth inhibition.
Tukenmez et al. (2019) [100]	Exopolysaccharides (EPSs) produced by *Lactobacillus spp*	Experimental study	Colon cancer	Induction of apoptosis via increased expression of Bax, caspase 3, and caspase 9, and decreased expression of Bcl-2 and Survivin	All EPSs contained mannose, glucose, and N-acetylglucosamine; some also contained arabinose or fructose.
Nozari et al. (2019) [101]	Cell wall protein fractions from *Lactobacillus paracase*	Experimental study	Colon cancer	Significant antiproliferative effects were observed, particularly with the F1 protein fraction	The cell wall proteins of *L. paracasei* can potentially be chemotherapeutic agents against colon cancer.
Sentürk et al. (2020) [102]	Secondary metabolites of *Lactobacillus plantarum*	Experimental study	Breast cancer	Reduced expression of antiapoptotic genes Increased expression of proapoptotic genes	High binding affinity with apoptotic genes, enhancing its therapeutic potential

(*Continued*)

TABLE 3.1
(Continued)

Study	Postbiotic Type	Study Design	Cancer Type Targeted	Mechanism of Action	Key Findings
Chen et al. (2020) [103]	Butyrate-producing bacteria (*Clostridium butyricum*)	Experimental study	Colorectal cancer	Inhibition of intestinal tumor development via suppression of Wnt/β-catenin signaling pathway Modulation of gut microbiota composition with increases in beneficial bacteria	Treatment resulted in decreased proliferation and increased apoptosis of intestinal tumor cells.
Kim et al. (2015) [104]	Cell-free extracts of *Lactobacillus casei*	In vitro study	Cervical cancer	Assessment of synergistic potential between *L. casei* extracts and anticancer drugs in inhibiting cancer cell growth	*L. casei* extract did not significantly affect the growth rate of cervical cancer cell lines.

3.9 IN VITRO AND IN VIVO STUDIES IN THE FIELD OF CANCER AND POSTBIOTICS

This chapter focuses on the importance of postbiotics for cancer treatment and prevention, focusing on their mechanisms and research findings from both in vitro and in vivo research [104].

3.9.1 Mechanisms of Action of Postbiotics in Cancer

Postbiotics work to fight cancer in a myriad of ways. They stimulate apoptosis within cancer cells via activation of mechanisms of mitochondria and the caspase enzyme. In addition, by regulating the immune systems, they increase the function of the NK cells and T-lymphocytes with cytotoxic properties. Anti-inflammatory properties decrease proinflammatory cytokines, such as TNF and IL-6, which can help reduce chronic inflammation that causes cancer. The gut microbiota of the postbiotics can also be affected by their composition, which leads to the creation of SCFAs that have protective properties. They also reduce the effects of oxidative stress by neutralizing ROS and limiting DNA destruction [105].

3.9.2 In Vitro Studies on Postbiotics in Cancer

Cytotoxic effects were evident against breast and colorectal cancer cells; postbiotics specifically inhibit their proliferation [106]. Studies also reveal postbiotics suppress cancer cell migration and invasion by regulating epithelial-to-mesenchymal transition (EMT) markers. In the case of colorectal cancer, postbiotics regulate the Wnt/β catenin signaling pathway that suppresses the expression of oncogenic genes. In addition, they reduce the DNA damage due to oxidative stress that occurs in cancerous lung cells by stimulating antioxidant enzymes [107].

3.9.3 In Vivo Studies on Postbiotics in Cancer

An association has been made between postbiotics and chemotherapy; specifically, postbiotics improve the efficacy of medications like 5-fluorouracil [108]. Postbiotics successfully reduced inflammation and tumor burden in colitis-associated cancer models through downregulating NF-kB signaling [109]. Immunomodulatory effects demonstrated through immunotherapy proved potency against cancerous cell growth with enhanced activation of dendritic cells and T lymphocytes that improved antitumor responses [110]. Postbiotics have anticancer potential while emphasizing further validation in clinical settings [111]. Postbiotic research faces several hurdles, such as variations in formulations and lack of standardization, translation from preclinical findings into clinical applications being complicated due to human complexity, and limitations with animal models also preventing accurate prediction [112]. Extensive clinical research is required for effective cancer treatments [113]. See Table 3.2 [114–130].

TABLE 3.2
Summary of Postbiotic Studies in Cancer

Category	Focus Area	Key Findings	References
Mechanisms of action	Apoptosis induction	Activation of mitochondrial pathways and caspase enzymes	[114]
	Immune modulation	Enhanced NK cell and cytotoxic T-lymphocyte activity	[115]
	Anti-inflammatory effects	Downregulation of proinflammatory cytokines like TNF-α and IL-6	[116]
	Gut microbiota modulation	Production of SCFAs with anticancer properties	[117]
	Oxidative stress reduction	Neutralization of ROS and prevention of DNA damage	[118]
In vitro studies	Cytotoxic effects	Selective inhibition of cancer cell proliferation	[119]
	Suppression of migration and invasion	Regulation of EMT markers in gastric cancer cells	[120]
	Wnt/β-catenin pathway modulation	Suppression of oncogenic gene expression in colorectal cancer	[121]
In vivo studies	Tumor growth suppression	Enhanced immune cell infiltration and NK cell activity in mice	[122]
	Synergistic effects with chemotherapy	Increased efficacy of 5-fluorouracil in colon cancer models	[123]
	Anti-inflammatory effects	Reduced tumor burden in colitis-associated cancer models	[124]
	Metastasis reduction in breast cancer	Modulation of gut microbiota and delayed tumor onset	[125]
	Immunomodulatory effects in melanoma	Enhanced dendritic cell and T-lymphocyte activation	[126]
Challenges	Standardization	Variability in postbiotic formulations	[127]
	Translational research	Difficulty in applying preclinical findings to human models	[128]
	Model limitations	Constraints of animal studies in predicting human responses	[129]
Future directions	Clinical trials	Validation of efficacy and safety in humans	[10]
	Personalized therapies	Development based on microbiome profiles	[130]

3.10 SAFETY CONSIDERATIONS AND REGULATORY APPROVAL OF POSTBIOTICS IN CANCER

As postbiotics move beyond experimental studies into possible clinical use, understanding safety issues and regulatory frameworks becomes critical to their growth and approval [131]. While many of the byproducts can be beneficial, too much accumulation could potentially induce toxicity or disrupt normal metabolic function—studies focusing on metabolic safety are vital in guaranteeing that postbiotics do not generate potentially hazardous materials during interaction with host systems [132]. Allergenicity testing involves measuring whether these allergenicity components provoke immune hypersensitivity reactions [133]. Postbiotics may interact with chemotherapy or radiation therapies. Pharmacokinetic and pharmacodynamic studies examine the efficacy of existing cancer therapies and increase the therapeutic ratio. These investigations utilize postbiotic-specific studies conducted alongside chemotherapy or radiation for better evaluation and possible interactions [134].

3.11 REGULATORY APPROVAL PATHWAYS

Regulators like the U.S. Food and Drug Administration (FDA) and European Food Safety Authority (EFSA) have created pathways for reviewing new therapies, including postbiotics. Postbiotics intended for therapeutic use must undergo rigorous evaluation under either the Biologics Control Act or Dietary Supplement Health and Education Act in the United States; in Europe, they must also pass assessments as novel foods requiring evidence of safety, effectiveness, and absence of toxicity [135]. In vitro studies help identify biomarkers of efficacy or potential side effects and biomarkers of efficacy or adverse events (BEATs) [136, 137]. The regulatory bodies require compliance with good manufacturing practices (GMP) in manufacturing postbiotics. Manufacturers should also prove batch-to-batch reproducibility and document the process steps in detail [138, 139].

3.12 CONCLUSION

Postbiotics may provide an effective and prevention-based strategy for the treatment of cancer. The combination of postbiotics and existing treatments like chemotherapy, radiation, and immunotherapy can result in synergistic benefits that can enhance the overall outcomes of treatment. Postbiotics can play a part in increasing the quality of life of patients with cancer by decreasing inflammation, improving gut health, and supporting immune function throughout the recovery process. Future research should concentrate on the possibilities of postbiotics for treating cancers. By addressing current challenges and using technological advancements to improve their effectiveness, postbiotics could be an essential element in the fight against cancer and will provide secure, efficient, and innovative solutions for cancer patients across the globe.

REFERENCES

1. Salminen, S.; Collado, M.C.; Endo, A., Hill, C.; Lebeer, S.; Quigley, E.M.M.; Sanders, M.E.; Shamir, R.; Swann, J.R.; Szajewska, H.; et al. The International Scientific Association of Probiotics and Prebiotics (ISAPP) consensus statement on the definition and scope of postbiotics. *Nat. Rev. Gastroenterol. Hepatol.* 2021, *18*, 649–667, doi:10.1038/s41575-021-00440-6
2. Kudra, A.; Kaźmierczak-Siedlecka, K.; Sobocki, B.K.; Muszyński, D.; Połom, J.; Carbone, L.; Marano, L.; Roviello, F.; Kalinowski, L.; Stachowska, E. Postbiotics in oncology: science or science fiction? *Front. Microbiol.* 2023, *14*, 1182547, doi:10.3389/fmicb.2023.1182547
3. Dahiya, D.; Nigam, P.S. Probiotics, Prebiotics, synbiotics, and fermented foods as potential biotics in nutrition improving health via microbiome-gut-brain axis. *Fermentation* 2022, *8*, doi:10.3390/fermentation8070303

4. Kesika, P.; Suganthy, N.; Sivamaruthi, B.S.; Chaiyasut, C. Role of gut-brain axis, gut microbial composition, and probiotic intervention in Alzheimer's disease. *Life Sci.* 2021, *264*, 118627, doi:10.1016/j.lfs.2020.118627
5. Manirarora, J.N.; Kosiewicz, M.M.; Alard, P. Feeding lactobacilli impacts lupus progression in (NZBxNZW)F1 lupus-prone mice by enhancing immunoregulation. *Autoimmunity* 2020, *53*, 323–332, doi:10.1080/08916934.2020.1777282
6. He, M.; Shi, B. Gut microbiota as a potential target of metabolic syndrome: the role of probiotics and prebiotics. *Cell Biosci.* 2017, *7*, 54, doi:10.1186/s13578-017-0183-1
7. Johnson, C.N.; Kogut, M.H.; Genovese, K.; He, H.; Kazemi, S.; Arsenault, R.J. Administration of a postbiotic causes immunomodulatory responses in broiler gut and reduces disease pathogenesis following challenge. *Microorganisms* 2019, *7*, doi:10.3390/microorganisms7080268
8. Mayorgas, A.; Dotti, I.; Salas, A. Microbial metabolites, postbiotics, and intestinal epithelial function. *Mol. Nutr. Food Res.* 2021, *65*, 2000188, doi:https://doi.org/10.1002/mnfr.202000188
9. Dameshghian, M.; Tafvizi, F.; Tajabadi Ebrahimi, M.; Hosseini Doust, R. Anticancer potential of postbiotic derived from *Lactobacillus brevis* and *Lactobacillus casei*: In vitro analysis of breast cancer cell line. *Probiotics and Antimicrobial Proteins* 2024, doi:10.1007/s12602-024-10288-2
10. Eladwy, R.A.; Alsherbiny, M.A.; Chang, D.; Fares, M.; Li, C.-G.; Bhuyan, D.J. The postbiotic sodium butyrate synergizes the antiproliferative effects of dexamethasone against the AGS gastric adenocarcinoma cells. *Frontiers in Nutrition* 2024, *11*, doi:10.3389/fnut.2024.1372982
11. Kuru-Yaşar, R.; Üstün-Aytekin, Ö. The crucial roles of diet, microbiota, and postbiotics in colorectal cancer. *Current Nutrition Reports* 2024, *13*, 126–151, doi:10.1007/s13668-024-00525-z
12. Elango, A.; Nesam, V.D.; Sukumar, P.; Lawrence, I.; Radhakrishnan, A. Postbiotic butyrate: role and its effects for being a potential drug and biomarker to pancreatic cancer. *Arch. Microbiol.* 2024, *206*, 156, doi:10.1007/s00203-024-03914-8
13. Silva, Y.P.; Bernardi, A.; Frozza, R.L. The role of short-chain fatty acids from gut microbiota in gut-brain communication. *Front. Endocrinol. (Lausanne)* 2020, *11*, 25, doi:10.3389/fendo.2020.00025
14. Ragavan, M.L.; Hemalatha, S. The functional roles of short chain fatty acids as postbiotics in human gut: future perspectives. *Food Sci. Biotechnol.* 2024, *33*, 275–285, doi:10.1007/s10068-023-01414-x
15. Parada Venegas, D.; De la Fuente, M.K.; Landskron, G.; Gonzalez, M.J.; Quera, R.; Dijkstra, G.; Harmsen, H.J.M.; Faber, K.N.; Hermoso, M.A. Short chain fatty acids (SCFAs)-mediated gut epithelial and immune regulation and its relevance for inflammatory bowel diseases. *Front. Immunol.* 2019, *10*, 277, doi:10.3389/fimmu.2019.00277
16. Xie, W.; Zhong, Y.S.; Li, X.J.; Kang, Y.K.; Peng, Q.Y.; Ying, H.Z. Postbiotics in colorectal cancer: intervention mechanisms and perspectives. *Front. Microbiol.* 2024, *15*, 1360225, doi:10.3389/fmicb.2024.1360225
17. Molujin, A.M.; Abbasiliasi, S.; Nurdin, A.; Lee, P.C.; Gansau, J.A.; Jawan, R. Bacteriocins as potential therapeutic approaches in the treatment of various cancers: a review of in vitro studies. *Cancers (Basel)* 2022, *14*, doi:10.3390/cancers14194758
18. Du, Y.; Liu, L.; Yan, W.; Li, Y.; Li, Y.; Cui, K.; Yu, P.; Gu, Z.; Zhang, W.; Feng, J.; et al. The anticancer mechanisms of exopolysaccharide from *Weissella cibaria* D-2 on colorectal cancer via apoptosis induction. *Sci. Rep.* 2023, *13*, 21117, doi:10.1038/s41598-023-47943-7
19. Sudaarsan, A.S.K.; Ghosh, A.R. Appraisal of postbiotics in cancer therapy. *Front. Pharmacol.* 2024, *15*, doi:10.3389/fphar.2024.1436021
20. Oh, B.S.; Choi, W.J.; Kim, J.S.; Ryu, S.W.; Yu, S.Y.; Lee, J.S.; Park, S.H.; Kang, S.W.; Lee, J.; Jung, W.Y.; et al. Cell-free supernatant of *Odoribacter splanchnicus* isolated from human feces exhibits anti-colorectal cancer activity. *Front. Microbiol.* 2021, *12*, 736343, doi:10.3389/fmicb.2021.736343
21. An, J.; Seok, H.; Ha, E.M. GABA-producing *Lactobacillus plantarum* inhibits metastatic properties and induces apoptosis of 5-FU-resistant colorectal cancer cells via GABA(B) receptor signaling. *J. Microbiol.* 2021, *59*, 202–216, doi:10.1007/s12275-021-0562-5
22. Ding, L.; Cao, J.; Lin, W.; Chen, H.; Xiong, X.; Ao, H.; Yu, M.; Lin, J.; Cui, Q. The roles of cyclin-dependent kinases in cell-cycle progression and therapeutic strategies in human breast cancer. *Int. J. Mol. Sci.* 2020, *21*, doi:10.3390/ijms21061960
23. Kim, S.; Kim, G.-H.; Cho, H. Postbiotics for cancer prevention and treatment. *Korean J. Microbiol.* 2021, *57*, 142–153, doi:10.7845/kjm.2021.1067

24. Hosseini, S.S.; Goudarzi, H.; Ghalavand, Z.; Hajikhani, B.; Rafeieiatani, Z.; Hakemi-Vala, M. Antiproliferative effects of cell wall, cytoplasmic extract of *Lactococcus* lactis and nisin through down-regulation of cyclin D1 on SW480 colorectal cancer cell line. *Iran J. Microbiol.* 2020, *12*, 424–430, doi:10.18502/ijm.v12i5.4603
25. Nazari, F.; Jafari, P.; Nomanpour, B.; Varmira, K.; Raissi, F. Inhibitory effects of postbiotic consisting sonication-killed *Bifidobacterium bifidum* on experimental triple negative breast neoplasm in mice: a preliminary study. *Iran J. Microbiol.* 2022, *14*, 689–697, doi:10.18502/ijm.v14i5.10964
26. Luo, S.; Li, Z.; Mao, L.; Chen, S.; Sun, S. Sodium butyrate induces autophagy in colorectal cancer cells through LKB1/AMPK signaling. *J. Physiol. Biochem.* 2019, *75*, 53–63, doi:10.1007/s13105-018-0651-z
27. Vrzáčková, N.; Ruml, T.; Zelenka, J. Postbiotics, metabolic signaling, and cancer. *Molecules* 2021, *26*, doi:10.3390/molecules26061528
28. Castriconi, R.; Carrega, P.; Dondero, A.; Bellora, F.; Casu, B.; Regis, S.; Ferlazzo, G.; Bottino, C. Molecular mechanisms directing migration and retention of natural killer cells in human tissues. *Front. Immunol.* 2018, *9*, 2324, doi:10.3389/fimmu.2018.02324
29. Jung, Y.J.; Kim, H.S.; Jaygal, G.; Cho, H.R.; Lee, K.B.; Song, I.B.; Kim, J.H.; Kwak, M.S.; Han, K.H.; Bae, M.J.; et al. Postbiotics enhance NK cell activation in stress-induced mice through gut microbiome regulation. *J. Microbiol. Biotechnol.* 2022, *32*, 612–620, doi:10.4014/jmb.2111.11027
30. Pérez, M.; Buey, B.; Corral, P.; Giraldos, D.; Latorre, E. Microbiota-derived short-chain fatty acids boost antitumoral natural killer cell activity. *J. Clin. Med.* 2024, *13*, 3885.
31. Ferrari, V.; Lo Cascio, A.; Melacarne, A.; Tanasković, N.; Mozzarelli, A.M.; Tiraboschi, L.; Lizier, M.; Salvi, M.; Braga, D.; Algieri, F.; et al. Sensitizing cancer cells to immune checkpoint inhibitors by microbiota-mediated upregulation of HLA class I. *Cancer Cell* 2023, *41*, 1717–1730.e1714, doi:https://doi.org/10.1016/j.ccell.2023.08.014
32. Lan, T.; Chen, L.; Wei, X. Inflammatory cytokines in cancer: comprehensive understanding and clinical progress in gene therapy. *Cells* 2021, *10*, doi:10.3390/cells10010100
33. Fang, H.; Rodrigues E.; Lacerda R.; Barra, N.G.; Kukje Zada, D.; Robin, N.; Mehra, A.; Schertzer, J.D. Postbiotic impact on host metabolism and immunity provides therapeutic potential in metabolic disease. *Endocr. Rev.* 2024, doi:10.1210/endrev/bnae025
34. Yeşilyurt, N.; Yılmaz, B.; Ağagündüz, D.; Capasso, R. Involvement of probiotics and postbiotics in the immune system modulation. *Biologics* 2021, *1*, 89–110.
35. Hong, N.; Ku, S.; Yuk, K.; Johnston, T.V.; Ji, G.E.; Park, M.S. Production of biologically active human interleukin-10 by *Bifidobacterium bifidum* BGN4. *Microbial Cell Factories* 2021, *20*, 16, doi:10.1186/s12934-020-01505-y
36. Ullah, A.; Jiao, W.; Shen, B. The role of proinflammatory cytokines and CXC chemokines (CXCL1–CXCL16) in the progression of prostate cancer: insights on their therapeutic management. *Cell. Mol Biol. Lett.* 2024, *29*, 73, doi:10.1186/s11658-024-00591-9
37. Li, D.; Wu, M. Pattern recognition receptors in health and diseases. *Signal Transduct. Targeted Ther.* 2021, *6*, 291, doi:10.1038/s41392-021-00687-0
38. Jastrząb, R.; Graczyk, D.; Siedlecki, P. Molecular and cellular mechanisms influenced by postbiotics. *Int. J. Mol. Sci.* 2021, *22*, doi:10.3390/ijms222413475
39. Woo, V.; Alenghat, T. Epigenetic regulation by gut microbiota. *Gut Microbes* 2022, *14*, 2022407, doi:10.1080/19490976.2021.2022407
40. Erfanian, N.; Safarpour, H.; Tavakoli, T.; Mahdiabadi, M.A.; Nasseri, S.; Namaei, M.H. Investigating the therapeutic potential of *Bifidobacterium breve* and *Lactobacillus rhamnosus* postbiotics through apoptosis induction in colorectal HT-29 cancer cells. *Iran J. Microbiol.* 2024, *16*, 68–78, doi:10.18502/ijm.v16i1.14873
41. Wang, Y.; Wang, Y.; Sun, T.; Xu, J. Bacteriocins in cancer treatment: Mechanisms and clinical potentials. *Biomolecules* 2024, *14*, doi:10.3390/biom14070831
42. Elena-Real, C.A.; Díaz-Quintana, A.; González-Arzola, K.; Velázquez-Campoy, A.; Orzáez, M.; López-Rivas, A.; Gil-Caballero, S.; De la Rosa, M.; Díaz-Moreno, I. Cytochrome c speeds up caspase cascade activation by blocking 14-3-3ε-dependent Apaf-1 inhibition. *Cell Death Dis.* 2018, *9*, 365, doi:10.1038/s41419-018-0408-1
43. Kim, Y.; Kim, H.J.; Ji, K. The proliferation inhibitory effect of postbiotics prepared from probiotics with antioxidant activity against HT-29 cells. *Appl. Sci.* 2022, *12*, 12519.

44. Nowak, A.; Zakłos-Szyda, M.; Rosicka-Kaczmarek, J.; Motyl, I. Anticancer potential of post-fermentation media and cell extracts of probiotic strains: an in vitro study. *Cancers (Basel)* 2022, *14*, 1853.
45. Kvakova, M.; Kamlarova, A.; Stofilova, J.; Benetinova, V.; Bertkova, I. Probiotics and postbiotics in colorectal cancer: prevention and complementary therapy. *World J. Gastroenterol.* 2022, *28*, 3370–3382, doi:10.3748/wjg.v28.i27.3370
46. Rezaie, N.; Aghamohammad, S.; Haj Agha Gholizadeh Khiavi, E.; Khatami, S.; Sohrabi, A.; Rohani, M. The comparative antioxidant and anti-inflammatory efficacy of postbiotics and probiotics through Nrf-2 and NF-kB pathways in DSS-induced colitis model. *Sci. Rep.* 2024, *14*, 11560, doi:10.1038/s41598-024-62441-0
47. Vrzáčková, N.; Ruml, T.; Zelenka, J. Postbiotics, metabolic signaling, and cancer. *Molecules* 2021, *26*, 1528.
48. Kerry, R.G.; Patra, J.K.; Gouda, S.; Park, Y.; Shin, H.S.; Das, G. Benefaction of probiotics for human health: a review. *J. Food Drug Anal.* 2018, *26*, 927–939, doi:10.1016/j.jfda.2018.01.002
49. Pasquereau-Kotula, E.; Martins, M.; Aymeric, L.; Dramsi, S. Significance of *Streptococcus gallolyticus* subsp. gallolyticus association with colorectal cancer. *Front. Microbiol.* 2018, *9*, 614, doi:10.3389/fmicb.2018.00614
50. Balendra, V.; Rosenfeld, R.; Amoroso, C.; Castagnone, C.; Rossino, M.G.; Garrone, O.; Ghidini, M. Postbiotics as adjuvant therapy in cancer care. *Nutrients* 2024, *16*, doi:10.3390/nu16152400
51. Wegh, C.A.M.; Geerlings, S.Y.; Knol, J.; Roeselers, G.; Belzer, C. Postbiotics and their potential applications in early life nutrition and beyond. *Int. J. Mol. Sci.* 2019, *20*, 4673.
52. Ashique, S.; Faruk, A.; Ahmad, F.J.; Khan, T.; Mishra, N. It is all about probiotics to control cervical cancer. *Prob Antimicrob Prot.* 2024, 16(3):979–992.
53. Zółkiewicz, J.; Marzec, A.; Ruszczyński, M.; Feleszko, W. Postbiotics—a step beyond pre- and probiotics. *Nutrients* 2020, *12*, doi:10.3390/nu12082189
54. Auger, S.; Kropp, C.; Borras-Nogues, E.; Chanput, W.; Andre-Leroux, G.; Gitton-Quent, O.; Benevides, L.; Breyner, N.; Azevedo, V.; Langella, P.; et al. Intraspecific diversity of microbial anti-inflammatory molecule (MAM) from *Faecalibacterium prausnitzii*. *Int J. Mol. Sci.* 2022, *23*, doi:10.3390/ijms23031705
55. Chen, M.; Ma, L.; Yu, H.; Huang, S.; Zhang, J.; Gong, J.; Yang, L.; Chen, L.; Luo, H.; Tian, L.; Wang, S. JK5G postbiotics attenuate immune-related adverse events in NSCLC patients by regulating gut microbiota: a randomized controlled trial in China. *Front. Oncol.* 2023 Aug 4, *13*, 1155592, doi: 10.3389/fonc.2023.1155592
56. Morita, A.; Ichihara, E.; Inoue, K.; Fujiwara, K.; Yokoyama, T.; Harada, D.; Ando, C.; Kano, H.; Oda, N.; Tamura, T.; Ochi, N.; Kawai, H.; Inoue, M.; Hara, N.; Fujimoto, N.; Ichikawa, H.; Oze, I.; Hotta, K.; Maeda, Y.; Kiura, K. Impacts of probiotics on the efficacies of immune checkpoint inhibitors with or without chemotherapy for patients with advanced non-small-cell lung cancer. *Int. J. Cancer.* 2024 May 1, *154*(9), 1607–1615, doi: 10.1002/ijc.34842
57. Mishra, N.; Garg, A.; Ashique, S.; Bhatt, S. Potential of postbiotics for the treatment of metabolic disorders. *Drug Discov Today.* 2024, 20:103921.
58. Kudra, A.; Kaźmierczak-Siedlecka, K.; Sobocki, B.K.; Muszyński, D.; Połom, J.; Carbone, L.; Marano, L.; Roviello, F.; Kalinowski, L.; Stachowska, E. Postbiotics in oncology: science or science fiction? *Front. Microbiol.* 2023 Aug 7, **14**:1182547, doi: 10.3389/fmicb.2023.1182547
59. Vrzáčková, N.; Ruml, T.; Zelenka, J. Postbiotics, metabolic signaling, and cancer. *Molecules* 2021 Mar 11, *26*(6), 1528, doi: 10.3390/molecules26061528
60. Souza, V.G.P.; Forder, A.; Pewarchuk, M.E.; Telkar, N.; de Araujo, R.P.; Stewart, G.L.; Vieira, J.; Reis, P.P.; Lam, W.L. The complex role of the microbiome in non-small cell lung cancer development and progression. *Cells* 2023 Dec *12*(24), 2801, doi: 10.3390/cells12242801
61. Wei, H.; Yue, Z.; Han, J.; Chen, P.; Xie, K.; Sun, Y.; Zhu, J. Oral compound probiotic supplements can improve the quality of life for patients with lung cancer during chemotherapy: a randomized placebo-controlled study. *Thorac. Cancer* 2024 Jan, *15*(2), 182–191, doi: 10.1111/1759-7714.15177
62. Cong, J.; Zhang, C.; Zhou, S.; Zhu, J.; Liang, C. A pilot study: favorable effects of *Clostridium butyricum* on intestinal microbiota for adjuvant therapy of lung cancer. *Cancers (Basel).* 2022 Jul 23, *14*(15), 3599.
63. Dameshghian, M.; Tafvizi, F.; Tajabadi Ebrahimi, M.; Hosseini, D.R. Anticancer potential of postbiotic derived from *Lactobacillus brevis* and *Lactobacillus casei*: In vitro analysis of breast cancer cell line. *Probiotics Antimicrob. Proteins.* 2024 May 17, doi: 10.1007/s12602-024-10288-2

64. Wasiak, J.; Głowacka, P.; Pudlarz, A.; Pieczonka, A.M.; Dzitko, K.; Szemraj, J.; Witusik-Perkowska, M. Lactic acid bacteria-derived postbiotics as adjunctive agents in breast cancer treatment to boost the antineoplastic effect of a conventional therapeutic comprising tamoxifen and a new drug candidate: an aziridine-hydrazide hydrazone derivative. *Molecules*. 2024 May 13, *29*(10), 2292, doi: 10.3390/molecules29102292
65. Kwa, M.; Plottel, C.S.; Blaser, M.J.; Adams, S. The intestinal microbiome and estrogen receptor-positive female breast cancer. *J. Natl. Cancer Inst.* 2016 Apr 22, *108*(8), djw029, doi: 10.1093/jnci/djw029
66. Nazari, F.; Jafari, P.; Nomanpour, B.; Varmira, K.; Raissi, F. Inhibitory effects of postbiotic consisting sonication-killed *Bifidobacterium bifidum* on experimental triple negative breast neoplasm in mice: a preliminary study. *Iran J. Microbiol.* 2022 Oct, *14*(5), 689–697, doi: 10.18502/ijm.v14i5.10964
67. Dameshghian, M.; Tafvizi, F.; Tajabadi Ebrahimi, M.; Hosseini, D.R. Anticancer potential of postbiotic derived from *Lactobacillus brevis* and *Lactobacillus casei*: In vitro analysis of breast cancer cell line. *Probiotics Antimicrob. Proteins* 2024 May 17, doi: 10.1007/s12602-024-10288-2
68. Nowak, A.; Zakłos-Szyda, M.; Rosicka-Kaczmarek, J.; Motyl, I. Anticancer potential of post-fermentation media and cell extracts of probiotic strains: an in vitro study. *Cancers (Basel)*. 2022 Apr 6, *14*(7), 1853, doi: 10.3390/cancers14071853
69. Cha, M.K.; Lee, D.K.; An, H.M.; Lee, S.W.; Shin, S.H.; Kwon, J.H.; Kim, K.J.; Ha, N.J. Antiviral activity of bifidobacterium adolescentis SPM1005-A on human papillomavirus type 16. *BMC Med.* 2012 Jul 12, *10*, 72. doi: 10.1186/1741-7015-10-72
70. Pawar, K.; Aranha, C. Lactobacilli metabolites restore E-cadherin and suppress MMP9 in cervical cancer cells. *Curr. Res. Toxicol.* 2022 Sep *21*, 3, 100088, doi: 10.1016/j.crtox.2022.100088
71. Rad, A.H.; Aghebati-Maleki, L.; Kafil, H.S.; Abbasi, A.. Molecular mechanisms of postbiotics in colorectal cancer prevention and treatment. *Crit. Rev. Food Sci. Nutr.* 2021, *61*(11), 1787–1803. doi: 10.1080/10408398.2020.1765310
72. Erfanian, N.; Safarpour, H.; Tavakoli, T.; Mahdiabadi, M.A.; Nasseri, S.; Namaei, M.H. Investigating the therapeutic potential of *Bifidobacterium breve* and *Lactobacillus rhamnosus* postbiotics through apoptosis induction in colorectal HT-29 cancer cells. *Iran J. Microbiol.* 2024 Feb, *16*(1), 68–78, doi: 10.18502/ijm.v16i1.14873
73. Ohara, T.; Mori, T. Antiproliferative effects of short-chain fatty acids on human colorectal cancer cells via gene expression inhibition. *Anticancer Res.* 2019 Sep, *39*(9), 4659–4666, doi: 10.21873/anticanres.13647
74. Li, X.; Xia, L.; Ouyang, X.; Suyila, Q.; Su, L.; Su, X. Bioactive peptides sensitize cells to anticancer effects of oxaliplatin in human colorectal cancer xenografts in nude mice. *Protein Pept. Lett.* 2019, *26*(7), 512–522, doi: 10.2174/0929866526666190405124955
75. Mohammed, S.; Çon, A.H. Postbiotic nanoparticles (postbiotics-NPs): a novel strategy for providing probiotics' health advantages through food consumption. *Food Sci. Biotechnol.* 2024 Jun 23, *33*(12), 2729–2736, doi: 10.1007/s10068-024-01629-6
76. Wang, P.; Wang, S.; Wang, D.; Li, Y.; Yip, R.C.S.; Chen, H. Postbiotics-peptidoglycan, lipoteichoic acid, exopolysaccharides, surface layer protein and pili proteins: structure, activity in wounds and their delivery systems. *Int. J. Biol. Macromol.* 2024 Aug, *274*(Pt 1), 133195, doi: 10.1016/j.ijbiomac.2024.133195
77. Eladwy, R.A.; Alsherbiny, M.A.; Chang, D.; Fares, M.; Li, C.G.; Bhuyan, D.J. The postbiotic sodium butyrate synergizes the antiproliferative effects of dexamethasone against the AGS gastric adenocarcinoma cells. *Front. Nutr.* 2024 Mar, *12*(11), 1372982, doi: 10.3389/fnut.2024.1372982
78. Elango, A.; Nesam, V.D.; Sukumar, P.; Lawrence, I.; Radhakrishnan, A. Postbiotic butyrate: role and its effects for being a potential drug and biomarker to pancreatic cancer. *Arch. Microbiol.* 2024 Mar 13, *206*(4), 156, doi: 10.1007/s00203-024-03914-8
79. Thoda, C.; Touraki, M. Probiotic-derived bioactive compounds in colorectal cancer treatment. *Microorganisms* 2023 Jul 27, *11*(8), 1898, doi: 10.3390/microorganisms11081898
80. Kubatka, P.; Mazurakova, A.; Samec, M.; Koklesova, L.; Zhai, K.; Al-Ishaq, R.; Kajo, K.; Biringer, K.; Vybohova, D.; Brockmueller, A.; Pec, M.; Shakibaei, M.; Giordano, F.A.; Büsselberg, D.; Golubnitschaja, O. Flavonoids against non-physiologic inflammation attributed to cancer initiation, development, and progression-3PM pathways. *EPMA J.* 2021 Oct 6, *12*(4), 559–587, doi: 10.1007/s13167-021-00257-y
81. Wei, W.C.; Shyur, L.F.; Yang, N.S. Cellular and molecular signaling as targets for cancer vaccine therapeutics. *Cells* 2022 May 9, *11*(9), 1590, doi: 10.3390/cells11091590

82. Xie, H.; Xu, X.; Chen, J.; Li, L.; Wang, J.; Fang, T.; Zhou, L.; Wang, H.; Zheng, S. Rational design of multifunctional small-molecule prodrugs for simultaneous suppression of cancer cell growth and metastasis in vitro and in vivo. *Chem. Commun. (Camb)*. 2016 Apr 25, *52*(32), 5601–5604, doi: 10.1039/c5cc 10367c
83. Mohammad, R.; Giri, A.; Goustin, A.S. Small-molecule inhibitors of Bcl-2 family proteins as therapeutic agents in cancer. *Recent Pat. Anticancer Drug Discov*. 2008 Jan, *3*(1), 20–30. doi: 10.2174/ 157489208783478676
84. Roy, S.R.; Li, G.; Hu, X.; Zhang, S.; Yukawa, S.; Du, E.; Zhang, Y. Integrin and heparan sulfate dual-targeting peptide assembly suppresses cancer metastasis. *ACS Appl. Mater. Interfaces*. 2020 Apr 29, *12*(17), 19277–19284. doi: 10.1021/acsami.0c02235
85. Tong, Y.; Guo, H.; Abbas, Z.; Zhang, J.; Wang, J.; Cheng, Q.; Peng, S.; Yang, T.; Bai, T.; Zhou, Y.; Li, J.; Wei, X.; Si, D.; Zhang, R. Optimizing postbiotic production through solid-state fermentation with *Bacillus amyloliquefaciens* J and *Lactiplantibacillus plantarum* SN4 enhances antibacterial, antioxidant, and anti-inflammatory activities. *Front. Microbiol*. 2023 Sep 7, *14*, 1229952, doi: 10.3389/fmicb.2023.1229952
86. Sharafi, H.; Divsalar, E.; Rezaei, Z.; Liu, S.Q.; Moradi, M. The potential of postbiotics as a novel approach in food packaging and biopreservation: a systematic review of the latest developments. *Crit. Rev. Food Sci. Nutr*. 2024, *64*(33), 12524–12554, doi: 10.1080/10408398.2023.2253909
87. Singh, V.; Vihal, S.; Rana, R.; Rathore, C. Nanocarriers for cannabinoid delivery: enhancing therapeutic potential. *Recent Adv. Drug Deliv. Formul*. 2024, *18*(4), 247–261, doi: 10.2174/01266738783003472407 18100814
88. Ruiz, P.S.L.; Serafini, M.R.; Alves, I.A.; Novoa, D.M.A. Recent progress in self-emulsifying drug delivery systems: a systematic patent review (2011-2020). *Crit. Rev. Ther. Drug Carrier Syst*. 2022, *39*(2), 1–77, doi: 10.1615/CritRevTherDrugCarrierSyst.2021038490
89. Fong, S.Y.; Brandl, M.; Bauer-Brandl, A. Phospholipid-based solid drug formulations for oral bioavailability enhancement: a meta-analysis. *Eur. J. Pharm. Sci*. 2015 Dec 1, *80*, 89–110, doi: 10.1016/ j.ejps.2015.08.005
90. Bloemberg, J.; Van Riel, L.; Dodou, D.; Breedveld, P. Focal therapy for localized cancer: a patent review. *Expert Rev. Med. Devices* 2021 Aug, *18*(8), 751–769, doi: 10.1080/17434440.2021.1943360
91. De Vasconcelos Cerqueira Braz, J.; Carvalho Nascimento Júnior, J.A.; Serafini, M.R. Terpenes with antitumor activity: a patent review. *Recent Pat. Anticancer Drug Discov*. 2020, *15*(4), 321–328, doi: 10. 2174/1574892815666201002162315
92. Ali Hazis, N.U.; Aneja, N.; Rajabalaya, R.; David, S.R. Systematic patent review of nanoparticles in drug delivery and cancer therapy in the last decade. *Recent Adv. Drug Deliv. Formul*. 2021, *15*(1), 59–74, doi: 10.2174/1872211314666210521105534
93. Cebada, J.; Perez-Santos, M.; Bandala, C.; Lara-Padilla, E.; Herrera-Camacho, I.; Rosas-Murrieta, N.H.; Millán-Pérez Peña, L.; Monjaraz, E.; Flores, A.; Anaya-Ruiz, M. OX40 agonists for cancer treatment: a patent review. *Expert Opin. Ther. Pat*. 2021 Jan, *31*(1), 81–90, doi: 10.1080/13543776.2021.1825688
94. Akhtar, N.; Mohammed, S.A.; Singh, V.; Abdellatif, A.A.; Mohammad, H.A.; Ahad, A.; Yusuf, M.; Khadri, H.; Naz, M.; Khan, O.; Rashid, M.; Khan, R.A. Liposome-based drug delivery of various anticancer agents of synthetic and natural product origin: a patent overview. *Pharm. Pat. Anal*. 2020 May, *9*(3), 87–116, doi: 10.4155/ppa-2019-0020
95. Eladwy, R.A.; Alsherbiny, M.A.; Chang, D.; Fares, M.; Li, C.G.; Bhuyan, D.J. The postbiotic sodium butyrate synergizes the antiproliferative effects of dexamethasone against the AGS gastric adenocarcinoma cells. *Front. Nutr*. 2024 Mar 12, *11*, 1372982, doi: 10.3389/fnut.2024.1372982
96. Riaz Rajoka, M.S.; Zhao, H.; Mehwish, H.M.; Li, N.; Lu, Y.; Lian, Z.; Shao, D.; Jin, M.; Li, Q.; Zhao, L.; Shi, J. Antitumor potential of cell free culture supernatant of *Lactobacillus rhamnosus* strains isolated from human breast milk. *Food Res. Int*. 2019 Sep, *123*, 286–297, doi: 10.1016/j.foodres.2019.05.002
97. Lee, J.E.; Lee, J.; Kim, J.H.; Cho, N.; Lee, S.H.; Park, S.B.; Koh, B.; Kang, D.; Kim, S.; Yoo, H.M. Characterization of the anticancer activity of the probiotic bacterium lactobacillus fermentum using 2D vs. 3D culture in colorectal cancer cells. *Biomolecules* 2019 Oct 1, *9*(10), 557, doi: 10.3390/biom9100557
98. Konishi, H.; Fujiya, M.; Tanaka, H.; Ueno, N.; Moriichi, K.; Sasajima, J.; Ikuta, K.; Akutsu, H.; Tanabe, H.; Kohgo, Y. Probiotic-derived ferrichrome inhibits colon cancer progression via JNK-mediated apoptosis. *Nat. Commun*. 2016 Aug 10, *7*, 12365, doi: 10.1038/ncomms12365
99. Bahmani, S.; Azarpira, N.; Moazamian, E. Anti-colon cancer activity of *Bifidobacterium* metabolites on colon cancer cell line SW742. *Turk. J. Gastroenterol*. 2019 Sep, *30*(9), 835–842, doi: 10.5152/ tjg.2019.18451

100. Tukenmez, U.; Aktas, B.; Aslim, B.; Yavuz, S. The relationship between the structural characteristics of lactobacilli-EPS and its ability to induce apoptosis in colon cancer cells in vitro. *Sci. Rep.* 2019 Jun 4, *9*(1), 8268, doi: 10.1038/s41598-019-44753-8
101. Nozari, S.; Faridvand, Y.; Etesami, A. Ahmad Khan Beiki, M.; Miresmaeili Mazrakhondi, S.A.; Abdolalizadeh, J. Potential anticancer effects of cell wall protein fractions from *Lactobacillus paracasei* on human intestinal Caco-2 cell line. *Lett. Appl. Microbiol.* 2019 Sep, *69*(3), 148–154, doi: 10.1111/lam.13198
102. Sentürk, M.; Ercan, F.; Yalcin, S. The secondary metabolites produced by *Lactobacillus plantarum* downregulate BCL-2 and BUFFY genes on breast cancer cell line and model organism *Drosophila melanogaster*: molecular docking approach. *Cancer Chemother. Pharmacol.* 2020 Jan, 85(1), 33–45, doi: 10.1007/s00280-019-03978-0
103. Chen, D.; Jin, D.; Huang, S.; Wu, J.; Xu, M.; Liu, T.; Dong, W.; Liu, X.; Wang, S.; Zhong, W.; Liu, Y.; Jiang, R.; Piao, M.; Wang, B.; Cao, H. *Clostridium butyricum*, a butyrate-producing probiotic, inhibits intestinal tumor development through modulating Wnt signaling and gut microbiota. *Cancer Lett.* 2020 Jan 28, 469, 456–467, doi: 10.1016/j.canlet.2019.11.019
104. Kim, S.N.; Lee, W.M.; Park, K.S.; Kim, J.B.; Han, D.J.; Bae, J. The effect of *Lactobacillus casei* extract on cervical cancer cell lines. *Contemp. Oncol. (Pozn).* 2015, 19(4), 306–312, doi: 10.5114/wo.2014.45292
105. Aguilar-Toalá, J.E.; Garcia-Varela, R.; Garcia, H.S.; Mata-Haro, V.; González-Córdova, A.F.; Vallejo-Cordoba, B.; Hernández-Mendoza, A. Postbiotics: an evolving term within the functional foods field. *Trends in Food Science & Technology* 2018, *75*, 105–114, doi: 10.1016/j.tifs.2018.03.009
106. Vrzáčková, N.; Ruml, T.; Zelenka, J. Postbiotics, metabolic signaling, and cancer. *Molecules* 2021, *26*(6), 1528, doi: 10.3390/molecules26061528
107. Florento, L.; Matias, R.; Tuaño, E.; Santiago, K.; Dela Cruz, F.; Tuazon, A. Comparison of cytotoxic activity of anticancer drugs against various human tumor cell lines using in vitro cell-based approach. *Int. J. Biomed. Sci.* 2012, *8*(1), 76–80.
108. Sun, L.; Xing, J.; Zhou, X.; Song, X.; Gao, S. Wnt/β-catenin signalling, epithelial-mesenchymal transition and crosslink signalling in colorectal cancer cells. *Biomed. Pharmacother.*, 2024, *175*, 116685. doi: 10.1016/j.biopha.2024.116685
109. Balendra, V.; Rosenfeld, R.; Amoroso, C.; Castagnone, C.; Rossino, M.G.; Garrone, O.; Ghidini, M. Postbiotics as adjuvant therapy in cancer care. *Nutrients*, 2024, *16*(15), 2400, doi: 10.3390/nu16152400
110. Chung, K.Y.; Kim, S.; Yoon, H.T.; Kwon, S.H.; Park, H.S.; Im, J.P.; Kim, J.S.; Kim, J.W.; Han, Y.M.; Koh, S.-J. Toll-like receptor 3 signaling attenuated colitis-associated cancer development in mice. *Sci. Rep.* 2024, *14*(1), 30308, doi: 10.1038/s41598-024-76954-1
111. Bleicher, R.J. Timing and delays in breast cancer evaluation and treatment. *Ann. Surg. Oncol.*, 2018, *25*(10), 2829–2838, https://doi.org/10.1245/s10434-018-6615-2
112. Saeidnia, S.; Manayi, A.; Abdollahi, M. From in vitro experiments to in vivo and clinical studies; pros and cons. *Curr. Drug Discov. Technol.*, 2015, *12*(4), 218–224. doi: 10.2174/157016381366616011409 3140
113. Kumar, A.; Green, K.M.; Rawat, M. A comprehensive overview of postbiotics with a special focus on discovery techniques and clinical applications. *Foods*, 2024, *13*(18), 2937, doi: 10.3390/foods13182937
114. Silverman, W.K.; Pettit, J.W.; Jaccard, J. Future directions in clinical trials and intention-to-treat analysis: fulfilling admirable intentions through the right questions. *J. Clin. Child Adol. Psychol.* 2024, *53*(5), 840–848, doi: 10.1080/15374416.2024.2384035
115. Redza-Dutordoir, M.; Averill-Bates, D.A. Activation of apoptosis signalling pathways by reactive oxygen species. *Biochimica et Biophysica Acta (BBA) - Molecular Cell Research*, 2016, *1863*(12), 2977–2992, doi: 10.1016/j.bbamcr.2016.09.012
116. Stenger, T.D.; Miller, J.S. Therapeutic approaches to enhance natural killer cell cytotoxicity. *Frontiers in Immunology* 2024, *15*, doi: 10.3389/fimmu.2024.1356666
117. Bhol, N.K.; Bhanjadeo, M.M.; Singh, A.K.; Dash, U.C.; Ojha, R.R.; Majhi, S.; Duttaroy, A.K.; Jena, A.B. The interplay between cytokines, inflammation, and antioxidants: mechanistic insights and therapeutic potentials of various antioxidants and anti-cytokine compounds. *Biomed. Pharmacother.* 2024, *178*, 117177, doi: 10.1016/j.biopha.2024.117177
118. Son, M.-Y.; Cho, H.-S. Anticancer effects of gut microbiota-derived short-chain fatty acids in cancers. *J. Microbiol. Biotechnol.* 2023, *33*(7), 849–856, doi: 10.4014/jmb.2301.01031
119. Hong, Y.; Boiti, A.; Vallone, D.; Foulkes, N.S. Reactive oxygen species signaling and oxidative stress: transcriptional regulation and evolution. *Antioxidants*, 2024, *13*(3), 312, doi: 10.3390/antiox13030312

120. Jara-Gutiérrez, C.; Mercado, L.; Paz-Araos, M.; Howard, C.; Parraga, M.; Escobar, C.; Mellado, M.; Madrid, A.; Montenegro, I.; Santana, P.; Murgas, P.; Jimenez-Jara, C.; González-Olivares, L.G.; Ahumada, M.; Villena, J. Oxidative stress promotes cytotoxicity in human cancer cell lines exposed to *Escallonia spp.* extracts. *BMC Complementary Medicine and Therapies* 2024, *24*(1), 38, doi: 10.1186/s12906-024-04341-4
121. Huang, L.; Wu, R.-L.; Xu, A.-M. Epithelial-mesenchymal transition in gastric cancer. *Am. J. Translat. Res.*, 2015, *7*(11), 2141–2158.
122. Zhao, H.; Ming, T.; Tang, S.; Ren, S.; Yang, H.; Liu, M.; Tao, Q.; Xu, H. Wnt signaling in colorectal cancer: Pathogenic role and therapeutic target. *Molecular Cancer*, 2022, *21*, 144, doi: 10.1186/s12943-022-01616-7
123. Melaiu, O.; Lucarini, V.; Cifaldi, L.; Fruci, D. Influence of the tumor microenvironment on NK cell function in solid tumors. *Front. Immunol.* 2020, *10*, doi: 10.3389/fimmu.2019.03038
124. Kamran, S.; Sinniah, A.; Alshawsh, M.A. Synergistic effects of 5-fluorouracil in combination with diosmetin in colorectal cancer cells. *Biol. Life Sci. Forum* 2021, 7(1), Article 1, doi: 10.3390/ECB2021-10276
125. Wang, T.; Huang, Y.; Jiang, P.; Yuan, X.; Long, Q.; Yan, X.; Huang, Y.; Wang, Z.; Li, C. Research progress on anti-inflammatory drugs for preventing colitis-associated colorectal cancer. *Int. Immunopharmacol.* 2025, *144*, 113583, doi: 10.1016/j.intimp.2024.113583
126. Zhang, J.; Xie, Q.; Huo, X.; Liu, Z.; Da, M.; Yuan, M.; Zhao, Y.; Shen, G. Impact of intestinal dysbiosis on breast cancer metastasis and progression. *Front. Oncol.* 2022, *12*, 1037831, doi: 10.3389/fonc.2022.1037831
127. Nanru, P. Immunomodulatory effects of immune cell-derived extracellular vesicles in melanoma. *Front. Immunol.* 2024, *15*, 1442573, doi: 10.3389/fimmu.2024.1442573
128. Mathur, H.; Linehan, K.; Flynn, J.; Byrne, N.; Dillon, P.; Conneely, M.; Grimaud, G.; Hill, C.; Stanton, C.; Ross, R.P. Emulsion-based postbiotic formulation is comparable to viable cells in eliciting a localized immune response in dairy cows with chronic mastitis. *Front. Microbiol.* 2022, *13*, 759649, doi: 10.3389/fmicb.2022.759649
129. Mahalmani, V.; Sinha, S.; Prakash, A.; Medhi, B. Translational research: bridging the gap between preclinical and clinical research. *Indian J. Pharmacol.* 2022, *54*(6), 393–396, doi: 10.4103/ijp.ijp_860_22
130. Barroca, N.C.B.; Della Santa, G.; Suchecki, D.; García-Cairasco, N.; Umeoka, E.H.De L. Challenges in the use of animal models and perspectives for a translational view of stress and psychopathologies. *Neurosci. Biobehav. Rev.* 2022, *140*, 104771 doi: 10.1016/j.neubiorev.2022.104771
131. Ratiner, K.; Ciocan, D.; Abdeen, S.K.; Elinav, E. Utilization of the microbiome in personalized medicine. *Nature Rev. Microbiol.* 2024, *22*(5), 291–308, doi: 10.1038/s41579-023-00998-9
132. Rafique, N.; Jan, S.Y.; Dar, A.H.; Dash, K.K.; Sarkar, A.; Shams, R.; Pandey, V.K.; Khan, S.A.; Amin, Q.A.; Hussain, S.Z. Promising bioactivities of postbiotics: a comprehensive review. *J. Agri. Food Res.* 2023, *14*, 100708, doi: 10.1016/j.jafr.2023.100708
133. Peluzio, M. Do C. G.; Martinez, J.A.; Milagro, F.I. Postbiotics: metabolites and mechanisms involved in microbiota-host interactions. *Trends Food Sci. Technol.* 2021, *108*, 11–26, doi: 10.1016/j.tifs.2020.12.004
134. Homayouni Rad, A.; Aghebati Maleki, L.; Samadi Kafil, H.; Fathi Zavoshti, H.; Abbasi, A. Postbiotics as promising tools for cancer adjuvant therapy. *Adv. Pharmaceut. Bull.* 2021, *11*(1), 1–5, doi: 10.34172/apb.2021.007
135. Owusu-Apenten, R.; Vieira, E. Food regulatory agencies. In R. Owusu-Apenten & E. R. Vieira (Eds.), *Elementary food science* (pp. 57–79), 2023, Springer International Publishing, doi: 10.1007/978-3-030-65433-7_3
136. Tubau-Juni, N.; Hontecillas, R.; Ehrich, M.; Leber, A.; Zoccoli-Rodriguez, V.; Bassaganya-Riera, J. Preclinical studies: efficacy and safety. In J. Bassaganya-Riera (Ed.), *Accelerated path to cures* (pp. 25–40). 2018, Springer International Publishing. doi: 10.1007/978-3-319-73238-1_3
137. Mosca, A.; Abreu, Y.; Abreu, A.T.; Gwee, K.A.; Ianiro, G.; Tack, J.; Nguyen, T.V.H.; Hill, C. The clinical evidence for postbiotics as microbial therapeutics. *Gut Microbes*, *14*(1), 2117508; doi: 10.1080/19490976.2022.2117508
138. WHO good manufacturing practices for biological products, Annex 2, TRS No 999. Retrieved January 11, 2025, from https://www.who.int/publications/m/item/annex-2-trs-no-999-WHO-gmp-for-biological-products
139. Liang, B.; Xing, D. The current and future perspectives of postbiotics. *Probiot. Antimicrob. Proteins* 2023, 1–18, https://doi.org/10.1007/s12602-023-10045-x.

4 Postbiotics
Beneficial Impact on Diabetes

Oyindamola John Samson
Olabisi Onabanjo University, Ago-Iwoye, Nigeria

Ayush Madan
People's University, Bhopal (MP), India

Ismail Babatunde Onajobi
Olabisi Onabanjo University, Ago-Iwoye, Nigeria

Mantosh Kumar Satapathy
Department of Physiology, Department of Pharmacology, College of Medicine, Taipei Medical University, Taipei, Taiwan.
Cathay General Hospital and Research Center, Taipei, Taiwan.
Graduate Institute of Biomedical Materials and Tissues Engineering, College of Biomedical Engineering, Taipei Medical University, Taipei, Taiwan.

4.1 INTRODUCTION

Diabetes mellitus, a chronic metabolic disorder characterized by persistent hyperglycemia, continues to pose a significant global health challenge. The rising prevalence of diabetes has necessitated the exploration of innovative and sustainable therapeutic strategies (Hossain et al., 2024). In recent years, the gut microbiome has emerged as a key player in metabolic health, with growing evidence suggesting that its modulation can significantly influence diabetes management. Among the various strategies to harness the gut microbiome, postbiotics—bioactive compounds produced by probiotics during fermentation—have garnered increasing attention for their potential health benefits, particularly in managing diabetes (Baars et al., 2024).

Postbiotics, which include metabolites such as short-chain fatty acids (SCFAs), vitamins, peptides, and exopolysaccharides, exert multiple physiological effects that contribute to their therapeutic potential (Bourebaba et al., 2022). Unlike live probiotics, postbiotics do not pose risks related to viability or colonization, making them a safer alternative for individuals with compromised health. This characteristic, combined with their stability and ease of incorporation into functional foods and supplements, positions postbiotics as an attractive option for addressing the multifaceted complications of diabetes (Ozma et al., 2024).

The beneficial impact of postbiotics on diabetes lies in their ability to influence key mechanisms underlying the condition. These include improving insulin sensitivity, reducing systemic inflammation, modulating gut microbiota composition, and enhancing intestinal barrier integrity (Peluzio et al., 2021). With these mechanisms, postbiotics can contribute to better glycemic control and reduce the risk of diabetes-related complications. Importantly, their natural origin and broad spectrum of bioactivities align with the growing preference for nonpharmacological and integrative approaches to diabetes management (Li et al., 2024).

DOI: 10.1201/9781003607779-4

As the burden of diabetes continues to grow, innovative solutions that leverage the gut microbiome's capabilities offer promising avenues for intervention (Yashas et al., 2025). Postbiotics represent a cutting-edge development in this field, with the potential to transform diabetes care and improve the quality of life for millions worldwide. This review describes the emerging role of postbiotics in diabetes management, highlighting their potential as a complementary strategy to conventional treatments. It emphasizes the mechanisms through which postbiotics exert their beneficial effects, supported by evidence from in vitro, animal, and human studies.

4.2 POSTBIOTICS

4.2.1 Definition and Classification

Postbiotics are bioactive compounds synthesized by probiotic microorganisms during fermentation processes, encompassing metabolites, cell components, or secretions that exert beneficial effects on host health (Isaac-Bamgboye et al., 2024). Unlike probiotics, which require viable microorganisms, postbiotics function independently of live microbes, offering increased safety, stability, and ease of application in health interventions. These compounds can be classified based on their origin (bacterial, fungal, or yeast-derived) or their functional properties, such as immunomodulatory, antimicrobial, or metabolic regulatory effects.

4.2.2 Types of Postbiotic Compounds

Postbiotics are a diverse group of bioactive compounds derived from probiotics during their metabolic activity. These compounds exert various health benefits by interacting with host cells and microbiota. Following are the key types of postbiotic compounds and their significance:

4.2.2.1 Short-Chain Fatty Acids (SCFAs)

SCFAs such as acetate, propionate, and butyrate are essential metabolites produced during the fermentation of dietary fibers by gut bacteria. SCFAs play a critical role in maintaining gut barrier integrity by promoting the production of tight junction proteins, which prevent the translocation of harmful pathogens and toxins. Additionally, SCFAs have potent anti-inflammatory effects, as they modulate the activity of immune cells and reduce pro-inflammatory cytokines, making them essential for managing conditions like inflammatory bowel disease (IBD) (Ashaolu et al., 2021).

Beyond gut health, SCFAs contribute significantly to metabolic regulation. Butyrate serves as a primary energy source for colonocytes, aiding in gut homeostasis, while propionate impacts hepatic gluconeogenesis and lipid metabolism. These functions make SCFAs beneficial in improving insulin sensitivity, reducing systemic inflammation, and aiding in diabetes and obesity management. Emerging evidence also highlights SCFAs' role in modulating appetite and energy balance via gut–brain signaling pathways (Salvi & Cowles, 2021).

4.2.2.2 Peptides and Proteins

Peptides and proteins derived from probiotics have shown immense potential due to their antimicrobial and immunomodulatory properties. Antimicrobial peptides, such as bacteriocins, inhibit pathogenic bacteria by disrupting their cellular membranes or interfering with metabolic pathways, offering a natural alternative to traditional antibiotics. This antimicrobial action is particularly valuable in addressing antibiotic-resistant pathogens (Jenab et al., 2020).

Immunomodulatory peptides regulate host immune responses by influencing cytokine production and the activity of immune cells like macrophages and T-cells. These peptides help balance the immune system, reducing chronic inflammation and preventing autoimmune disorders. Furthermore, certain peptides have been found to enhance glucose metabolism by improving insulin receptor sensitivity, making them promising candidates for managing metabolic disorders such as type 2 diabetes (Ashaolu et al., 2023).

4.2.2.3 Vitamins and Other Metabolites

Postbiotics also include essential vitamins like vitamin K_2 and B-group vitamins, synthesized by specific probiotic strains. Vitamin K_2 is vital for bone health and cardiovascular function, as it regulates calcium utilization and prevents arterial calcification. B-group vitamins, including B_{12} and folate, play critical roles in energy metabolism, red blood cell formation, and DNA synthesis, supporting overall health (Mayorgas et al., 2021).

Secondary metabolites such as exopolysaccharides, organic acids, and phenolic compounds further enhance the functional diversity of postbiotics. Exopolysaccharides improve gut barrier function by strengthening the mucus layer, while organic acids lower gut pH, creating an unfavorable environment for pathogens. Phenolic compounds exhibit potent antioxidant properties, neutralizing free radicals and reducing oxidative stress, thereby protecting against chronic diseases like cancer and cardiovascular disorders (Sadeghi et al., 2022).

4.2.3 Production and Characterization Methods

The production of postbiotics involves cultivating probiotics under specific conditions to optimize the yield of desired metabolites (Pihurov et al., 2021). Two primary methods are commonly employed:

1. *Fermentation*: This method uses substrates such as dairy products, plant extracts, or synthetic media to promote the metabolic activity of probiotics. Parameters like temperature, pH, and oxygen availability are carefully controlled to maximize the production of targeted compounds, such as SCFAs and peptides.
2. *Controlled bioreactor cultivation*: Bioreactors allow for precise control of environmental conditions, including nutrient supply, agitation, and microbial density. This approach ensures consistent and large-scale production of postbiotics, making it suitable for industrial applications.

Characterization techniques are crucial for identifying and quantifying postbiotic components. Gas chromatography is widely used for SCFA profiling, providing detailed insights into their composition and concentration. Spectrophotometry is employed for quantifying vitamins like vitamin K_2 and B-group vitamins, while mass spectrometry aids in the identification of peptides and proteins. These analytical methods help evaluate the bioactive properties of postbiotics, ensuring their efficacy and safety for therapeutic applications (Yang et al., 2022). The diverse types of postbiotic compounds, their health benefits, and the associated production and characterization methods, offering a comprehensive overview of their potential applications (Table 4.1). Postbiotics present a promising avenue for health improvement, particularly in addressing metabolic disorders like diabetes.

4.3 MECHANISMS OF ACTION IN DIABETES MANAGEMENT

Postbiotics play a pivotal role in diabetes management by targeting various physiological and molecular pathways. Their bioactive compounds, including SCFAs, peptides, and vitamins, contribute to improved glycemic control, enhanced insulin sensitivity, and reduced inflammation (Razi & Rashidinejad, 2021). Postbiotics, particularly SCFAs like butyrate and propionate, enhance insulin sensitivity by activating the G-protein-coupled receptors (GPR41 and GPR43) in adipose and liver tissues. This activation regulates glucose uptake and lipid metabolism, thereby reducing insulin resistance. SCFAs also stimulate the production of glucagon-like peptide-1 (GLP-1), a hormone that promotes insulin secretion and glucose homeostasis (Chen et al., 2024).

Postbiotics influence gut microbiota composition by promoting the growth of beneficial bacteria and suppressing pathogenic species. This modulation maintains a balanced gut ecosystem, reducing

TABLE 4.1
Types of Postbiotic Compounds and Their Functional Benefits

Type of Postbiotic Compound	Examples	Health Benefits	Production Method	Characterization Technique	References
Short-chain fatty acids (SCFAs)	Acetate, propionate, butyrate	Enhances gut barrier integrity, reduces inflammation, improves insulin sensitivity	Fermentation	Gas chromatography	Salvi & Cowles (2021)
Peptides and proteins	Bacteriocins, immunomodulatory peptides	Antimicrobial action, immune modulation, glucose metabolism regulation	Controlled bioreactor cultivation	Mass spectrometry	Ashaolu et al. (2023)
Vitamins and other metabolites	Vitamin K_2, B_{12}, exopolysaccharides	Supports bone and cardiovascular health, antioxidant activity, nutrient absorption	Fermentation	Spectrophotometry and mass spectrometry	Mayorgas et al. (2021)

dysbiosis—a condition linked to increased inflammation and insulin resistance in diabetes. By enhancing gut microbiota diversity, postbiotics indirectly support metabolic health and glucose regulation (Peluzio et al., 2021). Chronic low-grade inflammation is a hallmark of diabetes. Postbiotics, through their immunomodulatory properties, inhibit the release of pro-inflammatory cytokines such as TNF-α, IL-6, and IL-1β. They achieve this by interacting with immune cells, modulating nuclear factor-kappa B (NF-κB) signaling pathways, and reducing oxidative stress. This anti-inflammatory action alleviates the metabolic stress associated with diabetes (Cristofori et al., 2021).

Postbiotics strengthen the intestinal barrier by increasing the production of mucins and tight junction proteins. SCFAs, for example, nourish colonocytes and promote gut epithelial repair. This improvement in barrier integrity prevents the translocation of endotoxins such as lipopolysaccharides (LPSs) into the bloodstream, which are known to trigger inflammation and insulin resistance (Humam et al., 2021). Certain postbiotic compounds, such as peptides and organic acids, regulate lipid metabolism by inhibiting lipid peroxidation and modulating enzyme activity involved in fat storage. These actions help maintain a healthy lipid profile and reduce the risk of cardiovascular complications associated with diabetes. Additionally, postbiotics influence mitochondrial function, supporting energy metabolism and reducing cellular stress in key tissues (Glowacka et al., 2024). These interconnected pathways culminate postbiotics as a multifaceted approach to diabetes management (Table 4.2), addressing both its root causes and complications. Their natural origin and broad-spectrum efficacy make them a promising adjunct to traditional therapeutic strategies.

4.4 THERAPEUTIC APPLICATIONS OF POSTBIOTICS

Postbiotics hold significant promise in therapeutic applications due to their diverse bioactive properties. These compounds, derived from probiotic fermentation, offer a safe and effective approach to managing various health conditions. In diabetes management, postbiotics serve as adjuncts to conventional therapies by targeting key metabolic and inflammatory pathways, thereby improving glycemic control and reducing the risk of complications. Beyond diabetes, their therapeutic potential extends to other areas, including gastrointestinal disorders, immune modulation, and cardiovascular health (Fang et al., 2024).

TABLE 4.2
Postbiotic Mechanisms in the Management of Diabetes

Study	Mechanism	Postbiotic Components	Action in Diabetes Management
Kircher et al. (2022)	Improving insulin sensitivity	SCFAs (butyrate, propionate)	Activates GPRs, enhances GLP-1 production
Woelk and Snyder (2021)	Modulating gut microbiota	SCFAs, Exopolysaccharides	Promotes beneficial bacteria, reduces dysbiosis
Aron-Wisnewsky et al. (2020)	Reducing systemic inflammation	Peptides, SCFAs	Inhibits pro-inflammatory cytokines, modulates NF-κB
Zhang et al. (2023)	Enhancing intestinal barrier	SCFAs, Exopolysaccharides	Increases tight junction proteins, prevents LPS translocation
Yan and Horng (2020)	Regulating lipid metabolism	Organic acids, peptides	Reduces lipid peroxidation, improves mitochondrial function

One of the most prominent applications of postbiotics is in *metabolic disorders*, particularly type 2 diabetes. Postbiotics, such as SCFAs, improve insulin sensitivity, regulate glucose metabolism, and reduce systemic inflammation. These compounds also play a role in enhancing gut microbiota composition, which is critical in preventing and managing obesity-related insulin resistance. By supporting a balanced microbiome, postbiotics address one of the root causes of metabolic dysregulation (Bourebaba et al., 2022).

In *gastrointestinal health*, postbiotics strengthen intestinal barrier integrity and promote the growth of beneficial gut bacteria. This helps prevent leaky gut syndrome, reduce gut-related inflammation, and improve nutrient absorption. Postbiotics are particularly effective in conditions like irritable bowel syndrome (IBS), colitis, and other inflammatory bowel diseases (IBDs). Their antimicrobial properties also inhibit the growth of pathogenic microorganisms, reducing gastrointestinal infections (Liu et al., 2022).

Postbiotics also play a vital role in *immune system modulation*. When interacting with immune cells, these compounds can reduce pro-inflammatory cytokine production and enhance the body's ability to combat infections. They are being explored as potential interventions in autoimmune conditions and allergies, where immune responses need regulation. Their ability to modulate inflammation makes them suitable for managing chronic diseases with an inflammatory component, including cardiovascular conditions (Mehta et al., 2023).

Finally, in *cardiovascular health*, postbiotics contribute by regulating lipid metabolism, reducing oxidative stress, and lowering blood pressure. SCFAs and peptides derived from postbiotics improve lipid profiles by reducing LDL cholesterol and triglycerides while enhancing HDL cholesterol. This has a direct impact on reducing the risk of atherosclerosis and other cardiovascular complications, particularly in individuals with diabetes or metabolic syndrome (Flori et al., 2024).

A healthy intestine provides a balanced environment, supporting the gut microbiota, which initiates essential metabolic activities as shown in the sequential process of postbiotic production (Figure 4.1). Through bacterial fermentation, the microbiota breaks down nutrients, leading to the production of postbiotics. These bioactive compounds play a vital role in enhancing gut health, promoting a robust immune system, and improving overall well-being.

4.5 SUCCESSFUL CLINICAL EVIDENCE

Postbiotics have gained significant attention in clinical research due to their potential to address various health conditions effectively. Clinical studies provide evidence supporting their safety, efficacy, and therapeutic benefits in managing metabolic disorders, gastrointestinal health, and

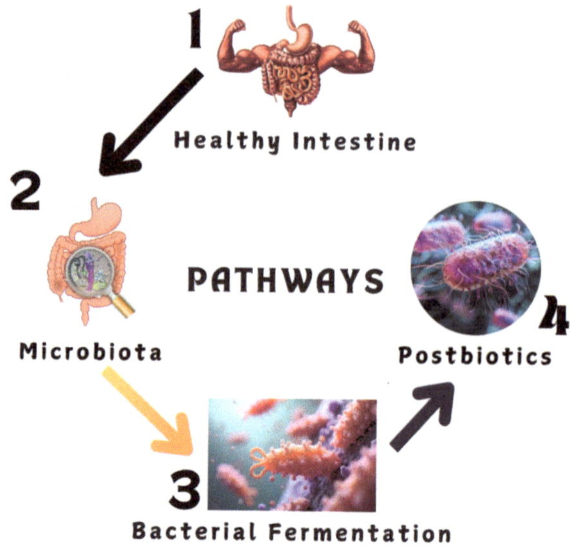

FIGURE 4.1 Pathways of postbiotic formation.

immune-related conditions. Several clinical trials have demonstrated the role of postbiotics in improving metabolic health, particularly in individuals with type 2 diabetes and obesity (Arellano-García et al., 2024). For instance, supplementation with butyrate and acetate—key SCFAs—has been shown to enhance insulin sensitivity, lower fasting blood glucose levels, and reduce inflammation. Studies have also reported reductions in HbA1c levels, a critical marker of long-term glycemic control, in patients treated with postbiotic-rich formulations (Bourebaba et al., 2022).

Postbiotics have proven effective in treating conditions such as IBS and IBDs. Clinical evidence suggests that postbiotic supplementation strengthens intestinal barrier integrity, reduces gut inflammation, and promotes the growth of beneficial bacteria. For example, a randomized controlled trial (RCT) involving patients with ulcerative colitis reported significant symptom relief and improved mucosal healing with postbiotic intervention (Martyniak et al., 2021).

Clinical studies also emphasize the immunomodulatory effects of postbiotics. In one trial, patients with allergic rhinitis showed reduced symptom severity and inflammatory markers after consuming postbiotic-enriched yogurt. Additionally, postbiotics have been investigated for their role in enhancing vaccine efficacy by promoting better immune responses, particularly in immunocompromised individuals (Yeşilyurt et al., 2021). Postbiotics are gaining recognition for their cardiovascular benefits. RCTs have highlighted improvements in lipid profiles, including reductions in LDL cholesterol and triglycerides, alongside increased HDL cholesterol levels. Some studies also report reduced systolic and diastolic blood pressure among participants consuming postbiotic formulations, suggesting their potential in managing hypertension and atherosclerosis (Park et al., 2023).

4.6 FUTURE OUTLOOK OF POSTBIOTICS

4.6.1 Emerging Research Directions

Emerging research in postbiotics is focused on elucidating their mechanisms of action at molecular and cellular levels. Advancements in omics technologies, such as genomics, transcriptomics, and metabolomics, are providing unprecedented insights into the bioactive compounds produced by probiotics. Researchers are exploring the synergistic effects of postbiotics with prebiotics and probiotics to maximize their therapeutic potential. Additionally, the role of postbiotics in modulating the gut–brain axis and their impact on neurological disorders, including depression and Alzheimer's disease, is gaining traction (Jastrząb et al., 2021).

Another promising area is the investigation of postbiotics' role in environmental sustainability. Studies are examining their application in reducing agricultural waste, improving soil health, and mitigating antibiotic resistance in livestock. These multifaceted benefits position postbiotics as a critical component of sustainable health and agricultural practices, warranting continued exploration (Checcucci et al., 2020).

4.6.2 Development of Novel Postbiotic Formulations

The development of novel postbiotic formulations is a priority to enhance their stability, bioavailability, and therapeutic efficacy. Encapsulation technologies, such as microencapsulation and nanotechnology-based delivery systems, are being employed to protect bioactive compounds from degradation during storage and gastrointestinal transit. These innovations ensure targeted release and prolonged activity of postbiotics, increasing their clinical applicability (Manocha et al., 2022).

Researchers are also investigating the use of plant-based substrates and alternative fermentation methods to produce postbiotics. These approaches aim to create formulations that are vegan-friendly, cost-effective, and sustainable. Additionally, combining postbiotics with complementary compounds, such as antioxidants or immune modulators, is being explored to develop multifunctional health products. Such advancements are paving the way for a new generation of postbiotic formulations tailored to diverse health needs (Abbaspour, 2024).

4.6.3 Potential for Personalized Medicine

The integration of postbiotics into personalized medicine represents a groundbreaking shift in healthcare. Advancements in microbiome profiling and precision diagnostics could lead to healthcare providers identifying individual-specific microbiota compositions and metabolic needs. This information allows for the design of customized postbiotic interventions to address specific health conditions, such as irritable bowel syndrome (IBS), obesity, or autoimmune disorders (Heniedy et al., 2024).

Furthermore, personalized postbiotic therapies hold promise for improving patient outcomes in complex diseases. For instance, tailored formulations targeting dysbiosis can restore microbial balance and enhance therapeutic efficacy in conditions like colorectal cancer. As precision medicine continues to evolve, postbiotics are poised to become a cornerstone in delivering targeted and effective treatments (Shukla et al., 2024).

4.6.4 Regulatory Considerations

The expanding market for postbiotics necessitates robust regulatory frameworks to ensure their safety, efficacy, and quality. Regulatory bodies, including the Food and Drug Administration (FDA) and European Food Safety Authority (EFSA), are working to establish standardized definitions and guidelines for postbiotic products. These frameworks are crucial for distinguishing postbiotics from probiotics and other microbiome-related interventions (Chugh et al., 2023).

Key regulatory challenges include validating the functional claims of postbiotics and ensuring transparency in labeling. Comprehensive safety assessments, including toxicity studies and clinical trials, are essential to gain regulatory approval and consumer trust. Collaborative efforts between researchers, industry stakeholders, and policymakers are critical for addressing these challenges and fostering the responsible development and commercialization of postbiotic products (Ong et al., 2023).

4.7 CLINICAL AND THERAPEUTIC APPLICATIONS

4.7.1 Guidelines for Clinical Use

Establishing clear guidelines for the clinical use of postbiotics is essential to ensure their safe and effective application. Clinicians must consider the specific type of postbiotic, its dosage, and the health condition being targeted. For example, SCFAs may be recommended for enhancing gut health and reducing inflammation, while bacteriocins could be employed as adjunct therapies for managing infections. Clinical guidelines should also emphasize the importance of sourcing high-quality postbiotic products that meet regulatory standards (Ailioaie & Litscher, 2021).

Timing and duration of postbiotic administration are equally critical. Guidelines suggest that postbiotics may be most effective when integrated into a comprehensive treatment plan, including diet modification and other therapeutic interventions. Additionally, clinicians should be aware of potential contraindications, such as in patients with severe immunodeficiency or allergies to specific compounds, to minimize risks and maximize benefits (Wegh et al., 2019).

4.7.2 Patient Selection Criteria

Identifying suitable candidates for postbiotic therapy requires careful evaluation of individual health profiles. Patients with conditions like IBS, IBD, or metabolic disorders such as diabetes may benefit significantly from postbiotic interventions. Selection criteria should include a thorough assessment of the patient's microbiome composition, as this can influence the efficacy of specific postbiotics (Maldonado et al., 2021).

Factors such as age, dietary habits, and underlying health conditions should also be considered during patient selection. For instance, elderly individuals with age-related dysbiosis may require tailored postbiotic formulations to restore microbial balance. Additionally, patients undergoing antibiotic therapy could benefit from postbiotics to mitigate the adverse effects on gut microbiota. Personalization of therapy based on these criteria ensures targeted and effective outcomes (Dong et al., 2024).

4.7.3 Monitoring and Assessment Protocols

Effective implementation of postbiotic therapies necessitates robust monitoring and assessment protocols. Clinicians should establish baseline health metrics, including gut microbiota composition, inflammatory markers, and metabolic parameters, before initiating treatment. Regular follow-ups should be conducted to track changes and evaluate the therapy's impact (Dong et al., 2024).

Monitoring tools such as stool analysis for microbiome profiling and blood tests for inflammatory and metabolic markers are essential. Patient-reported outcomes, including symptom improvement and quality of life, should also be incorporated into the assessment. These protocols enable clinicians to make data-driven adjustments to the therapy, ensuring sustained benefits and minimizing potential side effects (Franzosa et al., 2019).

4.7.4 Integration into Diabetes Care Plans

The integration of postbiotics into diabetes care plans offers a novel approach to improving glycemic control and overall metabolic health. Postbiotics like SCFAs and immunomodulatory peptides can enhance insulin sensitivity, reduce systemic inflammation, and promote gut health, all of which are crucial for managing diabetes. Incorporating postbiotics as part of a holistic care plan, alongside dietary modifications and pharmacological treatments, can optimize patient outcomes (Bourebaba et al., 2022).

Tailored postbiotic interventions can address specific challenges in diabetes management, such as reducing postprandial glucose spikes or mitigating the risk of cardiovascular complications.

Educational programs for patients and healthcare providers are essential to raise awareness about the benefits of postbiotics and their role in comprehensive diabetes care (Dong et al., 2024). By integrating postbiotics into standard care protocols, clinicians can offer patients a more personalized and effective treatment strategy.

4.8 CONCLUSION

Diabetes mellitus remains a significant global health concern, necessitating innovative approaches for effective management. Postbiotics, bioactive compounds derived from probiotic fermentation, have emerged as a promising therapeutic tool due to their ability to modulate gut microbiota and improve metabolic health. Their mechanisms of action—enhancing insulin sensitivity, reducing inflammation, and strengthening the gut barrier—offer substantial benefits for glycemic control and the prevention of diabetes complications. Unlike probiotics, postbiotics eliminate concerns regarding viability and colonization, making them safer and more stable for therapeutic use, particularly in vulnerable populations. Their ease of integration into functional foods and supplements further enhances their accessibility and appeal. The broad therapeutic applications of postbiotics make them a valuable tool in modern medicine, particularly for addressing chronic diseases. Their natural origin, safety, and multifunctional properties position them as a promising alternative or complement to conventional treatments. The growing body of clinical evidence highlights the therapeutic potential of postbiotics across multiple health domains. Their demonstrated safety and efficacy in clinical trials promotes their promise as a valuable addition to contemporary healthcare strategies. As research continues to uncover their full potential, postbiotics represent a sustainable, nonpharmacological intervention that aligns with modern integrative healthcare approaches. Postbiotics in diabetes care offer a transformative solution, providing an effective, natural strategy to improve outcomes and enhance the quality of life for diabetic individuals.

REFERENCES

Abbaspour, N. (2024). Fermentation's pivotal role in shaping the future of plant-based foods: An integrative review of fermentation processes and their impact on sensory and health benefits. *Applied Food Research*, 100468. https://doi.org/10.1016/j.afres.2024.100468

Ailioaie, L. M., & Litscher, G. (2021). Probiotics, photobiomodulation, and disease management: Controversies and challenges. *International Journal of Molecular Sciences*, *22*(9), 4942. https://doi.org/10.3390/ijms22094942

Arellano-García, L. I., Portillo, M. P., Martínez, J. A., Courtois, A., & Milton-Laskibar, I. (2024). Postbiotics for the management of obesity, insulin resistance/type 2 diabetes and NAFLD. Beyond microbial viability. *Critical Reviews in Food Science and Nutrition*, 1–24. https://doi.org/10.1080/10408398.2024.2437143

Aron-Wisnewsky, J., Warmbrunn, M. V., Nieuwdorp, M., & Clement, K. (2020). Nonalcoholic fatty liver disease: Modulating gut microbiota to improve severity?. *Gastroenterology*, *158*(7), 1881–1898. https://doi.org/10.1053/j.gastro.2020.01.049

Ashaolu, T. J., Ashaolu, J. O., & Adeyeye, S. A. O. (2021). Fermentation of prebiotics by human colonic microbiota in vitro and short-chain fatty acids production: A critical review. *Journal of Applied Microbiology*, *130*(3), 677–687. https://doi.org/10.1111/jam.14843

Ashaolu, T. J., Zarei, M., Agrawal, H., Kharazmi, M. S., & Jafari, S. M. (2023). A critical review on immunomodulatory peptides from plant sources; action mechanisms and recent advances. *Critical Reviews in Food Science and Nutrition*, *64*(20), 7220–7236. https://doi.org/10.1080/10408398.2023.2183380

Baars, D. P., Fondevila, M. F., Meijnikman, A. S., & Nieuwdorp, M. (2024). The central role of the gut microbiota in the pathophysiology and management of type 2 diabetes. *Cell Host & Microbe*, *32*(8), 1280–1300. https://doi.org/10.1016/j.chom.2024.07.017

Bourebaba, Y., Marycz, K., Mularczyk, M., & Bourebaba, L. (2022). Postbiotics as potential new therapeutic agents for metabolic disorders management. *Biomedicine & Pharmacotherapy*, *153*, 113138. https://doi.org/10.1016/j.biopha.2022.113138

Checcucci, A., Trevisi, P., Luise, D., Modesto, M., Blasioli, S., Braschi, I., & Mattarelli, P. (2020). Exploring the animal waste resistome: The spread of antimicrobial resistance genes through the use of livestock manure. *Frontiers in Microbiology*, *11*, 1416. https://doi.org/10.3389/fmicb.2020.01416

Chen, K., Wang, H., Yang, X., Tang, C., Hu, G., & Gao, Z. (2024). Targeting gut microbiota as a therapeutic target in T2DM: A review of multi-target interactions of probiotics, prebiotics, postbiotics, and synbiotics with the intestinal barrier. *Pharmacological Research*, 107483. https://doi.org/10.1016/j.phrs.2024.107483

Chugh, P., Misra, S., Dhar, M. S., & Raghuwanshi, S. (2023). Regulatory aspects relevant to probiotic products. In *Probiotics, prebiotics, synbiotics, and postbiotics: Human microbiome and human health* (pp. 513–534). Singapore: Springer Nature Singapore. https://doi.org/10.1007/978-981-99-1463-0_25

Cristofori, F., Dargenio, V. N., Dargenio, C., Miniello, V. L., Barone, M., & Francavilla, R. (2021). Anti-inflammatory and immunomodulatory effects of probiotics in gut inflammation: A door to the body. *Frontiers in Immunology*, *12*, 578386. https://doi.org/10.3389/fimmu.2021.578386

Dong, H. J., Brain, K., Olsson, M., Dragioti, E., Gerdle, B., & Ghafouri, B. (2024). Eating habits and the desire to eat healthier among patients with chronic pain: A registry-based study. *Scientific Reports*, *14*(1), 4705. https://doi.org/10.1038/s41598-024-55449-z

Fang, H., Rodrigues, E.; Lacerda, R., Barra, N. G., Kukje Zada, D., Robin, N., Mehra, A., & Schertzer, J. D. (2024). Postbiotic impact on host metabolism and immunity provides therapeutic potential in metabolic disease. *Endocrine Reviews*, bnae025. https://doi.org/10.1210/endrev/bnae025

Flori, L., Benedetti, G., Martelli, A., & Calderone, V. (2024). Microbiota alterations associated with vascular diseases: postbiotics as a next-generation magic bullet for gut–vascular axis. *Pharmacological Research*, 107334. https://doi.org/10.1016/j.phrs.2024.107334

Franzosa, E. A., Sirota-Madi, A., Avila-Pacheco, J. et al. (2019). Gut microbiome structure and metabolic activity in inflammatory bowel disease. *Nature Microbiology*, *4*(2), 293–305. https://doi.org/10.1038/s41564-018-0306-4

Głowacka, P., Oszajca, K., Pudlarz, A., Szemraj, J., & Witusik-Perkowska, M. (2024). Postbiotics as molecules targeting cellular events of aging brain—the role in pathogenesis, prophylaxis and treatment of neurodegenerative diseases. *Nutrients*, *16*(14), 2244. https://doi.org/10.3390/nu16142244

Heniedy, A. M., Mahdy, D. M., Elenien, W. I. A., Mourad, S., & El-Kadi, R. A. (2024). Postbiotics as a health-promoting technique: A review article on scientific and commercial interest. *Process Biochemistry*. https://doi.org/10.1016/j.procbio.2024.05.010

Hossain, M. J., Al-Mamun, M., & Islam, M. R. (2024). Diabetes mellitus, the fastest growing global public health concern: Early detection should be focused. *Health Science Reports*, *7*(3), e2004. https://doi.org/10.1002/hsr2.2004

Humam, A. M., Loh, T. C., Foo, H. L., Izuddin, W. I., Zulkifli, I., Samsudin, A. A., & Mustapha, N. M. (2021). Supplementation of postbiotic RI11 improves antioxidant enzyme activity, upregulated gut barrier genes, and reduced cytokine, acute phase protein, and heat shock protein 70 gene expression levels in heat-stressed broilers. *Poultry Science*, *100*(3), 100908. https://doi.org/10.1016/j.psj.2020.12.011

Isaac-Bamgboye, F. J., Mgbechidinma, C. L., Onyeaka, H., Isaac-Bamgboye, I. T., & Chukwugozie, D. C. (2024). Exploring the potential of postbiotics for food safety and human health improvement. *Journal of Nutrition and Metabolism*, *2024*(1), 1868161. https://doi.org/10.1155/2024/1868161

Jastrząb, R., Graczyk, D., & Siedlecki, P. (2021). Molecular and cellular mechanisms influenced by postbiotics. *International Journal of Molecular Sciences*, *22*(24), 13475. https://doi.org/10.3390/ijms222413475

Jenab, A., Roghanian, R., & Emtiazi, G. (2020). Bacterial natural compounds with anti-inflammatory and immunomodulatory properties (mini review). *Drug Design, Development and Therapy*, *14*, 3787–3801. https://doi.org/10.2147/DDDT.S261283

Kircher, B., Woltemate, S., Gutzki, F., Schlüter, D., Geffers, R., Bähre, H., & Vital, M. (2022). Predicting butyrate-and propionate-forming bacteria of gut microbiota from sequencing data. *Gut Microbes*, *14*(1), 2149019.

Li, X., Xia, Y., Song, X., Xiong, Z., Ai, L., & Wang, G. (2024). Probiotics intervention for type 2 diabetes mellitus therapy: A review from proposed mechanisms to future prospects. *Critical Reviews in Food Science and Nutrition*, 1–19. https://doi.org/10.1080/10408398.2024.2387765

Liu, Y., Wang, J., & Wu, C. (2022). Modulation of gut microbiota and immune system by probiotics, prebiotics, and post-biotics. *Frontiers in Nutrition*, *8*, 634897. https://doi.org/10.3389/fnut.2021.634897

Maldonado, R. R., Caram, A. L. A., de Oliveira, D. S., Kamimura, E. S., Mazalli, M. R., & Aguiar-Oliveira, E. (2021). The role of probiotics and prebiotics in the composition of the gut microbiota and their influence on inflammatory bowel disease, obesity, and diabetes. In *Microbiome–host interactions* (pp. 113–128). CRC Press.

Manocha, S., Dhiman, S., Grewal, A. S., & Guarve, K. (2022). Nanotechnology: An approach to overcome bioavailability challenges of nutraceuticals. *Journal of Drug Delivery Science and Technology*, *72*, 103418. https://doi.org/10.1016/j.jddst.2022.103418

Martyniak, A., Medyńska-Przęczek, A., Wędrychowicz, A., Skoczeń, S., & Tomasik, P. J. (2021). Prebiotics, probiotics, synbiotics, paraprobiotics and postbiotic compounds in IBD. *Biomolecules*, *11*(12), 1903. https://doi.org/10.3390/biom11121903

Mayorgas, A., Dotti, I., & Salas, A. (2021). Microbial metabolites, postbiotics, and intestinal epithelial function. *Molecular Nutrition & Food Research*, *65*(5), 2000188. https://doi.org/10.1002/mnfr.202000188

Mehta, J. P., Ayakar, S., & Singhal, R. S. (2023). The potential of paraprobiotics and postbiotics to modulate the immune system: A review. *Microbiological Research*, 127449. https://doi.org/10.1016/j.micres.2023.127449

Ong, K. J., Tejeda-Saldana, Y., Duffy, B., Holmes, D., Kukk, K., & Shatkin, J. A. (2023). Cultured meat safety research priorities: Regulatory and governmental perspectives. *Foods*, *12*(14), 2645. https://doi.org/10.3390/foods12142645

Ozma, M. A., Fadaee, M., Hosseini, H. M., Ataee, M. H., & Mirhosseini, S. A. (2024). A critical review of postbiotics as promising novel therapeutic agents for clostridial infections. *Probiotics and Antimicrobial Proteins*, 1–12. https://doi.org/10.1007/s12602-024-10406-0

Park, S. J., Sharma, A., & Lee, H. J. (2023). Postbiotics against obesity: perception and overview based on pre-clinical and clinical studies. *International Journal of Molecular Sciences*, *24*(7), 6414. https://doi.org/10.3390/ijms24076414

Peluzio, M. D. C. G., Martinez, J. A., & Milagro, F. I. (2021). Postbiotics: Metabolites and mechanisms involved in microbiota-host interactions. *Trends in Food Science & Technology*, *108*, 11–26. https://doi.org/10.1016/j.tifs.2020.12.004

Pihurov, M., Păcularu-Burada, B., Cotârleţ, M., Vasile, M. A., & Bahrim, G. E. (2021). Novel insights for metabiotics production by using artisanal probiotic cultures. *Microorganisms*, *9*(11), 2184. https://doi.org/10.3390/microorganisms9112184

Razi, S. M., & Rashidinejad, A. (2021). Bioactive compounds: chemistry, structure, and functionality. In *Spray drying encapsulation of bioactive materials* (pp. 1–46). CRC Press.

Sadeghi, A., Ebrahimi, M., Shahryari, S., Kharazmi, M. S., & Jafari, S. M. (2022). Food applications of probiotic yeasts; focusing on their techno-functional, postbiotic and protective capabilities. *Trends in Food Science & Technology*, *128*, 278–295. https://doi.org/10.1016/j.tifs.2022.08.018

Salvi, P. S., & Cowles, R. A. (2021). Butyrate and the intestinal epithelium: Modulation of proliferation and inflammation in homeostasis and disease. *Cells*, *10*(7), 1775. https://doi.org/10.3390/cells10071775

Shukla, V., Singh, S., Verma, S., Verma, S., Rizvi, A. A., & Abbas, M. (2024). Targeting the microbiome to improve human health with the approach of personalized medicine: Latest aspects and current updates. *Clinical Nutrition ESPEN*. https://doi.org/10.1016/j.clnesp.2024.08.005

Wegh, C. A., Geerlings, S. Y., Knol, J., Roeselers, G., & Belzer, C. (2019). Postbiotics and their potential applications in early life nutrition and beyond. *International Journal of Molecular Sciences*, *20*(19), 4673. https://doi.org/10.3390/ijms20194673

Woelk, C. H., & Snyder, A. (2021). Modulating gut microbiota to treat cancer. *Science*, *371*(6529), 573–574. https://doi.org/10.1126/science.abg2904

Yan, J., & Horng, T. (2020). Lipid metabolism in regulation of macrophage functions. *Trends in Cell Biology*, *30*(12), 979–989.

Yang, Y., Hu, N., Deng, J., & Yang, J. (2022). Electrochemical sensing for vitamins. *Chemosensors*, *10*(11), 494.

Yashas, N. S., Bhanuprakash, A., Devendra, D. N., & Narayanan, A. S. (2025). Gut microbiome engineering and diabetics. In *Human and animal microbiome engineering* (pp. 51–75). Academic Press. https://doi.org/10.1016/B978-0-443-22348-8.00003-9

Yeşilyurt, N., Yılmaz, B., Ağagündüz, D., & Capasso, R. (2021). Involvement of probiotics and postbiotics in the immune system modulation. *Biologics*, *1*(2), 89–110. https://doi.org/10.3390/biologics1020006

Zhang, Y., Zhu, X., Yu, X., Novák, P., Gui, Q., & Yin, K. (2023). Enhancing intestinal barrier efficiency: A novel metabolic diseases therapy. *Frontiers in Nutrition*, *10*, 1120168. https://doi.org/10.3389/fnut.2023.1120168

5 Postbiotics
In Gestational Diabetes Mellitus

Mrugendra Potdar
Shri Vile Parle Kelavani Mandal's Institute of Pharmacy, Dhule, Maharashtra, India

Monika Kaushik
Amity Institute of Pharmacy, Amity University Madhya Pradesh, Gwalior, Madhya Pradesh, India

Murugesan Vanangamudi
Institute of Pharmaceutical Research, GLA University, Mathura, Uttar Pradesh, India

Md Abubakar
National Institute of Pharmaceutical Education and Research, Hajipur, India

Mohammad Muztaba
Praduman Singh SPS Pharmacy College, Basti, Uttar Pradesh, India

Prathap Madeswaraguptha
Vignan's Foundation for Science, Technology and Research (Deemed to be University), Guntur, India

Vijayaraj Surendran
Dr. Kalam College of Pharmacy, Thanjavur, Tamil Nadu, India

5.1 INTRODUCTION

5.1.1 Overview of Gestational Diabetes Mellitus (GDM)

Diabetic hyperglycemia, resulting from insulin secretion or resistance, poses significant health risks, affecting vital organs such as the eyes, kidneys, brain, heart, and blood vessels (1). In China, the prevalence of diabetes and prediabetes is rising, with 9.7% of the population diagnosed as of 2007–2008 (2). Gestational diabetes mellitus (GDM) is characterized by hyperglycemia during pregnancy, with the American Diabetes Association noting that new diabetes cases can emerge in the second or third trimester (3). Concerns regarding pregnancy-related hyperglycemia are highlighted by statistics

DOI: 10.1201/9781003607779-5

showing that 16.2% of newborns in 2015 experienced perinatal hyperglycemia, with Southeast Asia reporting even higher rates (1). In Tianjin, 9.3% of women were diagnosed with GDM between 2010 and 2012, with various risk factors identified, including age, prepregnancy BMI, and family history of diabetes (4). Research indicates that 20–50% of women with GDM may develop type 2 diabetes mellitus (T2DM) within five years post-pregnancy.

Postbiotics, derived from probiotics, are emerging as beneficial compounds for health, including those from *Lactobacilli* and *Bifidobacterium* (5). The International Scientific Association for Probiotics and Prebiotics (ISAPP) defines postbiotics as inanimate microbes and their constituents that confer health benefits (6). According to Gomez-Arango et al. (7), gut microbiota changes greatly influenced obese women's first-trimester metabolic hormone changes. Compared to overweight women, pregnant women of normal weight had higher pathobionts (*E. coli*, *Staphylococcus*, and *Enterobacteriaceae*) and less *Bifidobacterium* and *Bacteroides* (8). Women with GDM showed different gut microbiota than adults with T2DM (9, 10). Changes in gut microbiota during pregnancy have been linked to metabolic hormone alterations, with differences noted between women with GDM and those with type 2 diabetes (11). This review explores the interactions between host and gut microbiota, the implications of gut dysbiosis, and the potential role of probiotics in supporting women post-GDM, emphasizing the need for further research in this area (8, 9).

5.1.2 Pathophysiology of GDM

GDM is a form of latent glucose intolerance that typically develops in the third trimester of pregnancy, affecting both the mother and fetus (10, 11). The exact origins of GDM remain unclear, but it is primarily linked to chronic insulin resistance and pancreatic β-cell dysfunction. Early detection through biomarkers is an area of ongoing research. The cellular microenvironment, feto-placental factors, and maternal genetic predisposition can obscure the diagnosis of GDM, making an understanding of cellular pathophysiology and risk factors essential for effective screening and treatment (12–14). Studies on animal models, such as Zucker obese rats and Sprague Dawley rats, highlight the importance of β-cell mass and proliferation in maintaining glucose homeostasis (15–17). Factors like epigenetic downregulation of key transcription factors and the effects of prolactin on β-cell growth have been observed. GDM can lead to hyperinsulinemic infants, as maternal blood glucose levels directly influence fetal growth and fat storage (18).

5.1.3 Importance of Gut Microbiota in GDM

Pregnancy-related glucose levels indicate the metabolic health of both the mother and her unborn child, and further study is required to reverse GDM's effects on both. Colonizing the host gut is "the second genome" gut microbiota (19, 20). The human gut microbiome contains 500–1000 microorganisms. Gut microbiota has 2,000,000 coding genes, 100 times more than the genome (21). The gut microbiome greatly affects host health (22). According to growing evidence, gut bacteria may affect several metabolic problems. Multiple studies have shown that T2DM patients have dysregulated gut microbiota with more harmful bacteria such *Clostridium hathewayi* and *Escherichia coli*. Metabolic illness T2DM involves peripheral insulin resistance and pancreatic β-cell dysfunction. Interestingly, the gut microbiome changed as diabetes advanced. The intervention group had less *Clostridium* species than the control group during prediabetes. This suggests complex gut–host interactions may govern glucose metabolism (23). 17 of 23 meta-analyses found that pregnant women with GDM had substantially decreased gut microbiota alpha and beta diversity (24). We don't know why the gut microbiota is linked to glucose metabolism disorders during pregnancy, but it's clear that it prevents GDM and improves pregnancy outcomes related to maternal hyperglycemia. Intervention trials showed that probiotic supplementation improved glucose metabolism in the GDM group and reduced GDM risk in women (25). This research examined how GDM affects children's metabolism via the gut microbiota and early-life interventions to improve it.

5.1.4 Definition and Concept of Postbiotics

Postbiotics, derived from the Greek terms "post" (after) and "bios" (life), refer to substances produced after microorganisms are dead or inactivated. They encompass a range of components, including microbe-produced proteins, peptides, and metabolites, which may enhance health benefits but are not essential for classification (26). A key aspect of postbiotics is that they must originate from known probiotic organisms, which have demonstrated health benefits. This requirement, however, poses challenges, as the health benefits of live probiotics do not necessarily translate to inactivated forms, potentially stifling innovation in the field (27).

The literature predominantly defines postbiotics in terms of microbially produced metabolites, with inactivated microbes also included. However, the definitions often lack clarity regarding health benefits, leading to questions about the efficacy of postbiotics without demonstrated advantages. Furthermore, some definitions fail to distinguish between postbiotics administered to a target host and those generated in situ by resident microorganisms, complicating their application in food and feed products. Additionally, restrictions on the delivery of postbiotics to specific body sites, such as the gut lumen, limit their potential uses (28–30).

5.2 CLASSIFICATION OF POSTBIOTICS RELEVANT TO GDM

During fermentation, probiotics produce postbiotics. GDM therapy with postbiotics has garnered attention recently (31). Researchers have found postbiotic substances generated by internal and exterior probiotic microorganisms. Figure 5.1 describes the impact of postbiotics in GDM management.

5.2.1 Short-Chain Fatty Acids

Gut microbiota fermentation of dietary fiber produces short-chain fatty acids (SCFAs) like acetate, propionate, and butyrate, which may serve as postbiotics for GDM (32). SCFAs enhance insulin sensitivity, promote glucose uptake, and reduce systemic inflammation by inhibiting NF-κB signaling. They also stimulate gut hormones GLP-1 and PYY, improving insulin production and

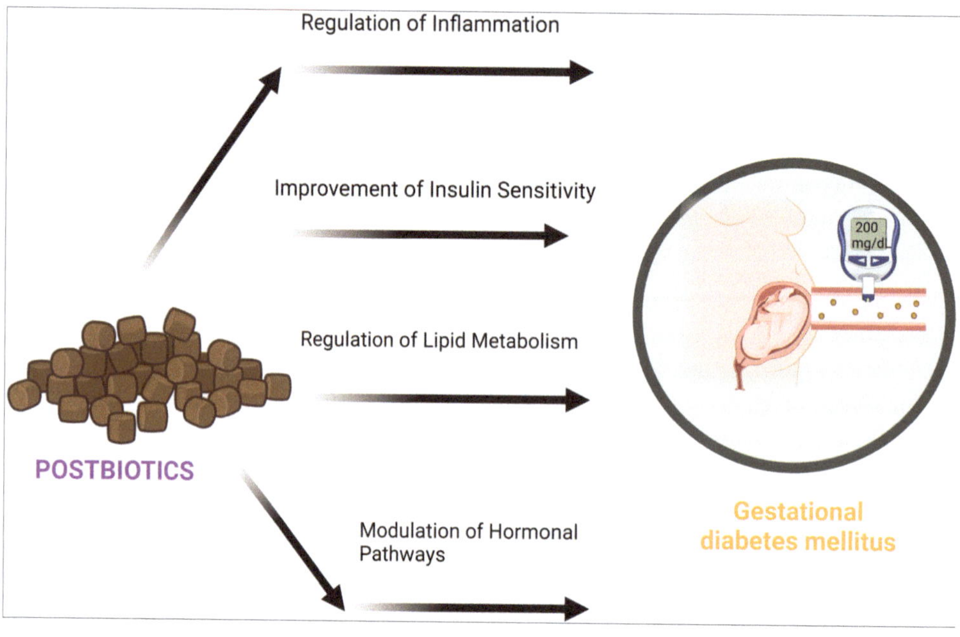

FIGURE 5.1 Postbiotics in GDM Management.

lowering blood glucose. Increasing SCFA synthesis through supplementation or high-fiber diets could effectively manage GDM and benefit maternal and fetal health (33).

5.2.2 Exopolysaccharides (EPS)

Probiotics that produce exopolysaccharides have considerable potential as postbiotics for GDM. Their gut microbiome changes boost beneficial bacteria, glucose metabolism, and inflammation reduction. EPS's antioxidant properties minimize oxidative stress, which is connected to GDM, and activate the AMP-activated protein kinase (AMPK) pathway to improve insulin sensitivity. SCFAs like butyrate enhance gut health and immune response modulation (34). EPS may improve maternal and fetal outcomes and reduce preterm delivery and macrosomia by addressing oxidative stress, inflammation, and glucose dysregulation. Preclinical and a few human investigations have showed promising outcomes, but further clinical trials are needed to evaluate EPS's safety, efficacy, and optimal use in GDM therapy (35).

5.2.3 Bacteriocins and Bacteria-Derived Enzymes

Bacteriocins are antibacterial peptides produced by lactic acid bacteria. Bacteriocins reduce harmful bacteria and increase beneficial microorganisms, improving gut health and systemic inflammation (36). Systemic inflammation, which causes insulin resistance during pregnancy, is reduced. It regulates SCFA synthesis, enhancing glucose metabolism and insulin sensitivity. Bacterial amylase inhibitors limit carbohydrate glucose release, preventing postprandial glucose increases. Acetate and butyrate-producing enzymes regulate metabolic pathways and inflammation to increase insulin sensitivity. Bacterial enzymes such superoxide dismutase reduce oxidative stress, which causes GDM problems (37).

5.2.4 Vitamins

Vitamin K, produced by gut bacteria, is vital for blood clotting, wound healing, and bone health. B vitamins, synthesized by bacteria like *Lactobacillus*, support energy production, DNA synthesis, and neurological function, with folate being crucial for cell division, particularly during pregnancy (38). Gut microbes also activate vitamin D (39). Postbiotic vitamins enhance nutrient absorption, bolster immune defense, and promote gut health, potentially preventing chronic conditions such as obesity and diabetes. Gut dysbiosis can lead to metabolic syndromes and other diseases. Dietary changes, probiotics, and postbiotics can improve glucose metabolism and reduce gestational diabetes risk, with multi-omics techniques providing insights into microbiota–host interactions (39).

5.2.5 Antimicrobial Peptides (AMPs)

Vitamin K, produced by gut bacteria, is essential for blood clotting, wound healing, and bone health. B vitamins, including B_2, B_6, B_{12}, and folate, support energy production and various bodily functions. Folate is particularly important for cell division during pregnancy (39, 40). Gut microbes also activate vitamin D. Postbiotic vitamins enhance nutrient absorption, bolster immune defenses, and reduce inflammation, helping prevent chronic diseases like diabetes and obesity. Gut dysbiosis can lead to metabolic syndromes, but restoring gut health can improve glucose metabolism and reduce gestational diabetes risk (41).

5.2.6 Cell Wall Components

Beneficial bacteria like *Lactobacillus* and *Bifidobacterium* produce cell wall components (peptidoglycans, teichoic acid, etc.) that act as postbiotics and have a number of health-promoting qualities (42). These constituents, which interact with the host immune system and other biological processes, include polysaccharides, lipoteichoic acids, and peptidoglycans. Cell wall components are not impacted by intestinal habitats or storage conditions, in contrast to live probiotics. Derivatives of nonviable bacteria lower the risk of infection in people with weakened immune systems. To produce certain biological effects, some components interact with host receptors (43).

5.2.7 Postbiotic Signaling Molecules

Bacterial signaling molecules known as quorum-sensing (QS) molecules, in particular Autoinducer-2 (AI-2), control population-wide behaviors in response to cell density. These compounds are becoming more well-known as postbiotics because of their capacity to affect host physiology and microbial ecosystems (44). AI-2 is special because it bridges interspecies communication by acting as a universal communication signal for several bacterial species (45). The types of postbiotics and their application in treating GDM are given in Table 5.1.

TABLE 5.1
Types of Postbiotic and Their Application in Treating GDM

S. No	Type of Postbiotics	Example	Mechanism	Applications in GDM
1	SCFAs	Acetate, propionate, butyrate	Enhance insulin sensitivity by activating G-protein-coupled receptors (GPCRs); improve gut barrier integrity.	• Improve glucose uptake • Reduce insulin resistance • Lower systemic inflammation contributing to GDM
2	Exopolysaccharides (EPS)	From *Lactobacillus* and *Bifidobacterium*	Modulate immune responses; reduce oxidative stress; improve gut microbiota balance	• Reduce oxidative stress • Alleviate inflammation • Support fetal development and maternal health
3	Organic acids	Lactate, pyruvate	Regulate glycolysis and energy metabolism; strengthen gut epithelial barrier	• Enhance glycolysis and glucose homeostasis (46)
4	Bacteriocins	Nisin, reuterin	Suppress pathogenic bacteria; promote probiotic colonization	Enhance gut health; indirectly improve glucose metabolism
5	Bacterial-derived Enzymes	Bile salt hydrolase, lactase	Alter bile acid pools; improve insulin secretion and lipid metabolism (18)	• Optimize lipid metabolism • Improve glycemic control
6	Phenolic compounds	Derived from fermented foods	Exhibit antioxidant properties; modulate insulin signaling pathways	• Mitigate oxidative damage • Enhance insulin signaling pathways (47)
7	Vitamins and amino acids	Folate, riboflavin, tryptophan metabolites	Act as cofactors in metabolic processes; reduce inflammation and oxidative stress	• Improve maternal and fetal glucose metabolism • Enhance maternal energy production (48)
8	Antimicrobial Peptides	Bacteriocins, reuterin	Disrupting the cell membrane, inhibit cell wall synthesis	• Improve gut health • Reduce intestinal inflammation
9	Cell wall components	Peptidoglycans, lipoteichoic acids	Interact with toll-like receptors (TLRs) to mediate anti-inflammatory effects	• Lower insulin resistance by reducing inflammation • Strengthen maternal immune function
10	Postbiotic signaling molecules	Quorum-sensing molecules like AI-2	• Regulate microbial community behavior • Modulate host metabolism	• Support gut microbiota diversity • Improve glucose metabolism through microbiome–host communication

5.3 POSTBIOTICS AND DIETARY INTERVENTIONS IN GDM MANAGEMENT

5.3.1 Dietary Sources of Postbiotics

5.3.1.1 Organic Acids

Postbiotics play a crucial role in preventing food spoilage by blocking the growth of bacteria. Organic acids, such as acetic and citric acids, have antimicrobial properties and can reduce the virulence of bacteria by making the environment acidic (49). These acids balance bacterial cell membranes by reducing cell pH and maintaining membrane integrity. They can also prevent or control the acidification of the cytoplasm and energy production (50). Studies have shown that certain strains of *Lactobacillus plantarum* can kill harmful bacteria using organic acids, such as lactic and acetic acids. This research suggests that new antibacterial agents for the food industry could be created by mixing organic acids (51).

5.3.1.2 Bacteriocin

Bacteriocins are antimicrobial peptides produced by *Archaebacteria* and *Eubacteria*, utilized in fermented foods for their antibacterial properties (49, 52). They combat gastrointestinal infections, withstand heat and pH changes, and inhibit food-spoiling bacteria by acidifying membranes and creating pores. Additionally, they damage pathogenic bacteria's membranes and proteins, making them effective in preventing infections and spoilage (53, 54).

5.3.1.3 Essential Fatty Acids

Antibacterial fatty acids, produced by attaching a hydrophilic carboxylic group to a carbon chain, can kill bacteria and inhibit their growth (55). A 2020 study found that these fatty acids can lyse *Klebsiella oxytoca*'s cell wall, indicating potential as alternatives to traditional antibiotics (56).

5.3.1.4 Peptides

Peptides produced by microbes inhibit bacterial growth by disrupting macromolecule synthesis and degrading membranes. They are categorized into ribosomal and nonribosomal antimicrobial peptides (57). Ribosomal peptides damage microbial membranes, while nonribosomal peptides are synthesized by nonribosomal peptide synthetases. These peptides can create holes in membranes, acidify them, and damage internal components. Examples include *E. coli* Nissle 1917 peptides used against *Salmonella* in turkeys and *Bacillus subtilis* peptides tested against *E. coli*, highlighting their food preservation potential (58).

5.3.1.5 The Hydroxyl Radicals

Hydrogen peroxides (H_2O_2), produced by catalase-negative bacteria during aerobic growth, generates hydroxyl radicals that exhibit antibacterial properties (58). Factors such as bacterial strains, temperature, and pH influence H_2O_2 concentration and its effectiveness against bacteria. Research on probiotics like *Bifidobacterium longum* and *Lactobacillus* species shows that H_2O_2 can inhibit *Staphylococcus aureus* growth (59). This suggests that postbiotic compounds like hydrogen peroxide could serve as alternatives to antibiotics for combating infections and food spoilage.

5.3.1.6 The Vitamins

Probiotic bacteria, particularly in dairy products, significantly contribute to vitamin production in the gut, which helps combat harmful microorganisms. Research shows that while probiotics like *Lactobacillus plantarum* are beneficial, vitamin C exhibits a stronger antimicrobial effect by disrupting bacterial membranes. Postbiotics chemicals antimicrobial activities protect and prolong food shelf life (60).

5.3.2 Probiotic Versus Postbiotic Approaches

The human gut microbiota is a diverse ecosystem essential for regulating intestinal inflammation, metabolic balance, and immune development (61). Dysbiosis, or imbalances in this microbiota, can contribute to various health issues, including obesity, type 2 diabetes, inflammatory bowel disease, neurological disorders, cardiovascular diseases, and certain cancers (62). Probiotics, which are beneficial microorganisms, can modify gut microbiota, enhance the intestinal barrier, and support immune function (63). They also have a role in the gut–brain axis, influencing the nervous system (64). Postbiotics, comprising inanimate bacteria or their components, offer health benefits such as anti-inflammatory and antioxidant effects. They include proteins, short-chain fatty acids, and other bioactive metabolites (65, 66). Postbiotics are considered safer than probiotics for individuals with compromised immune systems or serious health conditions. Both probiotics and postbiotics contribute positively to gut health and overall well-being, highlighting their importance in maintaining balanced microbiota (67).

5.3.3 Postbiotic Supplementation in Pregnancy

Postbiotic supplementation during pregnancy may provide benefits similar to probiotics, such as regulating gut microbiota and potentially lowering the risk of complications like pre-eclampsia and GDM (68). However, it is crucial to consult a healthcare provider before taking any supplements during pregnancy. Pregnancy is a critical period for future health, as even minor reductions in maternal glucose levels can decrease the risk of metabolic syndrome (MS) later in life (69). Effective postpartum monitoring of those with GDM is vital, as it can lead to type 2 diabetes and MS (70). Understanding the interactions between gut bacteria and the host may improve health outcomes for women with GDM (71). Probiotic supplements may help individuals with metabolic issues maintain diverse gut flora and stable metabolism. Postbiotics, metabolites produced by microbes, have shown potential in preventing hypertension in children that is linked to maternal diets (72). However, the impact of additional postbiotics on MS remains unclear due to the variety of bioactive compounds produced (73). There is insufficient evidence to establish a consensus on the effectiveness of symbiotic or probiotic supplements for GDM, highlighting the need for long-term, high-quality trials to determine optimal dosages and safety (74). Further research is necessary to understand the mechanisms and safety of probiotics for mothers and children.

5.3.4 Effect on Maternal Weight Management

Postbiotics, derived from probiotics, can help manage maternal obesity during pregnancy and postnatally. They impact inflammatory processes, gastrointestinal health, and metabolic rate, which are all involved in controlling obesity (75, 76). Obesity is a risk factor for multiple metabolic disorders, and dysbiosis can lead to weight gain and metabolic disorders. Postbiotics, including bacterial peptides, cell wall components, and SCFAs, help restore gut microbial equilibrium, allowing proper digestion and nutrient absorption (77, 78). They also increase levels of GLP-1 and PYY, curbing food cravings and making individuals feel fuller for longer (79).

Postbiotics reduce chronic low-grade inflammation, contributing to obesity and metabolic dysfunction (77). Decreased inflammation leads to enhanced lipid metabolism and avoidance of fat buildup, reducing the risk of long-term weight retention and promoting a healthy rise in weight during pregnancy and a drop in weight after delivery. Postbiotics may also affect children's health, with better metabolic health in mothers and better prenatal programming potentially reducing the prevalence of metabolic problems and obesity in children (80).

Lactobacillus gasseri supplements have been shown to help overweight and obese people lose weight by reducing abdominal fat (77). Probiotics from the *Lactobacillus paracasei* family may alleviate metabolic disorders and obesity. Animal studies have shown that probiotics can help regulate obesity in rats by reducing insulin receptor substrate-1, JNK phosphorylation, and TLR4 activation. Probiotics also enhance the diversity of Firmicutes and *Actinobacteria* while suppressing the abundance of Bacteroidetes (81). In obese mice, probiotic therapy preserves Bacteroidetes, boosts Actinobacteria, and reduces Firmicutes (82). The release of anorexigenic peptides like GLP-1 and

PYY improves glucose tolerance and energy utilization. However, the findings of these studies remain inconsistent (83).

5.4 MECHANISMS OF POSTBIOTICS IN GDM

5.4.1 Overview of Postbiotics

Consuming postbiotics, inanimate bacteria or their components, may be healthy. In probiotic cell-free supernatants, metabolic byproducts include vitamins, enzymes, organic acids, secreted proteins, amino acids, SCFAs, peptides, and biosurfactants. Substances called postbiotics. In numerous human trials, like Nataraj et al. (2020), taking a supplement with pure microbial metabolites like butyrate, alone or with inulin, improves inflammation and cardiometabolic performance (84). Experimental investigations and clinical trials have shown that probiotics, prebiotics, synbiotics, and postbiotics (PPSP) reduce obesity, type 2 diabetes, and other metabolic illnesses. However, a pregnant patient may complicate matters. Probiotics with pasteurization are called postbiotics. *Lactobacillus* and *Bifidobacterium* probiotics dominate postbiotic research. *Lactobacillus plantarum* L-14 postbiotic exopolysaccharide suppresses immature cell differentiation into mature adipocytes and regulates body weight gain and lipid profiles in mice by upregulating the TLR2–AMPK signaling pathway (85). Targeting NOD2 and IRF4 reduces obesity-related insulin resistance. According to Cavallari et al. (2020), postbiotics like muropeptide and NOD2 may improve insulin sensitivity and inflammation; however, NOD1 may worsen metabolic disorders (86). More specified chemical makeup, longer shelf life, and safe dosing are postbiotics' advantages over probiotics. They are widely distributed, easily absorbed and digested, and give probiotic-like benefits without bioavailability, instability, or resistance gene transfer (87).

5.4.2 Interaction Between Gut Microbiota and Metabolic Health in GDM

Hyperglycemia caused by glucose resistance and associated with alterations in the gut flora define GDM, a metabolic condition that manifests during pregnancy (88). In GDM, opportunistic taxa like *Enterococcus* flourish while beneficial taxa like *Lactobacillus* and *Bifidobacterium* decrease. These changes, which manifest during the first trimester, might be early markers of the risk of GDM. Additionally, GDM impairs the metabolic outputs of microorganisms, especially those that produce short-chain fatty acids and bile acids. The host's ability to control inflammation, insulin sensitivity, and glucose is all dependent on these metabolites.

The maternal GDM status has an impact on babies' gut flora, which may have implications for future generations' health. The emergence of multi-omics approaches is revealing new information on the intricate microbiota–host interactions in GDM via metabolites (89).

Disease and host immunity are influenced by the gut microbiome, and metabolic syndromes such as obesity, type 2 diabetes, allergy problems, some types of cancer, and neurological illnesses are associated with gut dysbiosis (90). An imbalance in the gut microbiota known as gut dysbiosis may lead to an excess of reactive oxygen species (ROS), which can weaken the gut barrier, trigger the immune system, alter metabolic pathways, and result in inflammation. By reestablishing healthy gut microbiota by dietary modifications, probiotics, or postbiotics, glucose metabolism may be improved and the risk of GDM decreased.

5.4.3 Postbiotics in the Regulation of Inflammation in GDM

The gut microbiota may develop a metabolic axis with the host microorganisms by using the host's nutrition to create microbial metabolites. Nutritional metabolism and immunological homeostasis are both impacted by this axis, which in turn influences the host's metabolism as a whole (91). Tissue inflammation, obesity, blood glucose, insulin, and endocrine metabolism are affected by gut bacteria.

Combating metabolic illness typically involves prebiotics or probiotics. Without specificity, prebiotics may harm bacterial communities (92). In order to affect human immunity or metabolism, probiotics need live bacteria to colonize. Postbiotics, created from bacterial components and compounds, may change host immunometabolism without affecting colonization efficiency or microbiota (89). In this article, we discuss the impact of certain postbiotics on metabolic illness and their processes. Certain immunological responses activated by bacterial cell wall components such as lipopolysaccharides, muropeptides, lipoteichoic acids, and flagellin affect host metabolism in context (93). Certain lipopolysaccharides and muropeptide postbiotics may sensitize or resist insulin, affecting host metabolism. SCF, bile acids, lactate, glycerol, succinate, ethanolamine, and ethanol are host metabolism substrates. Immunomodulation or hormone mimicry by postbiotics may feed host metabolic pathways or impact endocrine metabolism. It is important to include the postbiotic impact on host–microbe interactions when studying metabolic inflammation and disease (94).

5.4.4 Postbiotic Effects on Endocrine Signaling (Insulin and Glucagon)

According to research by Wu et al. (2023), a healthy endocrine system is maintained by a slender body, but as weight increases, it leads to a hormonal imbalance and disrupts normal physiology (95). A certain weight/adiposity range is maintained via the regulation of hunger, body composition, and glucose homeostasis by hormones produced by endocrine, adipose, and neuroendocrine cells; this range is affected by variables such as calorie intake, meal composition, and lifestyle (95). Primarily, hyperglycemia during pregnancy may be managed with insulin treatment, which activates the insulin signaling cascade by binding to the insulin receptor in trophoblast membranes. Placental levels of active insulin mediators in gestating diabetic mothers may affect fetal metabolism. According to research out of Spain, some fatty acid transporters are elevated, and placental fatty acid levels are enhanced when exogenous insulin treatment promotes the phosphorylation of placental insulin mediators. Due to the potential risks of excessive fat transfer to the fetus via the placenta, it is important to monitor maternal blood lipid levels while using insulin for glycemic management. Insulin resistance is a main symptom of every kind of diabetes. Evidence linking insulin resistance to an aberrant inflammatory state in the body is mounting (96).

5.4.5 Gut–Pancreatic Axis and Its Role in GDM

Gut bacteria regulate islet function and endocrine metabolism via the complicated gut–pancreatic axis (97). It has gut flora, hormones, amino acids, and angiotensin-converting enzyme (ACE) (98). GDM may result from gut bacteria composition affecting insulin production, blood sugar control, insulin sensitivity, and pancreatic beta cell function. Distorted gut–pancreas communication and reduced glucose tolerance during pregnancy may cause GDM (99). Dysbiotic gut microbiomes during pregnancy change bacteria's metabolism of carbohydrates, lipids, and amino acids, impacting energy balance and glucose homeostasis. Reduced insulin sensitivity and β-cell activity may lead to hyperglycemia and GDM, affecting hormone production, substrate accessibility, and nutrition. New GDM markers and treatment targets may be found by studying dysregulated gut bacteria metabolism during pregnancy (88).

Food and chemicals metabolized by the GM alter the host's metabolism and inflammation (100). Bile acids, SCFAs, and GLP-1 modulate endocrine responses by changing pancreatic hormones, increasing hepatic glycogen manufacturing, and increasing energy expenditure. The gut–liver–pancreas axis (GLPA) and other normal endocrine and metabolic activities may be affected by GM dysbiosis-related microbial imbalance (101).

5.5 CLINICAL EVIDENCE AND RESEARCH ON POSTBIOTICS IN GDM

5.5.1 Animal Studies on Postbiotics and GDM

Probiotic BiLaEn is used to study how GDM influences mother metabolic health and child islet development in C57BL/6J mice. GDM was induced in high-fat-fed female mice treated with low or high BiLaEn doses. Ordinary mice ate a conventional low-fat diet. Possibly a safe GDM treatment, BiLaEn decreased maternal metabolic stress and enhanced islet development in GDM children (102).

Research examined how postbiotics benefit animal growth, immunology, and gastrointestinal health. Yeast and lactic acid bacteria (LAB) postbiotics supported milk production, feed efficiency, and immunological responses while conserving rumen microbiota and lowering heat stress inflammation. By producing SCFA and promoting gut bacteria, LAB metabolites enhanced monogastric animal development, diarrhea, and intestinal health. Post-weaning lambs' intestinal barrier, antioxidant enzyme activity, and inflammation improved with postbiotic feed. Pig postbiotics improved food intake, diarrhea, and gut microbiomes (103).

Gestational diabetes patients' oxidative stress and inflammation were investigated with probiotics. *Lactobacillus acidophilus* and *Bifidobacterium* strains inhibit malondialdehyde (MDA), glutathione reductase (GSHR), and glutathione peroxidase (GPx) and increase superoxide dismuthase (SOD). Nuclear factor kappa B (NF-κB) and mitogen-activated protein kinase (MAPK) pathways were modulated by probiotics in animal trials, leading to enhanced inflammatory biomarkers and decreased TNF-α and interleukin levels (104).

In a C57BL/6J mouse model, GDM was examined on the immune system, newborn gastrointestinal microbiota, and intrauterine microbial exposure. In GDM animals' umbilical cord blood, LPSs were higher, suggesting greater microbial exposure. Children of mothers with GDM have dysbiosis due to reduced gut microbiota diversity and increased Firmicutes/Bacteroidetes ratio. In GDM progeny, *Lactobacillus* colonization reduced quicker than *E. coli*, showing probiotic fragility. In memory T cells, microbial antigen exposure increased cytokines, suggesting greater adaptive immunity. These data show that maternal health affects offspring microbiota and immune systems, enabling GDM prevention research (105).

5.5.2 Human Clinical Trials Investigating Postbiotics in GDM

Many clinical studies have examined the effectiveness of postbiotics in managing GDM. Research shows that probiotic-derived postbiotics improve insulin sensitivity and glucose control. SCFAs and exopolysaccharides (EPS) have been shown to be effective postbiotics in tests at HaEmek Medical Center in Israel and Hvidovre University Hospital in Denmark. This research is investigating how postbiotics and prebiotics synergistically improve gut flora and metabolic health in pregnant women. These studies assess pregnancy-related dosage, timing, and safety. More study is needed to determine treatment recommendations, although preliminary findings suggest reducing fasting plasma glucose and increasing maternal glycemic control (106).

5.5.3 Evidence on Postbiotic Supplementation and Insulin Resistance

The systematic review "Probiotics Supplementation and Insulin Resistance" examines how probiotics improve insulin sensitivity in people and animals. Novel therapies are needed due to rising obesity, insulin resistance, and type 2 diabetes. Living bacteria called probiotics improve gut health, endotoxemia, and insulin sensitivity. PRISMA standards were used to evaluate *Lactobacillus*, *Bifidobacterium*, and *Akkermansia* strains on insulin resistance from 1990 to 2020. Animal studies demonstrated that probiotics improved insulin sensitivity, glucose metabolism, lipid profiles, and inflammation in obese rats fed high-fat diets. Some bacteria, such *Lactobacillus*, *Bifidobacterium*,

Clostridium, and *Akkermansia*, improve metabolic health. Probiotics increased short-chain fatty acid synthesis, tight junction protein expression, and metabolic endotoxemia and inflammation reduction. They sometimes worked like metformin and vildagliptin. Five of seven human clinical studies demonstrated insulin resistance benefits, while two did not (107). Differences in sample size, technique, and probiotic dosage limit direct comparisons in human studies. Animal studies show that probiotics improve intestinal barrier function, immunological responses, and inflammatory markers. *Akkermansia muciniphila* and *Bifidobacterium breve* may reduce obesity-related insulin resistance. Human study methodology is varied, and the long-term effects are unclear (108, 109).

5.6 CHALLENGES IN UTILIZING POSTBIOTICS FOR GDM

The pathophysiology of GDM is complex and includes genetic predisposition, insulin resistance, and hormonal changes. Targeted postbiotic therapies are challenging due to the lack of understanding of the interactions between these variables and gut flora. Postbiotics contain a variety of bioactive substances, but standardization and efficacy evaluation are hindered by variations in composition and production techniques. Postbiotics' safety profile during pregnancy is still being studied, and thorough preclinical and clinical validation is necessary due to potential negative effects on mother and fetal outcomes. Postbiotics lack a consistent definition and regulatory structure, affecting their practical use and eroded patient and healthcare provider confidence. The efficiency of postbiotics is influenced by host factors like food, genetic background, and baseline microbiota composition.

5.7 CONCLUSION

In gestational diabetes mellitus, hormonal, genetic, and environmental factors may impact the mother and baby. Due to improved knowledge of gut bacteria's metabolic function, postbiotics may cure GDM. Their stability and safety eliminate worries about swallowing live germs while pregnant. To integrate postbiotics into therapeutic practice, one must understand their mechanisms of action, protect susceptible populations, and overcome individual response variability. Using peptides, SCFAs, and cell wall components, postbiotics increase insulin sensitivity, modulate inflammatory pathways, and restore the intestinal barrier. They may improve GDM by affecting gut flora and metabolic balance. More research is required to determine how these chemicals impact glucose metabolism and maternal health. Safety is paramount while utilizing postbiotics during pregnancy. Therapeutic therapies must be proven to avoid unanticipated effects on mother and fetus. Safety parameters must be established via preclinical and clinical studies. Researchers and regulatory authorities must collaborate to standardize postbiotic practices. Postbiotic clinical usage is complicated by the lack of a consensual definition and regulatory framework.

REFERENCES

1. Law KP, Zhang H. The Pathogenesis and Pathophysiology of Gestational Diabetes Mellitus: Deductions from a Three-Part Longitudinal Metabolomics Study in China. *Clinica Chimica Acta; International Journal of Clinical Chemistry*. 2017;468:60–70.
2. Gao J, Xu B, Zhang X, Cui Y, Deng L, Shi Z, et al. Association Between Serum Bile Acid Profiles and Gestational Diabetes Mellitus: A Targeted Metabolomics Study. *Clinica Chimica Acta; International Journal of Clinical Chemistry*. 2016;459:63–72.
3. Han Y, Sun H, Zhang A, Yan G, Wang XJ. Chinmedomics, a New Strategy for Evaluating the Therapeutic Efficacy of Herbal Medicines. *Pharmacology & Therapeutics*. 2020;216:107680.
4. Christian LM, Porter K. Longitudinal Changes in Serum Proinflammatory Markers Across Pregnancy and Postpartum: Effects of Maternal Body Mass Index. *Cytokine*. 2014;70(2):134–40.
5. Xia JF, Liang QL, Hu P, Wang YM, Li P, Luo GA. Correlations of Six Related Purine Metabolites and Diabetic Nephropathy in Chinese Type 2 Diabetic Patients. *Clinical Biochemistry*. 2009;42(3):215–20.

6. American Diabetes Association. 2. Classification and diagnosis of diabetes. *Diabetes Care*. 2015;38 Suppl:S8–16.
7. Gomez-Arango LF, Barrett HL, McIntyre HD, Callaway LK, Morrison M, Dekker Nitert M. Connections Between the Gut Microbiome and Metabolic Hormones in Early Pregnancy in Overweight and Obese Women. *Diabetes*. 2016;65(8):2214–23.
8. Santacruz A, Collado MC, García-Valdés L, Segura MT, Martín-Lagos JA, Anjos T, et al. Gut Microbiota Composition is Associated with Body Weight, Weight Gain and Biochemical Parameters in Pregnant Women. *The British Journal of Nutrition*. 2010;104(1):83–92.
9. Crusell MKW, Hansen TH, Nielsen T, Allin KH, Rühlemann MC, Damm P, et al. Gestational Diabetes Is Associated with Change in the Gut Microbiota Composition in Third Trimester of Pregnancy and Postpartum. *Microbiome*. 2018;6(1):89.
10. Ferrocino I, Ponzo V, Gambino R, Zarovska A, Leone F, Monzeglio C, et al. Changes in the Gut Microbiota Composition During Pregnancy in Patients with Gestational Diabetes Mellitus (GDM). *Scientific Reports*. 2018;8(1):12216.
11. Liu L, Yin T, Zhang X, Sun L, Yin Y. Temporal and Spatial Variation of the Human Placental Microbiota During Pregnancy. *American Journal of Reproductive Immunology* 2024;92(6):e70023.
12. Sharma AK, Singh S, Singh H, Mahajan D, Kolli P, Mandadapu G, et al. Deep Insight of the Pathophysiology of Gestational Diabetes Mellitus. *Cells*. 2022;11(17):2672.
13. Zhou T, Huang L, Wang M, Chen D, Chen Z, Jiang SW. A Critical Review of Proteomic Studies in Gestational Diabetes Mellitus. *Journal of Diabetes Research*. 2020;2020:6450352.
14. Defronzo RA. From the Triumvirate to the Ominous Octet: A New Paradigm for the Treatment of Type 2 Diabetes Mellitus. *Diabetes*. 2009;58(4):773–95.
15. Eizirik DL, Pasquali L, Cnop M. Pancreatic β-cells in Type 1 and Type 2 Diabetes Mellitus: Different Pathways to Failure. *Nature Reviews Endocrinology*. 2020;16(7):349–62.
16. Delghingaro-Augusto V, Nolan CJ, Gupta D, Jetton TL, Latour MG, Peshavaria M, et al. Islet Beta Cell Failure in the 60% Pancreatectomised Obese Hyperlipidaemic Zucker Fatty Rat: Severe Dysfunction with Altered Glycerolipid Metabolism Without Steatosis or a Falling Beta Cell Mass. *Diabetologia*. 2009;52(6):1122–32.
17. Auffret J, Freemark M, Carré N, Mathieu Y, Tourrel-Cuzin C, Lombès M, et al. Defective Prolactin Signaling Impairs Pancreatic B-Cell Development During the Perinatal Period. *American Journal of Physiology Endocrinology and Metabolism*. 2013;305(10):E1309–18.
18. Henquin JC, Rahier J. Pancreatic Alpha Cell Mass in European Subjects with Type 2 Diabetes. *Diabetologia*. 2011;54(7):1720–5.
19. Ren Y, Zeng Y, Wu Y, Yu J, Zhang Q, Xiao X. The Role of Gut Microbiota in Gestational Diabetes Mellitus Affecting Intergenerational Glucose Metabolism: Possible Mechanisms and Interventions. *Nutrients*. 2023;15(21):4551.
20. Li D, Wang P, Wang P, Hu X, Chen F. The Gut Microbiota: A Treasure for Human Health. *Biotechnology Advances*. 2016;34(7):1210–24.
21. Almeida A, Mitchell AL, Boland M, Forster SC, Gloor GB, Tarkowska A, et al. A New Genomic Blueprint of the Human Gut Microbiota. *Nature*. 2019;568(7753):499–504.
22. Preethy S, Ranganathan N, Raghavan K, Dedeepiya VD, Ikewaki N, Abraham SJK. Integrating the Synergy of the Gut Microbiome into Regenerative Medicine: Relevance to Neurological Disorders. *Journal of Alzheimer's Disease: JAD*. 2022;87(4):1451–60.
23. Zhou Z, Sun B, Yu D, Zhu C. Gut Microbiota: An Important Player in Type 2 Diabetes Mellitus. *Frontiers in Cellular and Infection Microbiology*. 2022;12:834485.
24. Medici Dualib P, Ogassavara J, Mattar R, Mariko Koga da Silva E, Atala Dib S, de Almeida Pititto B. Gut Microbiota and Gestational Diabetes Mellitus: A Systematic Review. *Diabetes Research and Clinical Practice*. 2021;180:109078.
25. Kamińska K, Stenclik D, Błażejewska W, Bogdański P, Moszak M. Probiotics in the Prevention and Treatment of Gestational Diabetes Mellitus (GDM): A Review. *Nutrients*. 2022;14(20):4303.
26. Vinderola G, Sanders ME, Salminen S. *The Concept of Postbiotics. Foods* (Basel, Switzerland). 2022;11(8).
27. Binda S, Hill C, Johansen E, Obis D, Pot B, Sanders ME, et al. Criteria to Qualify Microorganisms as "Probiotic" in Foods and Dietary Supplements. *Frontiers in Microbiology*. 2020;11:1662.
28. Tsilingiri K, Rescigno M. Postbiotics: What Else? *Beneficial Microbes*. 2013;4(1):101–7.

29. Aguilar-Toalá JE, Arioli S, Behare P, Belzer C, Berni Canani R, Chatel JM, et al. Postbiotics—When Simplification Fails to Clarify. *Nature Reviews Gastroenterology & Hepatology*. 2021;18(11):825–6.
30. Salminen S, Collado MC, Endo A, Hill C, Lebeer S, Quigley EMM, et al. Reply to: Postbiotics—When Simplification Fails to Clarify. *Nature Reviews Gastroenterology & Hepatology*. 2021;18(11):827–8.
31. Wegh CAM, Geerlings SY, Knol J, Roeselers G, Belzer C. Postbiotics and Their Potential Applications in Early Life Nutrition and Beyond. *International Journal of Molecular Sciences*. 2019;20(19):4673.
32. Fusco W, Lorenzo MB, Cintoni M, Porcari S, Rinninella E, Kaitsas F, et al. Short-Chain Fatty-Acid-Producing Bacteria: Key Components of the Human Gut Microbiota. *Nutrients*. 2023;15(9):2211.
33. He J, Zhang P, Shen L, Niu L, Tan Y, Chen L, et al. Short-Chain Fatty Acids and Their Association with Signalling Pathways in Inflammation, *Glucose and Lipid Metabolism. International Journal of Molecular Sciences*. 2020;21(17):6356.
34. de Mendonça E, Fragoso MBT, de Oliveira JM, Xavier JA, Goulart MOF, de Oliveira ACM. Gestational Diabetes Mellitus: *The Crosslink Among Inflammation, Nitroxidative Stress, Intestinal Microbiota and Alternative Therapies. Antioxidants* (Basel, Switzerland). 2022;11(1).
35. Aguilar-Toalá JE, Garcia-Varela R, Garcia HS, Mata-Haro V, González-Córdova AF, Vallejo-Cordoba B, et al. Postbiotics: An Evolving Term Within the Functional Foods Field. *Trends in Food Science & Technology*. 2018;75:105–14.
36. Hernández-González JC, Martínez-Tapia A, Lazcano-Hernández G, García-Pérez BE, Castrejón-Jiménez NS. Bacteriocins from Lactic Acid Bacteria. A Powerful Alternative as Antimicrobials, Probiotics, and Immunomodulators in Veterinary Medicine. *Animals: An Open Access Journal from MDPI*. 2021;11(4).
37. Tang R, Li L. Modulation of Short-Chain Fatty Acids as Potential Therapy Method for Type 2 Diabetes Mellitus. *The Canadian Journal of Infectious Diseases & Medical Microbiology = Journal Canadien Des Maladies Infectieuses et De La Microbiologie Medicale*. 2021;2021:6632266.
38. Tardy AL, Pouteau E, Marquez D, Yilmaz C, Scholey A. Vitamins and Minerals for Energy, Fatigue and Cognition: A Narrative Review of the Biochemical and Clinical Evidence. *Nutrients*. 2020;12(1):228.
39. Thorakkattu P, Khanashyam AC, Shah K, Babu KS, Mundanat AS, Deliephan A, et al. Postbiotics: Current Trends in Food and Pharmaceutical Industry. *Foods* (Basel, Switzerland). 2022;11(19).
40. Gubatan J, Holman DR, Puntasecca CJ, Polevoi D, Rubin SJ, Rogalla S. Antimicrobial Peptides and the Gut Microbiome in Inflammatory Bowel Disease. *World Journal of Gastroenterology*. 2021;27(43):7402–22.
41. Fu Y, Li S, Xiao Y, Liu G, Fang J. A Metabolite Perspective on the Involvement of the Gut Microbiota in Type 2 Diabetes. *International Journal of Molecular Sciences*. 2023;24(19).
42. Nataraj BH, Ali SA, Behare PV, Yadav H. Postbiotics-Parabiotics: The New Horizons in Microbial Biotherapy and Functional Foods. *Microbial Cell Factories*. 2020;19(1):168.
43. Zheng Z, Diaz-Arévalo D, Guan H, Zeng M. Noninvasive Vaccination Against Infectious Diseases. *Human Vaccines & Immunotherapeutics*. 2018;14(7):1717–33.
44. Salman MK, Abuqwider J, Mauriello G. Anti-Quorum Sensing Activity of Probiotics: The Mechanism and Role in Food and Gut Health. *Microorganisms*. 2023;11(3):793.
45. Azimi S, Klementiev AD, Whiteley M, Diggle SP. Bacterial Quorum Sensing During Infection. *Annual Review of Microbiology*. 2020;74:201–19.
46. Rafique N, Jan SY, Dar AH, Dash KK, Sarkar A, Shams R, et al. Promising Bioactivities of Postbiotics: A Comprehensive Review. *Journal of Agriculture and Food Research*. 2023;14:100708.
47. İncili GK, Karatepe P, Akgöl M, Güngören A, Koluman A, İlhak O, et al. Characterization of Lactic Acid Bacteria Postbiotics, Evaluation In-Vitro Antibacterial Effect, Microbial and Chemical Quality on Chicken Drumsticks. *Food Microbiology*. 2022;104:104001.
48. Peluzio MdCG, Martinez JA, Milagro FI. Postbiotics: Metabolites and Mechanisms Involved in Microbiota–Host Interactions. *Trends in Food Science & Technology*. 2021;108:11–26.
49. Mishra B, Mishra AK, Mohanta YK, Yadavalli R, Agrawal DC, Reddy HP, et al. Postbiotics: The New Horizons of Microbial Functional Bioactive Compounds in Food Preservation and Security. *Food Production, Processing and Nutrition*. 2024;6(1):28.
50. Chang HM, Foo HL, Loh TC, Lim ETC, Abdul Mutalib NE. Comparative Studies of Inhibitory and Antioxidant Activities, and Organic Acids Compositions of Postbiotics Produced by Probiotic *Lactiplantibacillus plantarum* Strains Isolated from Malaysian Foods. *Frontiers in Veterinary Science*. 2020;7:602280.
51. Hu CH, Ren LQ, Zhou Y, Ye BC. Characterization of Antimicrobial Activity of Three *Lactobacillus Plantarum* Strains Isolated from Chinese Traditional Dairy Food. *Food science & Nutrition*. 2019;7(6): 1997–2005.

52. Ozma MA, Abbasi A, Akrami S, Lahouty M, Shahbazi N, Ganbarov K, et al. Postbiotics as the Key Mediators of the Gut Microbiota–Host Interactions. *Le Infezioni in Medicina*. 2022;30(2):180–93.
53. Wang S, Han X, Zhang L, Zhang Y, Li H, Jiao Y. Whole Peptidoglycan Extracts from the *Lactobacillus paracasei* subsp. Paracasei M5 Strain Exert Anticancer Activity In Vitro. *BioMed Research International*. 2018;2018:2871710.
54. Kim SW, Ha YJ, Bang KH, Lee S, Yeo JH, Yang HS, et al. Potential of Bacteriocins from *Lactobacillus Taiwanensis* for Producing Bacterial Ghosts as a Next Generation Vaccine. *Toxins*. 2020;12(7):432.
55. Patil S, Sawant S, Hauff K, Hampp G. Validated Postbiotic Screening Confirms Presence of Physiologically-Active Metabolites, Such as Short-Chain Fatty Acids, Amino Acids and Vitamins in Hylak® Forte. *Probiotics and Antimicrobial Proteins*. 2019;11(4):1124–31.
56. Higashi B, Mariano TB, de Abreu Filho BA, Gonçalves RAC, de Oliveira AJB. Effects of Fructans and Probiotics on the Inhibition of *Klebsiella oxytoca* and the Production of Short-Chain Fatty Acids Assessed by NMR Spectroscopy. *Carbohydrate Polymers*. 2020;248:116832.
57. Forkus B, Ritter S, Vlysidis M, Geldart K, Kaznessis YN. Antimicrobial Probiotics Reduce *Salmonella enterica* in Turkey Gastrointestinal Tracts. *Scientific Reports*. 2017;7:40695.
58. Li HY, Zhou DD, Gan RY, Huang SY, Zhao CN, Shang A, et al. Effects and Mechanisms of Probiotics, Prebiotics, Synbiotics, and Postbiotics on Metabolic Diseases Targeting Gut Microbiota: A Narrative Review. *Nutrients*. 2021;13(9):3211.
59. Żółkiewicz J, Marzec A, Ruszczyński M, Feleszko W. Postbiotics: A Step Beyond Pre- and Probiotics. *Nutrients*. 2020;12(8):2189.
60. Górska A, Przystupski D, Niemczura MJ, Kulbacka J. Probiotic Bacteria: A Promising Tool in Cancer Prevention and Therapy. *Current Microbiology*. 2019;76(8):939–49.
61. Valdes AM, Walter J, Segal E, Spector TD. Role of the Gut Microbiota in Nutrition and Health. *BMJ* (Clinical research ed). 2018;361:k2179.
62. Gul S, Durante-Mangoni E. Unraveling the Puzzle: Health Benefits of Probiotics—A Comprehensive Review. *Journal of Clinical Medicine*. 2024;13(5):1436.
63. Hill C, Guarner F, Reid G, Gibson GR, Merenstein DJ, Pot B, et al. Expert Consensus Document. The International Scientific Association for Probiotics and Prebiotics Consensus Statement on the Scope and Appropriate Use of the Term Brobiotic. *Nature Reviews Gastroenterology & Hepatology*. 2014;11(8):506–14.
64. Sanders ME, Guarner F, Guerrant R, Holt PR, Quigley EM, Sartor RB, et al. An Update on the Use and Investigation of Probiotics in Health and Disease. *Gut*. 2013;62(5):787–96.
65. Mörkl S, Butler MI, Holl A, Cryan JF, Dinan TG. Probiotics and the Microbiota–Gut–Brain Axis: Focus on Psychiatry. *Current Nutrition Reports*. 2020;9(3):171–82.
66. Salminen S, Collado MC, Endo A, Hill C, Lebeer S, Quigley EMM, et al. The International Scientific Association of Probiotics and Prebiotics (ISAPP) Consensus Statement on the Definition and Scope of Postbiotics. *Nature Reviews Gastroenterology & Hepatology*. 2021;18(9):649–67.
67. Teame T, Wang A, Xie M, Zhang Z, Yang Y, Ding Q, et al. Paraprobiotics and Postbiotics of Probiotic Lactobacilli, Their Positive Effects on the Host and Action Mechanisms: A Review. *Frontiers in Nutrition*. 2020;7:570344.
68. Obuchowska A, Gorczyca K, Standyło A, Obuchowska K, Kimber-Trojnar Z, Wierzchowska-Opoka M, et al. Effects of Probiotic Supplementation During Pregnancy on the Future Maternal Risk of Metabolic Syndrome. *International Journal of Molecular Sciences*. 2022;23(15):8253.
69. Xu Y, Zhang M, Zhang J, Sun Z, Ran L, Ban Y, et al. Differential Intestinal and Oral Microbiota Features Associated with Gestational Diabetes and Maternal Inflammation. *American Journal of Physiology Endocrinology and Metabolism*. 2020;319(2):E247–e53.
70. Mokkala K, Paulin N, Houttu N, Koivuniemi E, Pellonperä O, Khan S, et al. Metagenomics Analysis of Gut Microbiota in Response to Diet Intervention and Gestational Diabetes in Overweight and Obese Women: A Randomised, Double-Blind, Placebo-Controlled Clinical Trial. *Gut*. 2021;70(2): 309–18.
71. Brantsaeter AL, Myhre R, Haugen M, Myking S, Sengpiel V, Magnus P, et al. Intake of Probiotic Food and Risk of Preeclampsia in Primiparous Women: the Norwegian Mother and Child Cohort Study. *American Journal of Epidemiology*. 2011;174(7):807–15.
72. Rangaswami J, Naranjo M, McCullough PA. Preeclampsia as a Form of Type 5 Cardiorenal Syndrome: An Underrecognized Entity in Women's Cardiovascular Health. *Cardiorenal Medicine*. 2018;8(2):160–72.

73. Erez O, Romero R, Jung E, Chaemsaithong P, Bosco M, Suksai M, et al. Preeclampsia and Eclampsia: The Conceptual Evolution of a Syndrome. *American Journal of Obstetrics and Gynecology*. 2022;226(2s): S786–s803.
74. Davidson SJ, Barrett HL, Price SA, Callaway LK, Dekker Nitert M. Probiotics for Preventing Gestational Diabetes. *The Cochrane Database of Systematic Reviews*. 2021;4(4):Cd009951.
75. Maslennikov R, Ivashkin V, Efremova I, Poluektova E, Shirokova E. Gut-Liver Axis in Cirrhosis: Are Hemodynamic Changes a Missing Link? *World Journal of Clinical Cases*. 2021;9(31):9320–32.
76. Wiciński M, Gębalski J, Gołębiewski J, Malinowski B. Probiotics for the Treatment of Overweight and Obesity in Humans: A Review of Clinical Trials. *Microorganisms*. 2020;8(8):1148.
77. Bagarolli RA, Tobar N, Oliveira AG, Araújo TG, Carvalho BM, Rocha GZ, et al. Probiotics Modulate Gut Microbiota and Improve Insulin Sensitivity in DIO Mice. *The Journal of Nutritional Biochemistry*. 2017;50:16–25.
78. Kennedy A, Martinez K, Schmidt S, Mandrup S, LaPoint K, McIntosh M. Antiobesity Mechanisms of Action of Conjugated Linoleic Acid. *The Journal of Nutritional Biochemistry*. 2010;21(3):171–9.
79. Álvarez-Arraño V, Martín-Peláez S. Effects of Probiotics and Synbiotics on Weight Loss in Subjects with Overweight or Obesity: A Systematic Review. *Nutrients*. 2021;13(10):3627.
80. Ilmonen J, Isolauri E, Poussa T, Laitinen K. Impact of Dietary Counselling and Probiotic Intervention on Maternal Anthropometric Measurements During and After Pregnancy: A Randomized Placebo-Controlled Trial. *Clinical Nutrition (Edinburgh, Scotland)*. 2011;30(2):156–64.
81. Fawzy El-Sayed KM, Klingebiel P, Dörfer CE. Toll-like Receptor Expression Profile of Human Dental Pulp Stem/Progenitor Cells. *Journal of Endodontics*. 2016;42(3):413–7.
82. Cho I, Yamanishi S, Cox L, Methé BA, Zavadil J, Li K, et al. Antibiotics in Early Life Alter the Murine Colonic Microbiome and Adiposity. *Nature*. 2012;488(7413):621–6.
83. Mazloom K, Siddiqi I, Covasa M. Probiotics: How Effective Are They in the Fight Against Obesity? *Nutrients*. 2019;11(2):258.
84. Nataraj BH, Ali SA, Behare PV, Yadav H. (2020). Postbiotics-Parabiotics: The New Horizons in Microbial Biotherapy and Functional Foods. *Microbial Cell Factories*, 19, 1–22.
85. Huang Y-H, Tain Y-L, Hsu C-N. (2022). Maternal Supplementation of Probiotics, Prebiotics or Postbiotics to Prevent Offspring Metabolic Syndrome: The Gap Between Preclinical Results and Clinical Translation. *International Journal of Molecular Sciences*, 23(17), 10173. https://www.mdpi.com/1422-0067/23/17/10173
86. Cavallari JF, Barra NG, Foley KP, Lee A, Duggan BM, Henriksbo BD, Anhê, FF, Ashkar AA, Schertzer JD. (2020). Postbiotics for NOD2 Require Nonhematopoietic RIPK2 to Improve Blood Glucose and Metabolic Inflammation in Mice. *American Journal of Physiology-Endocrinology and Metabolism*, 318(4), E579–E585.
87. Chen Q, Li X-J, Xie W, Su Z-A, Qin G-M, Yu C-H. (2024). Postbiotics: Emerging Therapeutic Approach in Diabetic Retinopathy. *Frontiers in Microbiology*, 15, 1359949.
88. Ren Y, Zeng Y, Wu Y, Yu J, Zhang Q, Xiao X. The Role of Gut Microbiota in Gestational Diabetes Mellitus Affecting Intergenerational Glucose Metabolism: Possible Mechanisms and Interventions. *Nutrients*. 2023 Oct 27;15(21):4551.
89. Sun H, Saeedi P, Karuranga S, Pinkepank M, Ogurtsova K, Duncan BB, Stein C, Basit A, Chan JC, Mbanya JC, Pavkov ME. IDF Diabetes Atlas: Global, Regional and Country-Level Diabetes Prevalence Estimates for 2021 and Projections for 2045. *Diabetes Research and Clinical Practice*. 2022 Jan 1;183:109119.
90. Quan L-H, Zhang C, Dong M, Jiang J, Xu H, Yan C, Liu X, Zhou H, Zhang H, Chen L. (2020). Myristoleic Acid Produced by Enterococci Reduces Obesity Through Brown Adipose Tissue Activation. *Gut*, 69(7), 1239–1247.
91. Caissutti C, Berghella V. Scientific Evidence for Different Options for GDM Screening and Management: Controversies and Review of the Literature. *BioMed Research International*. 2017;2017(1):2746471.
92. Wang X, Liu H, Li Y, Huang S, Zhang L, Cao C, et al. Altered Gut Bacterial and Metabolic Signatures and Their Interaction in Gestational Diabetes Mellitus. *Gut Microbes*. 2020;12(1):1840765.
93. Alipour M, Zaidi D, Valcheva R, Jovel J, Martínez I, Sergi C, et al. Mucosal Barrier Depletion and Loss of Bacterial Diversity Are Primary Abnormalities in Paediatric Ulcerative Colitis. *Journal of Crohn's and Colitis*. 2016;10(4):462–71.

94. Dalby MJ, Ross AW, Walker AW, Morgan PJ. Dietary Uncoupling of Gut Microbiota and Energy Harvesting from Obesity and Glucose Tolerance in Mice. *Cell Reports*. 2017;21(6):1521–33.
95. Wu W, Chen Z, Han J, Qian L, Wang W, Lei J, Wang H. (2023). Endocrine, Genetic, and Microbiome Nexus of Obesity and Potential Role of Postbiotics: A Narrative Review. *Eating and Weight Disorders-Studies on Anorexia, Bulimia and Obesity*, 28(1), 84.
96. Kuang Y-S, Lu J-H, Li S-H, Li J-H, Yuan M-Y, He J-R, et al. Connections Between the Human Gut Microbiome and Gestational Diabetes Mellitus. *Gigascience*. 2017;6(8):gix058.
97. Yang S, Cao J, Sun C, Yuan L. The Regulation Role of the Gut–Islets Axis in Diabetes. *Diabetes, Metabolic Syndrome and Obesity*. 2024 Dec 31:1415–23.
98. Holst JJ. The Incretin System in Healthy Humans: The Role of GIP and GLP-1. *Metabolism*. 2019 Jul 1;96:46–55.
99. Lavergne A, Tarifeño-Saldivia E, Pirson J, Reuter AS, Flasse L, Manfroid I, Voz ML, Peers B. Pancreatic and Intestinal Endocrine Cells in Zebrafish Share Common Transcriptomic Signatures and Regulatory Programmes. *BMC Biology*. 2020 Dec;18:1–9.
100. Marroncini G, Naldi L, Martinelli S, Amedei A. Gut–Liver–Pancreas Axis Crosstalk in Health and Disease: From the Role of Microbial Metabolites to Innovative Microbiota Manipulating Strategies. *Biomedicines*. 2024 Jun 24;12(7):1398.
101. Thomas S, Izard J, Walsh E, Batich K, Chongsathidkiet P, Clarke G, Sela DA, Muller AJ, Mullin JM, Albert K, Gilligan JP. The Host Microbiome Regulates and Maintains Human Health: A Primer and Perspective for Non-Microbiologists. *Cancer Research*. 2017 Apr 15;77(8):1783–812.
102. Hu R, Zou L, Wang L, Xu C, Qi M, Yang Z, et al. Probiotics Alleviate Maternal Metabolic Disorders and Offspring-Islet Abnormalities in Gestational Diabetic Mice. *Journal of Functional Foods*. 2022;99:105300.
103. Zhao X, Liu S, Li S, Jiang W, Wang J, Xiao J, et al. Unlocking the Power of Postbiotics: A Revolutionary Approach to Nutrition for Humans and Animals. *Cell Metabolism*. 2024;36(4):725–44.
104. Hajifaraji M, Jahanjou F, Abbasalizadeh F, Aghamohammadzadeh N, Abbasi MM, Dolatkhah N. Effect of Probiotic Supplements in Women with Gestational Diabetes Mellitus on Inflammation and Oxidative Stress Biomarkers: a Randomized Clinical Trial. *Asia Pacific Journal of Clinical Nutrition*. 2018;27(3):581–91.
105. Liu J, Chen Y, Laurent I, Yang P, Xiao X, Li X. Gestational Diabetes Exacerbates Intrauterine Microbial Exposure Induced Intestinal Microbiota Change in Offspring Contributing to Increased Immune Response. *Nutrition & Diabetes*. 2024;14(1):87.
106. Salles BIM, Cioffi D, Ferreira SRG. Probiotics Supplementation and Insulin Resistance: A Systematic Review. *Diabetology & Metabolic Syndrome*. 2020;12(1):98.
107. Nagpal J, Rawat S, Gupta L, Negi A, Oraon DS. A Stepped Wedge Cluster Randomized Trial to Evaluate the Effectiveness of a Community Leader-Driven Kit-Based Diabetes Self-Management Education Approach in Improving Diabetes Control and Care: Study Protocol for the Delhi Diabetes INTervention Trial (DEDINTT). *Trials*. 2023;24(1):673.
108. Çetinkaya Özdemir S, Küçüktürkmen Paşa B, Metin T, Dinçer B, Sert H. The Effect of Probiotic and Synbiotic Use on Glycemic Control in Women with Gestational Diabetes: A Systematic Review and Meta-Analysis. *Diabetes Research and Clinical Practice*. 2022;194:110162.

6 Postbiotics
Beneficial Impact on Obesity

Satyajit Mohanty, Tuhin Mukherjee, Nikita Nayak and Ashok Pattnaik
Birla Institute of Technology, Mesra, Ranchi, Jharkhand, India

6.1 INTRODUCTION

Obesity, defined as a BMI $\geq$30 kg/m^2, is characterized by unhealthy fat buildup in adipose tissue due to energy imbalance. Due to genetic, physiological, and environmental variables like high-calorie diets and sedentary lifestyles, excessive fat accumulation disturbs glucose, protein, and lipid metabolism. Emerging evidence links gut microbiome dysbiosis to obesity and oxidative stress [1]. Weight problems can lead to hypertension, type 2 diabetes, cardiovascular disease, nonalcoholic fatty liver disease, musculoskeletal diseases, and cancer. Since 1975, obesity has nearly tripled worldwide, with 650 million obese and 1.9 billion overweight adults in 2019. One in five persons will be obese by 2025 [2]. Childhood obesity is a growing concern, particularly in the United States, with over 330 million children aged 5–19 and 40 million under 5 classified as overweight or obese in 2016. In South Korea, where obesity is defined as a BMI $\geq$25 kg/m^2, prevalence has increased from 31.3% in 2009 to 37.1% in 2021 and is expected to rise further by 2030. This underscores the urgent need for effective prevention and treatment strategies [3].

Adipose tissue is categorized into white (WAT), brown (BAT), and beige adipose tissue. WAT stores energy as triglycerides produces adipokines and acts as an endocrine organ, with adipocytes characterized by large lipid droplets and few mitochondria. In contrast, BAT contains multilocular lipid droplets, numerous mitochondria, and uncoupling protein-1 (UCP-1), facilitating energy expenditure via nonshivering thermogenesis, offering protection against obesity [4]. Obese individuals exhibit lower BAT activity than those with normal BMI. Beige adipocytes, which emerge in WAT through cold exposure or other stimuli, possess thermogenic properties similar to BAT and hold potential for obesity treatment. Transcriptional regulators like PRDM16, PPARγ, PGC-1α, and UCP-1 promote browning of WAT and are key targets for obesity management [5].

Metabolic syndrome encompasses central obesity, insulin resistance, hyperglycemia, dyslipidemia, and hypertension, significantly increasing the risk of diabetes, atherosclerotic cardiac disease, and cerebrovascular complications. Visceral adiposity, chronic inflammation, and neurohormonal activation are central to metabolic syndrome development, often triggered by excessive caloric intake and sedentary habits [6]. Weight loss remains the first-line intervention, with a 5–10% reduction in body weight improving insulin sensitivity, lipid profiles, and inflammatory markers. Dysbiosis in the gastrointestinal microbiome, characterized by reduced Bacteroidetes and increased Firmicutes proportions, has been linked to obesity and insulin resistance. Gut microbiota influences metabolism through bile acid regulation, short-chain fatty acid production, and metabolic endotoxemia, contributing to systemic inflammation and insulin resistance [7]. Postbiotics are soluble, nonliving metabolites such as vitamins, peptidoglycans, and polysaccharides, which directly or indirectly benefit the host's health. These substances can influence metabolic processes, enhance insulin secretion, and improve insulin sensitivity. For example, the postbiotic muramyl dipeptide reduces inflammation and enhances insulin signaling in conditions like metabolic endotoxemia, hyperglycemia, and obesity through a mechanism involving the NOD2 pathway [8].

 DOI: 10.1201/9781003607779-6

6.2 OVERVIEW OF POSTBIOTICS

The Greek words "post" (after) and "bios" (life) form "postbiotic." Billions of bacteria, viruses, fungi, and archaea make up the human gut microbiome. Gut microbes interact with the immune system and get nutrients from the host to impact host physiology. Low-molecular-weight metabolites are essential for bacteria's growth, reproduction, intercellular communication, and environmental stress resistance [9]. Postbiotics are bioactive components produced by live bacteria or released after lysis. Postbiotic preparations must have specific microbial compositions, with or without metabolites. Fermented foods often use unidentified microbe cultures, whereas postbiotics do not [10]. Beneficial gut bacteria produce postbiotics via metabolizing prebiotics or probiotic components or degrading bacterial cell walls. Heat, enzymatic reactions, ultrasonication, ultraviolet treatment, centrifugation, and ultrafiltration are common postbiotic extraction methods [11, 12].

Postbiotics have many advantages over probiotics. Probiotics, living bacteria that enhance health in large doses, are unstable during storage, processing, and gastrointestinal transit. Probiotics may be unsafe for immunocompromised and elderly patients. Postbiotics are stable, easy to store and carry, and transfer active components to the gut, making them suitable for resource-constrained areas. Postbiotics can improve probiotic function and safety and stability. However, other studies have found emesis, stomach distension, and dehydration, especially in children under 5 [13, 14].

6.3 POSTBIOTICS PRODUCTION AND CHARACTERIZATION TECHNIQUE

Fermentation is a major postbiotic source, as microbial cells use prebiotic substrates to produce postbiotic molecules. These chemicals have antibacterial, antioxidant, and anticancer effects, improving food matrix functionality. Laboratory methods for postbiotic production include heat inactivation, formalin treatment, UV exposure, high-pressure processing, dehydration, ionizing radiation, CO_2 supercritical fluid extraction, pH adjustment, sonication, and omics technologies [15]. Electroconductive heating, pulsed electric fields, and high-intensity ultrasound also inactivate probiotic microorganisms. These methods work for bench-scale production, but further study is needed to develop cost-effective, time-saving, and novel commercial-scale postbiotic manufacturing methods that preserve their functional qualities. Postbiotic production can cause bacterial cell viability loss due to mechanical damage, genetic material denaturation, cell membrane rupture, or physiological changes. Raman, atomic force microscopy, proton nuclear magnetic resonance, flow cytometry, infrared, and GC-MS, LC-MS characterize postbiotics [16, 17].

Bacillus, *Lactobacillus*, *Faecalibacterium*, *Streptococcus*, and *Bifidobacterium* create postbiotics. Probiotic yeast *Saccharomyces boulardii* has postbiotic potential, according to recent investigations. Postbiotics can imitate probiotic functions without strict production or storage conditions, making them useful in resource-limited contexts. Postbiotics reduce live microbe exposure, making them ideal for youngsters with weak immune systems and intestinal barriers. Despite increased interest, a full evaluation of postbiotic antiobesity mechanisms is lacking. Postbiotics reduce obesity by increasing and regulating lipid metabolism and energy expenditure, inhibiting adipogenesis and adipocyte differentiation, suppressing food intake, reducing lipid absorption, and modulating gut dysbiosis [18, 19] (Figure 6.1).

6.4 VARIOUS TYPES OF POSTBIOTICS AND THEIR THERAPEUTIC USES

Metabolic illnesses are becoming more common due to inadequate nutrition, sedentary habits, and lack of physical activity, presenting a considerable global public health concern. The makeup of intestinal microbiota profoundly affects the etiology of metabolic diseases, with imbalances or reduced bacterial diversity associated with their onset. Postbiotics, originating from microbial metabolism, include molecules such as peptidoglycans, exopolysaccharides, teichoic acids,

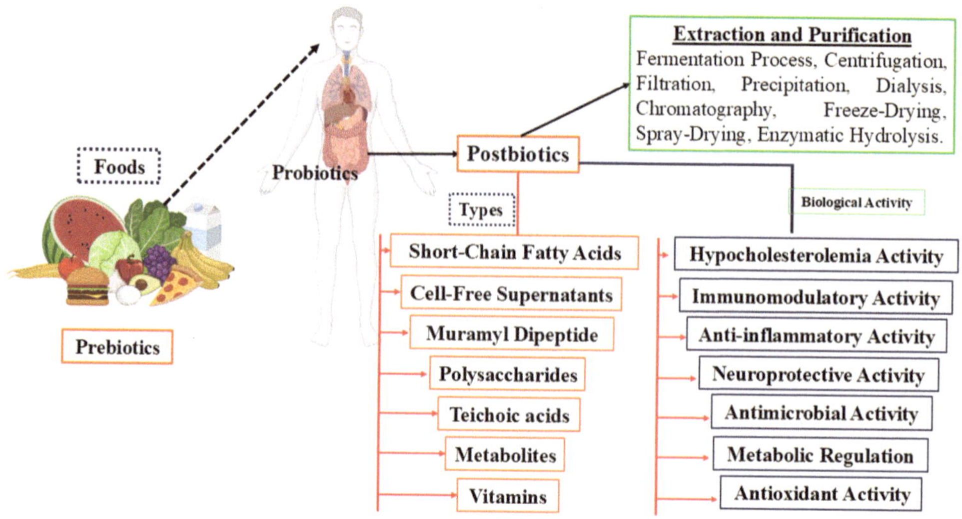

FIGURE 6.1 Production and characterization of postbiotics.

short-chain fatty acids, organic acids, and cell-free supernatants [20]. Postbiotics demonstrate several advantages, including antiobesity, anti-inflammatory, antibacterial, immunomodulatory, antioxidant, anticarcinogenic, and antihypertensive properties. Pathways facilitate these advantages, including lipid metabolism management, intestinal dysbiosis rectification, and immune system modulation (Table 6.1).

TABLE 6.1
Pharmacological Activity of Postbiotics in Obesity

Postbiotic Component	Source	Mechanism of Action	Effect on Obesity	Reference
Muramyl dipeptide	Gram-Ve, Gram +Ve	Activates NOD2 and RIPK2 pathways lowers adipose inflammation, improves glucose tolerance, reduces hepatic insulin resistance	Reduces adipose inflammation and improves glucose metabolism without altering microbiome composition	[21–23]
Surface layer proteins (SLPs)	*Lactobacillus kefiri*, *Leuconostoc mesenteroides*, plant sources	Modulates inflammation, targets adipogenesis, induces apoptosis in adipocytes	Improves systemic inflammation, reduces lipid accumulation, and prevents obesity in HFD models	[24–28]
Teichoic acid	*Bifidobacterium animalis*, *Lactobacillus plantarum*, *L. paracasei*	Activates TLR2/p38-AMPK pathways, modulates NF-κB signaling, reduces pro-inflammatory cytokines like IL-6, TNF-α, and IL-1β	Reduces fat accumulation, inflammation, and metabolic dysfunction; modulates immune response	[29–34]
Polysaccharides	*Bacillus licheniformis*, *Bacillus* sp.	Stimulates antioxidant pathways, enhances insulin sensitivity, reduces blood glucose levels	Inhibits hyperglycemia, reduces oxidative stress, and improves peripheral insulin sensitivity	[35–40]
Vitamin B_{12}	*Lactobacillus reuteri*, *L. plantarum*	Reduces homocysteine, alleviates oxidative stress, and modulates inflammation	Prevents obesity-associated oxidative stress and inflammation	[41–48]

6.4.1 Muramyl Dipeptide

Peptidoglycan, an essential part of bacterial cell walls, comprises muropeptides that are recognized by NOD proteins: NOD1 identifies diaminopimelic acid (meso-DAP), whereas NOD2 recognizes muramyl dipeptide (MDP). MDP, present in both Gram-positive and Gram-negative bacteria, has demonstrated encouraging postbiotic effects. MDP supplementation mitigated adipose inflammation and enhanced glucose tolerance in obese mice through NOD2 and IRF4, irrespective of weight reduction or alterations in the microbiota [21]. Moreover, MDP reduced hepatic insulin resistance in diet-induced obesity, with RIPK2 activation recognized as a crucial downstream mechanism for NOD2-mediated enhancements in glucose metabolism. Recent investigations demonstrated sex-dependent glucose-lowering effects of MDP, with adipocyte IRF4 playing a crucial role in male mice experiencing endotoxemia and obesity produced by a high-fat diet (HFD). Conversely, female mice exhibited enhancements in glucose levels irrespective of IRF4. Notwithstanding this, both genders exhibited reduced expression of inflammatory markers following MDP therapy. These findings underscore MDP's potential as a postbiotic therapy for obesity and associated metabolic diseases [22, 23].

6.4.2 Short-Chain Fatty Acids

Short-chain fatty acids (SCFAs) are volatile fatty acids created by intestinal microorganisms anaerobically fermenting nondigestible polysaccharides, mostly dietary fibers. Fiber increases SCFA production, with acetate, propionate, and butyrate dominating in a 60:20:20 ratio. SCFAs affect the intestinal environment to promote good bacteria like *Firmicutes*, which produce butyrate, and *Bacteroidetes*, which produce acetate and propionate [24]. SCFAs regulate lipid and glucose metabolism, reduce inflammation, prevent obesity and diabetes, decrease hunger and fat storage, and maintain intestinal barrier integrity. They control cardiovascular health, immune system responses, and adipose tissue remodeling. Specific strategies involve inhibiting histone deacetylase (HDAC), resulting in hyperacetylation [25].

SCFAs are anti-inflammatory and anticancer. Extracellular SCFA metabolites from *E. coli* KUB-36 exhibited more lethal effects on MCF breast cancer cells than colon or leukemia cells, reducing macrophage production [26]. SCFA advantages have been shown in clinical trials. For 24 weeks, overweight adults using inulin-propionate ester enhanced GLP-1 and PYY secretion, decreased appetite, caloric intake, intra-abdominal adipose tissue, and intra-hepatocellular lipid content. SCFAs activate GPR41 and GPR43, which increase gut hormone release and lower food intake and appetite, aiding weight loss. SCFAs lower blood pressure, heart hypertrophy, and fibrosis. Butyric acid reduced mice's angiotensin II-induced hypertension. Hypertension treatments may target SCFA receptors such GPR41, GPR43, GPR109a, and Olf78 (OR51E2 in humans). Therefore, SCFA administration may prevent cardiovascular disease and promote metabolic health [27, 28].

6.4.3 Teichoic Acids

Gram-positive bacteria cell walls include anionic glycopolymers called teichoic acids (TAs). Phosphodiester-linked polyol repeat units and ribitol polymers regulate cell structure, cell division, antibiotic resistance, host–bacterial interactions, and metabolic processes [29]. Glycolipids anchor lipoteichoic acids (LTAs) to bacterial membranes. Wall teichoic acids (WTAs) are covalently bonded to peptidoglycan and have a phosphodiester-linked polyol main chain and disaccharide linkage unit [30].

Several investigations have shown LTAs' immunomodulatory effects [31]. LTA from Lactobacillus plantarum inhibits JNK, ERK, and p38 kinase phosphorylation, as well as IκBα and IκBβ degradation, reducing LPS-induced TNF-α production in endotoxin shock-induced mice. LTA from *L.*

paracasei strain D3-5 treated high-fat diet-induced metabolic dysfunction, improved physical and cognitive abilities, and reduced inflammation. Increased phospho-p38-AMPK levels, TLR2 activation, and NF-κB reduction led to decreased pro-inflammatory cytokines such as IL-6, TNF-α, and IL-1β [32].

Further research shows that LTA from *L. rhamnosus* GG strain dose-dependently activates NF-κB in intestinal HEK293T and Caco-2 cells via TLR2/6 contact. According to another study, LTA from *L. plantarum* K8 inhibits TNF-α and IFN-γ by blocking p65, p38, STAT1/2, and JAK2 phosphorylation. In HaCaT cFells, complement component 3 mRNA and protein expression is reduced. As C3 is crucial for immune response pathways stimulated by cytokines like IL-1, IL-2, IL-6, TNF-α, and IFN-γ, LTA has immunomodulatory potential. Many studies show that LTAs reduce inflammation and intestinal tissue damage, although some imply the opposite [33]. Teichoic acids' intriguing anti-inflammatory, immunomodulatory, and anti-obesogenic activities require further research to determine their therapeutic potential and negative effects [34].

6.4.4 Polysaccharides

Glycans, or polysaccharides, are abundant macromolecules that include intracellular and structural components like LPS, peptidoglycan, and CPS, which may be produced as exopolysaccharides (EPS) during bacterial development. These chemicals are found on the cell surface or discharged into the environment and have physiochemical and biological properties related to structural support or energy storage [35]. EPS from *Lactobacillus plantarum* L-14 improved glucose absorption in adipocytes via the AS160 pathway, preventing and treating obesity and metabolic diseases. EPS from L-14 reduced lipid accumulation during early adipogenic differentiation by regulating AMPK downstream targets like ACC, SCD1, SREBP-1c, and FAS without affecting cell viability [36].

Bacillus subtilis exopolysaccharides (BSEPS) from *Bacillus subtilis* has also been found to treat hyperglycemia and dyslipidemia in diabetic animals. BSEPS increases high-density lipoprotein (HDL) and decreases blood glucose, troponin, total cholesterol, very high-density lipoprotein (VHDL), low-density lipoprotein (LDL), and triglycerides while increasing insulin levels, suggesting it may treat hyperglycemia and dyslipidemia. Levan, a *Bacillus licheniformis* polysaccharide, reduced hyperglycemia and oxidative stress in diabetic rats in another study. Levan supplementation lowered blood glucose by 110%, possibly by enhancing peripheral insulin sensitivity, activating pancreatic islets, and boosting antioxidants [37, 38]. In insulin-resistant HepG2 cells, EPS H31-2 from *Lactobacillus plantarum* H31 reduced glucose in the supernatant and increased glucose absorption via the AMPK/PI3K/Akt signaling pathway. EPS H31-2 increased glycometabolism genes such Akt-2, GLUT-4, and AMPK. Research suggests that EPS H31-2 may reduce blood glucose levels in type 2 diabetes mellitus (T2DM) via suppressing pancreatic α-amylase activity, making it a possible treatment option [39, 40].

6.4.5 Cell-Free Supernatants

Lactic acid bacteria and yeasts release cell-free supernatants (CFS), which contain biomolecules and metabolites such organic acids, di-acetylene, carbon dioxide, and bacteriocin-like compounds that support health and homeostasis. CFS composition depends on growing media nutrients. CFSs have antioxidant, anti-inflammatory, antitumoral, antibacterial, and biofilm formation suppression capabilities, making them useful in clinical applications [49]. Several *Lactobacillus* strain–derived intracellular fractions have been demonstrated to increase glutathione, a nonenzymatic antioxidant that helps maintain intracellular redox equilibrium. These nonenzymatic postbiotics may also scavenge ROS and reactive nitrogen species, which may explain their antioxidant properties [50]. Researchers found that *Bifidobacterium longum* and *Lactobacillus acidophilus* CFS prevented linoleic acid peroxidation and scavenged DPPH free radicals. Also, *Saccharomyces* species, including *Saccharomyces boulardii*, which is good for gut health, produce antioxidant mixtures that neutralize intracellular ROS and other free radicals, reducing oxidative stress and modulating mitochondrial dysfunction and redox imbalances [51].

Lactobacillus strains like *L. casei* and *L. acidophilus* have anti-inflammatory effects on neutrophils, intestinal epithelial cells, and macrophages by reducing pro-inflammatory cytokines like TNF-α and promoting IL-10 release, in addition to antioxidant effects [52]. On the other hand, *L. lactis*, *L. reuteri*, and *Saccharomyces boulardii* supernatants significantly reduced PGE-2 production in HT-29 human intestinal epithelial cells stimulated by LPS [53]. Additionally, *L. lactis* derivatives reduced IL-8 production, while *L. reuteri* supernatants reduced NO, COX-2, Hsp70, TNF-α, and IL-6 in macrophages. *S. boulardii* has anti-inflammatory capabilities due to its low molecular weight soluble components that inhibit NF-κB activation and IL-8 production in monocytes and intestinal epithelial cells [54, 55].

6.4.6 Metabolites

Urolithins are bioactive chemicals produced by gut microbes on ellagitannins and ellagic acid. Ingestion hydrolyses ellagitannins to ellagic acid, which converts the gut microbiota into urolithins C, A, D, and B. The gut's main metabolites, Uro-A and Uro-B, are highly active [56, 57]. An in vivo investigation examined gut dysbiosis in rats fed high-fat diets with Uro-A and Uro-B supplements. Both urolithins independently increased *Parabacteroides* and decreased desulfovibrionaceae and coriobacteriaceae, which are related to body weight growth. This modulation reduced adiposity. Molecular findings imply that Uro-A stimulates white adipose tissue browning and brown adipose tissue thermogenesis, increasing thyroid hormone-dependent energy expenditure [58, 59].

Another study demonstrated that acetate supplementation in drinking water stimulated thermogenic markers and mitochondrial biogenesis in BAT in mice. This effect was mediated by activating G-protein-coupled receptor 43 in brown adipocytes, which promoted mitochondrial biogenesis and enhanced the cells' energy-expending capacity [60]. Additionally, acetate induced adipogenesis and mitochondrial biogenesis in immortalized brown adipocyte cells (IM-BAT). In mice fed a high-fat diet (HFD), acetate supplementation led to the expression of browning markers and was associated with reduced inflammation and improved gut microbiota balance, characterized by an increase in *Bacteroidetes* and a decrease in *Firmicutes*. These findings suggest that acetate may have potential as a preventive strategy against obesity [61].

6.4.7 Vitamins

Vitamin shortages and biological abnormalities are linked to these molecules' vital physiological functions. Immune regulation, bone health, and antioxidant activity require vitamins A, D, and E. Mammals cannot biosynthesize these chemicals; hence, they must be eaten or supplemented. Gut commensals and certain probiotics help the body synthesize folates and B-complex vitamins needed for one-carbon metabolism and amino acid and nucleoside synthesis [41]. Poor diets, whether due to quality, imbalance, or starvation, reduce vitamin availability. This deficit impairs metabolic and enzymatic pathways, affecting DNA repair, cell division, energy production, and immunological response. As a fat-soluble coenzyme, vitamin K deficiency can affect blood coagulation. In addition, B-complex deficits can cause metabolic problems [42].

Excess ROS can damage lipids and proteins, creating cellular dysfunctions that lead to irreversible diseases including diabetes and microvascular and cardiovascular problems. Enzymes and nonenzymes like vitamins C and E neutralize ROS. For T2DM patients, folate and vitamin B_{12} deficiencies can increase oxidative stress. They decrease plasma homocysteine and operate in numerous metabolic pathways to reduce oxidative stress (OS), improve endothelial function, and prevent apoptosis [43]. Folic acid lowers blood glucose and increases SOD and CAT in experimentally induced diabetes. Folate may reduce SOD and CAT in diabetic rats by decreasing superoxide radicals. Folate's antioxidant properties may reduce NOS-produced superoxides. *Lactobacillus helveticus* CD6's intracellular folate antioxidant defends against free radicals. Deficient B_{12} and GSH are associated with oxidative stress-related diseases like T2DM. In B_{12}-deficient rats or diabetic people, TNF-α activity rises and epidermal growth factor drops, suggesting vitamin B_{12} may mitigate inflammation-induced oxidative stress via modulating cytokine and growth factor expression [44, 45].

6.5 CURRENT CHALLENGES AND FUTURE PERSPECTIVES

The biotics business is growing rapidly, yet consumers know little about postbiotics compared to probiotics. Postbiotic commercialization, mainly in the United States, Japan, and Europe, will likely accelerate due to a growing understanding of their benefits. Postbiotics have unique health benefits, but probiotics, prebiotics, and postbiotics are all important. Postbiotics do not require live bacteria, eliminating storage, processing, and viability difficulties [46]. They affect the oral cavity, skin, genitourinary tract, nasopharynx, and digestive tract. Postbiotics are stable, easy to process, and can be used in food, drinks, healthcare, cosmetics, and nutraceuticals. Due to customer demand and health awareness, postbiotic applications have moved from juices, smoothies, and meal replacements to bread and snacks. Probiotics have been extensively examined, but postbiotics are promising and have fewer disadvantages. Postbiotics and para-probiotics reduce obesity in cell cultures, animal models, and limited human research [47].

In animal studies, postbiotics boost fatty acid oxidation, thermogenic biomarker expression, and browning indicators while lowering food intake, weight gain, and lipid buildup. Oral postbiotics may be digested by gut bacteria before reaching the large intestine, reducing their potency. Bioavailability and targeted effect evaluation may benefit from in vitro gut models. According to limited clinical investigations, postbiotics increase humans' energy expenditure, lipolysis, and metabolic control. Despite promising results, knowledge gaps persist. Postbiotics don't grow in the gut like probiotics, limiting their effects after dosing. More studies are needed to find and quantify postbiotic biomarkers to determine appropriate doses and understand their involvement in disease prevention and therapy. Postbiotic components, including LTA and LPS, can modulate the immune system and cause excessive inflammation. To ensure their safety, dosage and administration must be studied [48].

6.6 CONCLUSION

The rising prevalence of obesity and related metabolic conditions, such as insulin resistance and diabetes mellitus, presents a major public health concern. Postbiotics, isolated from probiotics via secretion or bacterial lysis, are prospective biotherapeutics for various issues. Since postbiotics are nonliving, they are more stable, safe, and easy to process, store, and transport than probiotics. Postbiotics, produced by intestinal microbiota, have antiobesity, antioxidant, immunomodulatory, hypocholesterolemia, antidiabetic, and anti-inflammatory properties, making them useful for managing obesity and metabolic disorders. According to preclinical and clinical research, postbiotics stimulate thermogenesis, increase lipid oxidation, decrease adipocyte fat storage, and modulate obesity-related gene expression. Postbiotics can increase browning indicators, thermogenic proteins, and fatty acid oxidation while blocking fat formation pathways like PPARγ and SREBP-1c. Additionally, postbiotics reduce inflammation by lowering IL-6 and NF-κB levels and promoting death in adipocytes. SCFAs, a type of postbiotic, activate GRP-41 and GRP-43 receptors, which release intestinal hormones including GLP-1 and PYY, enhancing lipolysis and lowering weight gain. Most of the evidence comes from animal research on rats, mice, pigs, and nematodes, although a few randomized, controlled clinical trials have shown postbiotics work in humans. These SCFA and heat-inactivated probiotic trials in overweight and obese participants show that postbiotics can reduce weight gain, modulate metabolic pathways, and increase energy expenditure. Postbiotics also reduce antibiotic resistance from probiotic virulence factors, improving safety.

REFERENCES

1. Broughton, D. E., & Moley, K. H. (2017). Obesity and female infertility: Potential mediators of obesity's impact. *Fertility and Sterility*, 107(4), 840–847. https://doi.org/10.1016/j.fertnstert.2017.01.017
2. Faccioli, N., Poitou, C., Clément, K., & Dubern, B. (2023). Current treatments for patients with genetic obesity. *Journal of Clinical Research in Pediatric Endocrinology*, 15(2), 108–119. https://doi.org/10.4274/jcrpe.galenos.2023.2023-3-2
3. Conway, B., & Rene, A. (2004). Obesity as a disease: No lightweight matter. *Obesity Reviews: An Official Journal of the International Association for the Study of Obesity*, 5(3), 145–151. https://doi.org/10.1111/j.1467-789X.2004.00144.x
4. Ahmed, B., Sultana, R., & Greene, M. W. (2021). Adipose tissue and insulin resistance in obese. *Biomedicine & Pharmacotherapy = Biomedecine & pharmacotherapie*, 137, 111315. https://doi.org/10.1016/j.biopha.2021.111315
5. Cheng, L., Wang, J., Dai, H., Duan, Y., An, Y., Shi, L., Lv, Y., Li, H., Wang, C., Ma, Q., Li, Y., Li, P., Du, H., & Zhao, B. (2021). Brown and beige adipose tissue: A novel therapeutic strategy for obesity and type 2 diabetes mellitus. *Adipocyte*, 10(1), 48–65. https://doi.org/10.1080/21623945.2020.1870060
6. Litwin, M., & Kułaga, Z. (2021). Obesity, metabolic syndrome, and primary hypertension. *Pediatric Nephrology (Berlin, Germany)*, 36(4), 825–837. https://doi.org/10.1007/s00467-020-04579-3
7. Barko, P. C., McMichael, M. A., Swanson, K. S., & Williams, D. A. (2018). The gastrointestinal microbiome: A review. *Journal of Veterinary Internal Medicine*, 32(1), 9–25. https://doi.org/10.1111/jvim.14875
8. Vallianou, N., Stratigou, T., Christodoulatos, G. S., Tsigalou, C., & Dalamaga, M. (2020). Probiotics, prebiotics, synbiotics, postbiotics, and obesity: Current evidence, controversies, and perspectives. *Current Obesity Reports*, 9(3), 179–192. https://doi.org/10.1007/s13679-020-00379-w
9. Li, H. Y., Zhou, D. D., Gan, R. Y., Huang, S. Y., Zhao, C. N., Shang, A., Xu, X. Y., & Li, H. B. (2021). Effects and mechanisms of probiotics, prebiotics, synbiotics, and postbiotics on metabolic diseases targeting gut microbiota: A narrative review. *Nutrients*, 13(9), 3211. https://doi.org/10.3390/nu13093211
10. Oh, K. K., Gupta, H., Min, B. H., Ganesan, R., Sharma, S. P., Won, S. M., Jeong, J. J., Lee, S. B., Cha, M. G., Kwon, G. H., Jeong, M. K., Hyun, J. Y., Eom, J. A., Park, H. J., Yoon, S. J., Choi, M. R., Kim, D. J., & Suk, K. T. (2022). Elucidation of prebiotics, probiotics, postbiotics, and target from gut microbiota to alleviate obesity via network pharmacology study. *Cells*, 11(18), 2903. https://doi.org/10.3390/cells11182903
11. Zhao, X., Liu, S., Li, S., Jiang, W., Wang, J., Xiao, J., Chen, T., Ma, J., Khan, M. Z., Wang, W., Li, M., Li, S., & Cao, Z. (2024). Unlocking the power of postbiotics: A revolutionary approach to nutrition for humans and animals. *Cell Metabolism*, 36(4), 725–744. https://doi.org/10.1016/j.cmet.2024.03.004
12. Youn, H. Y., Seo, K. H., Kim, H. J., Kim, Y. S., & Kim, H. (2022). Effect of postbiotics derived from kefir lactic acid bacteria-mediated bioconversion of citrus pomace extract and whey on high-fat diet-induced obesity and gut dysbiosis. *Food Research International (Ottawa, Ont.)*, 162(Pt A), 111930. https://doi.org/10.1016/j.foodres.2022.111930
13. Rawal, S., & Ali, S. A. (2023). Probiotics and postbiotics play a role in maintaining dermal health. *Food & Function*, 14(9), 3966–3981. https://doi.org/10.1039/d3fo00152k
14. Yoshioka, K., Kakihana, K., Doki, N., & Ohashi, K. (2017). Gut microbiota and acute graft-versus-host disease. *Pharmacological Research*, 122, 90–95. https://doi.org/10.1016/j.phrs.2017.05.028
15. Thorakkattu, P., Khanashyam, A. C., Shah, K., Babu, K. S., Mundanat, A. S., Deliephan, A., Deokar, G. S., Santivarangkna, C., & Nirmal, N. P. (2022). Postbiotics: Current trends in food and pharmaceutical industry. *Foods (Basel, Switzerland)*, 11(19), 3094. https://doi.org/10.3390/foods11193094
16. Zhu, M., Liu, X., Ye, Y., Yan, X., Cheng, Y., Zhao, L., Chen, F., & Ling, Z. (2022). Gut microbiota: A novel therapeutic target for Parkinson's disease. *Frontiers in Immunology*, 13, 937555. https://doi.org/10.3389/fimmu.2022.937555
17. Sabahi, S., Homayouni Rad, A., Aghebati-Maleki, L., Sangtarash, N., Ozma, M. A., Karimi, A., Hosseini, H., & Abbasi, A. (2023). Postbiotics as the new frontier in food and pharmaceutical research. *Critical Reviews in Food Science and Nutrition*, 63(26), 8375–8402. https://doi.org/10.1080/10408398.2022.2056727
18. Park, S. J., Sharma, A., & Lee, H. J. (2023). Postbiotics against obesity: Perception and overview based on pre-clinical and clinical studies. *International Journal of Molecular Sciences*, 24(7), 6414. https://doi.org/10.3390/ijms24076414

19. Mishra, N., Garg, A., Ashique, S., & Bhatt, S. (2024). Potential of postbiotics for the treatment of metabolic disorders. *Drug Discovery Today*, 29(4), 103921. https://doi.org/10.1016/j.drudis.2024.103921
20. Liu, L., Wang, H., Chen, X., Zhang, Y., Zhang, H., & Xie, P. (2023). Gut microbiota and its metabolites in depression: From pathogenesis to treatment. *EBioMedicine*, 90, 104527. https://doi.org/10.1016/j.ebiom.2023.104527
21. Cavallari, J. F., Fullerton, M. D., Duggan, B. M., Foley, K. P., Denou, E., Smith, B. K., Desjardins, E. M., Henriksbo, B. D., Kim, K. J., Tuinema, B. R., Stearns, J. C., Prescott, D., Rosenstiel, P., Coombes, B. K., Steinberg, G. R., & Schertzer, J. D. (2017). Muramyl dipeptide-based postbiotics mitigate obesity-induced insulin resistance via IRF4. *Cell Metabolism*, 25(5), 1063–1074.e3. https://doi.org/10.1016/j.cmet.2017.03.021
22. Williams, L., Alshehri, A., Robichaud, B., Cudmore, A., & Gagnon, J. (2020). The role of the bacterial muramyl dipeptide in the regulation of GLP-1 and glycemia. *International Journal of Molecular Sciences*, 21(15), 5252. https://doi.org/10.3390/ijms21155252
23. Williams, L., Alshehri, A., Robichaud, B., Cudmore, A., & Gagnon, J. (2020). The role of the bacterial muramyl dipeptide in the regulation of GLP-1 and glycemia. *International Journal of Molecular Sciences*, 21(15), 5252. https://doi.org/10.3390/ijms21155252
24. Cuevas-Sierra, A., Ramos-Lopez, O., Riezu-Boj, J. I., Milagro, F. I., & Martinez, J. A. (2019). Diet, gut microbiota, and obesity: Links with host genetics and epigenetics and potential applications. *Advances in Nutrition (Bethesda, Md.)*, 10(suppl_1), S17–S30. https://doi.org/10.1093/advances/nmy078
25. Jiang, S., Zhai, H., Li, D., Huang, J., Zhang, H., Li, Z., Zhang, W., & Xu, G. (2016). AMPK-dependent regulation of GLP1 expression in L-like cells. *Journal of Molecular Endocrinology*, 57(3), 151–160. https://doi.org/10.1530/JME-16-0099
26. Budai, M. M., Varga, A., Milesz, S., Tőzsér, J., & Benkő, S. (2013). Aloe vera downregulates LPS-induced inflammatory cytokine production and expression of NLRP3 inflammasome in human macrophages. *Molecular Immunology*, 56(4), 471–479. https://doi.org/10.1016/j.molimm.2013.05.005
27. Schwager, J., Richard, N., Widmer, F., & Raederstorff, D. (2017). Resveratrol distinctively modulates the inflammatory profiles of immune and endothelial cells. *BMC Complementary and Alternative Medicine*, 17(1), 309. https://doi.org/10.1186/s12906-017-1823-z
28. Ang, Z., & Ding, J. L. (2016). GPR41 and GPR43 in obesity and inflammation - protective or causative?. *Frontiers in Immunology*, 7, 28. https://doi.org/10.3389/fimmu.2016.00028
29. Balaguer, F., Enrique, M., Llopis, S., Barrena, M., Navarro, V., Álvarez, B., Chenoll, E., Ramón, D., Tortajada, M., & Martorell, P. (2022). Lipoteichoic acid from *Bifidobacterium animalis* subsp. lactis BPL1: A novel postbiotic that reduces fat deposition via IGF-1 pathway. *Microbial Biotechnology*, 15(3), 805–816. https://doi.org/10.1111/1751-7915.13769
30. Loo, T. M., Kamachi, F., Watanabe, Y., Yoshimoto, S., Kanda, H., Arai, Y., Nakajima-Takagi, Y., Iwama, A., Koga, T., Sugimoto, Y., Ozawa, T., Nakamura, M., Kumagai, M., Watashi, K., Taketo, M. M., Aoki, T., Narumiya, S., Oshima, M., Arita, M., Hara, E., … Ohtani, N. (2017). Gut microbiota promotes obesity-associated liver cancer through PGE2-mediated suppression of antitumor immunity. *Cancer Discovery*, 7(5), 522–538. https://doi.org/10.1158/2159-8290.CD-16-0932
31. Kim, H. G., Kim, N. R., Gim, M. G., Lee, J. M., Lee, S. Y., Ko, M. Y., Kim, J. Y., Han, S. H., & Chung, D. K. (2008). Lipoteichoic acid isolated from Lactobacillus plantarum inhibits lipopolysaccharide-induced TNF-alpha production in THP-1 cells and endotoxin shock in mice. *Journal of Immunology*, 180(4), 2553–2561. https://doi.org/10.4049/jimmunol.180.4.2553
32. Kim, H. G., Lee, S. Y., Kim, N. R., Ko, M. Y., Lee, J. M., Yi, T. H., Chung, S. K., & Chung, D. K. (2008). Inhibitory effects of *Lactobacillus plantarum* lipoteichoic acid (LTA) on *Staphylococcus aureus* LTA-induced tumor necrosis factor-alpha production. *Journal of Microbiology and Biotechnology*, 18(6), 1191–1196.
33. Asif, A., Afzaal, M., Shahid, H., Saeed, F., Ahmed, A., Shah, Y. A., Ejaz, A., Ghani, S., Ateeq, H., & Khan, M. R. (2023). Probing the functional and therapeutic properties of postbiotics in relation to their industrial application. *Food Science & Nutrition*, 11(8), 4472–4484. https://doi.org/10.1002/fsn3.3465
34. Asif, A., Afzaal, M., Shahid, H., Saeed, F., Ahmed, A., Shah, Y. A., Ejaz, A., Ghani, S., Ateeq, H., & Khan, M. R. (2023). Probing the functional and therapeutic properties of postbiotics in relation to their industrial application. *Food Science & Nutrition*, 11(8), 4472–4484. https://doi.org/10.1002/fsn3.3465
35. Cuevas-Sierra, A., Ramos-Lopez, O., Riezu-Boj, J. I., Milagro, F. I., & Martinez, J. A. (2019). Diet, gut microbiota, and obesity: Links with host genetics and epigenetics and potential applications. *Advances in Nutrition (Bethesda, Md.)*, 10(suppl_1), S17–S30. https://doi.org/10.1093/advances/nmy078

36. Tsai, Y. L., Lin, T. L., Chang, C. J., Wu, T. R., Lai, W. F., Lu, C. C., & Lai, H. C. (2019). Probiotics, prebiotics and amelioration of diseases. *Journal of Biomedical Science*, 26(1), 3. https://doi.org/10.1186/s12929-018-0493-6
37. Zhao, Y., Zheng, Y., Xie, K., Hou, Y., Liu, Q., Jiang, Y., Zhang, Y., & Man, C. (2023). Combating obesity: Harnessing the synergy of postbiotics and prebiotics for enhanced lipid excretion and microbiota regulation. *Nutrients*, 15(23), 4971. https://doi.org/10.3390/nu15234971
38. Luzzi, A., Briata, I. M., Di Napoli, I., Giugliano, S., Di Sabatino, A., Rescigno, M., & Cena, H. (2024). Prebiotics, probiotics, synbiotics and postbiotics to adolescents in metabolic syndrome. *Clinical Nutrition (Edinburgh, Scotland)*, 43(6), 1433–1446. https://doi.org/10.1016/j.clnu.2024.04.032
39. Park, M., Joung, M., Park, J. H., Ha, S. K., & Park, H. Y. (2022). Role of postbiotics in diet-induced metabolic disorders. *Nutrients*, 14(18), 3701. https://doi.org/10.3390/nu14183701
40. Rodrigues, E., Lacerda, R., Fang, H., Robin, N., Bhatwa, A., Marko, D. M., & Schertzer, J. D. (2023). Microbiota and NOD-like receptors balance inflammation and metabolism during obesity and diabetes. *Biomedical Journal*, 46(5), 100610. https://doi.org/10.1016/j.bj.2023.100610
41. Mishra, N., Garg, A., Ashique, S., & Bhatt, S. (2024). Potential of postbiotics for the treatment of metabolic disorders. *Drug Discovery Today*, 29(4), 103921. https://doi.org/10.1016/j.drudis.2024.103921
42. Fiore, G., Pascuzzi, M. C., Di Profio, E., Corsello, A., Agostinelli, M., La Mendola, A., Milanta, C., Campoy, C., Calcaterra, V., Zuccotti, G., & Verduci, E. (2023). Bioactive compounds in childhood obesity and associated metabolic complications: Current evidence, controversies and perspectives. *Pharmacological Research*, 187, 106599. https://doi.org/10.1016/j.phrs.2022.106599
43. Mostafavi Abdolmaleky, H., & Zhou, J. R. (2024). Gut microbiota dysbiosis, oxidative stress, inflammation, and epigenetic alterations in metabolic diseases. *Antioxidants (Basel, Switzerland)*, 13(8), 985. https://doi.org/10.3390/antiox13080985
44. Kumareswaran, A., Ekeuku, S. O., Mohamed, N., Muhammad, N., Hanafiah, A., Pang, K. L., Wong, S. K., Chew, D. C. H., & Chin, K. Y. (2023). The effects of tocotrienol on gut microbiota: A scoping review. *Life (Basel, Switzerland)*, 13(9), 1882. https://doi.org/10.3390/life13091882
45. Karamshetty, V., Acharya, J. D., Ghaskadbi, S., & Goel, P. (2016). Mathematical modeling of glutathione status in type 2 diabetics with vitamin B12 deficiency. *Frontiers in Cell and Developmental Biology*, 4, 16. https://doi.org/10.3389/fcell.2016.00016
46. Thorakkattu, P., Khanashyam, A. C., Shah, K., Babu, K. S., Mundanat, A. S., Deliephan, A., Deokar, G. S., Santivarangkna, C., & Nirmal, N. P. (2022). Postbiotics: Current trends in food and pharmaceutical industry. *Foods (Basel, Switzerland)*, 11(19), 3094. https://doi.org/10.3390/foods11193094
47. Duarte, M., Oliveira, A. L., Oliveira, C., Pintado, M., Amaro, A., & Madureira, A. R. (2022). Current postbiotics in the cosmetic market: An update and development opportunities. *Applied Microbiology and Biotechnology*, 106(18), 5879–5891. https://doi.org/10.1007/s00253-022-12116-5
48. Maftei, N. M., Raileanu, C. R., Balta, A. A., Ambrose, L., Boev, M., Marin, D. B., & Lisa, E. L. (2024). The potential impact of probiotics on human health: An update on their health-promoting properties. *Microorganisms*, 12(2), 234. https://doi.org/10.3390/microorganisms12020234
49. Seo, K. H., Gyu Lee, H., Young Eor, J., Jin Jeon, H., Yokoyama, W., & Kim, H. (2022). Effects of kefir lactic acid bacteria-derived postbiotic components on high fat diet-induced gut microbiota and obesity. *Food Research International (Ottawa, Ont.)*, 157, 111445. https://doi.org/10.1016/j.foodres.2022.111445
50. Sørensen, H. M., Rochfort, K. D., Maye, S., MacLeod, G., Brabazon, D., Loscher, C., & Freeland, B. (2022). Exopolysaccharides of lactic acid bacteria: Production, purification and health benefits towards functional food. *Nutrients*, 14(14), 2938. https://doi.org/10.3390/nu14142938
51. Shin, H. H., Kim, J. H., Jung, Y. J., Kwak, M. S., Sung, M. H., & Imm, J. Y. (2024). Postbiotic potential of *Bacillus velezensis* KMU01 cell-free supernatant for the alleviation of obesity in mice. *Heliyon*, 10(5), e25263. https://doi.org/10.1016/j.heliyon.2024.e25263
52. Wu, Z. Q., Chen, X. M., Ma, H. Q., Li, K., Wang, Y. L., & Li, Z. J. (2023). *Akkermansia muciniphila* cell-free supernatant improves glucose and lipid metabolisms in *Caenorhabditis elegans*. *Nutrients*, 15(7), 1725. https://doi.org/10.3390/nu15071725
53. Ghaderi, F., Sotoodehnejadnematalahi, F., Hajebrahimi, Z., Fateh, A., & Siadat, S. D. (2022). Effects of active, inactive, and derivatives of *Akkermansia muciniphila* on the expression of the endocannabinoid system and PPARs genes. *Scientific Reports*, 12(1), 10031. https://doi.org/10.1038/s41598-022-13840-8
54. Kim, E., Lee, H. G., Han, S., Seo, K. H., & Kim, H. (2021). Effect of surface layer proteins derived from paraprobiotic kefir lactic acid bacteria on inflammation and high-fat diet-induced obesity. *Journal of Agricultural and Food Chemistry*, 69(50), 15157–15164. https://doi.org/10.1021/acs.jafc.1c05037

55. Xu, H., Wang, J., Cai, J., Feng, W., Wang, Y., Liu, Q., & Cai, L. (2019). Protective effect of *Lactobacillus rhamnosus* GG and its supernatant against myocardial dysfunction in obese mice exposed to intermittent hypoxia is associated with the activation of Nrf2 pathway. *International Journal of Biological Sciences*, 15(11), 2471–2483. https://doi.org/10.7150/ijbs.36465
56. Eslami, M., Pakmehr, A., Pourghazi, F., Kami, A., Ejtahed, H. S., Mohajeri-Tehrani, M., Hasani-Ranjbar, S., & Larijani, B. (2024). The anti-obesity effects of postbiotics: A systematic review of pre-clinical and clinical studies. *Clinical Nutrition ESPEN*, 64, 370–389. https://doi.org/10.1016/j.clnesp.2024.10.153
57. Kayser, E., He, F., Nixon, S., Howard-Varona, A., Lamelas, A., Martinez-Blanch, J., Chenoll, E., Davenport, G. M., & de Godoy, M. R. C. (2024). Effects of supplementation of live and heat-treated *Bifidobacterium animalis* subspecies lactis CECT 8145 on glycemic and insulinemic response, fecal microbiota, systemic biomarkers of inflammation, and white blood cell gene expression of adult dogs. *Journal of Animal Science*, 102, skae291. https://doi.org/10.1093/jas/skae291
58. Lima, M., & Paulino, L. C. (2024). Oral postbiotics as a therapeutic strategy for atopic dermatitis: A systematic review of randomized controlled trials. *Journal of the American Nutrition Association*, 43(2), 139–146. https://doi.org/10.1080/27697061.2023.2232021
59. Oh, K. K., Gupta, H., Min, B. H., Ganesan, R., Sharma, S. P., Won, S. M., Jeong, J. J., Lee, S. B., Cha, M. G., Kwon, G. H., Jeong, M. K., Hyun, J. Y., Eom, J. A., Park, H. J., Yoon, S. J., Choi, M. R., Kim, D. J., & Suk, K. T. (2022). Elucidation of prebiotics, probiotics, postbiotics, and target from gut microbiota to alleviate obesity via network pharmacology study. *Cells*, 11(18), 2903. https://doi.org/10.3390/cells11182903
60. García-Carrizo, F., Cannon, B., Nedergaard, J., Picó, C., Dols, A., Rodríguez, A. M., & Palou, A. (2020). Regulation of thermogenic capacity in brown and white adipocytes by the prebiotic high-esterified pectin and its postbiotic acetate. *International Journal of Obesity (2005)*, 44(3), 715–726. https://doi.org/10.1038/s41366-019-0445-6
61. Yau, W. W., Singh, B. K., Lesmana, R., Zhou, J., Sinha, R. A., Wong, K. A., Wu, Y., Bay, B. H., Sugii, S., Sun, L., & Yen, P. M. (2019). Thyroid hormone (T3) stimulates brown adipose tissue activation via mitochondrial biogenesis and MTOR-mediated mitophagy. *Autophagy*, 15(1), 131–150. https://doi.org/10.1080/15548627.2018.1511263

7 Postbiotics
Beneficial Impact on Mitochondrial Disorders

Shakira Ghazanfar
National institute of Genomics and Agriculture Biotechnology (NIGAB), National Agriculture Research Center, Islamabad, Pakistan

7.1 INTRODUCTION

7.1.1 History and Definition of Postbiotics

The term "postbiotic" is derived from the Greek words "post," which means after, and "*bios*," which means life. The "biotic" family (probiotics, prebiotics, synbiotics, and postbiotics) revolves around microbes or their product. Hence, the term "postbiotic" suitably refers to components derived from dead, inactive, or inanimate microbes (1). In 2004 the term "postbiotics" was used to define the bioactive products secreted by probiotic microorganisms (2).

In 2021, the International Scientific Association for Probiotics and Prebiotics (ISAPP) proposed the following extensive definition of postbiotics: "synthesis of inanimate or inactive microbes or their product that give health benefits to their consumers" (3). These are bioactive elements produced by healthy bacteria in our gut, called probiotic bacteria, which feed on different types of prebiotic food like fibers in our colon. This is the reason numerous health benefits of prebiotics and probiotics come from the secretion of postbiotics (4). These are soluble, cell-free supernatant (CFS) and nonviable components secreted by living microbes or released due to bacterial lysis, like short-chain fatty acids (SCFAs), enzymes, teichoic acids, peptides, cell wall peptides, polysaccharides, peptidoglycan-derived muropeptides, organic acids, and cell-surface proteins, that exert biological effects on the hosts (5, 6). Or they can be substances produced by microbes, released from food or microbial components, including nonviable cells that give health benefits when consumed in adequate amounts (6).

The concept of postbiotics is compared with prebiotics and probiotics. The Food and Agriculture Organization of the United Nations–World Health Organization (FAO-WHO) defined probiotics as "live microbes that give health benefits to its consumer when administered in sufficient amounts" (7). The recent survey of literature review by Nataraj and coauthors suggests that postbiotics have multiple pharmacodynamic properties over live microbes (8); these are as follows.

1. There is no risk of bacterial relocation from the intestinal lumen to blood capillaries along with vulnerable, harmful, and immunocompromised factors.
2. There is no chance of getting and transfer of different antibiotic resistance genes.
3. Postbiotics are more natural to extract, transport, and standardize and are easy to store.
4. Loss of viability due to cell lysis can exert further beneficial effects.
5. There are greater and more direct interactions of every released postbiotic with the epithelial cells of the gut.

In 1955, Glenn Gibson and Marcel Roberfroid introduced the concept of prebiotics defined as "a non-digestible food ingredient that beneficially affects the host by selectively stimulating the growth

DOI: 10.1201/9781003607779-7

and/or activity of one or a limited number of bacteria in the colon, and thus improves host health" (9). A synbiotic is a synergetic mixture of probiotics and prebiotics, which provide beneficial effects to the host by improving the number and colonization of live healthful microbes in the gastrointestinal tract (7). Postbiotics can be used as an alternate option for people who have trouble taking probiotics due to their safety concerns. They increase the effectiveness of active microbes or modify them into functional elements. They also possess anti-inflammatory, antiobesity, immunomodulatory, and cholesterol-lowering characteristics (10).

7.1.2 Production of Postbiotics

Postbiotics are produced by various bacterial strains like *Bifidobacterium*, *Lactobacillus*, *Faecalibacterium*, *Streptococcus*, *Saccharomyces*, and *Eubacterium*. These are also produced by fungi. Various fermented foods like yogurt, pickled vegetables, sauerkraut, and kombucha contain these microbes. During fermentation, these probiotic bacteria produce a variety of postbiotics or bioactive metabolites that have antioxidant, anti-inflammatory, and immunomodulatory activities (11). Postbiotics are also released due to cell lysis of complex bacterial cultures in food or the intestine. Postbiotics in the food include soluble factors, such as metabolic by-products released during bacterial growth in the medium, and these postbiotics don't undergo post-propagation treatment while postbiotics released due to cell lysis undergo specific fragmentation techniques like chemical, thermal, enzymatic, or solvent extraction (6). The quantity and kind of postbiotic compounds depend upon the type of bacterial culture, culture medium, and food processing methods like irradiation, high pressure, heat, and sonication. Microbes involved in the process of fermentation respond in different ways to these methods; for instance some proteins released from bacteria are inactivated or denatured by heat, while irradiation can cause mutations in the genome. The molecular mechanisms of postbiotic action are mediated by interaction between host and microbial products (12).

7.2 CLASSIFICATION OF POSTBIOTICS

The postbiotics are classified by various factors, including the type of microbes or the availability of different substrates or food like dietary fibers or prebiotics (2). There are many classes of postbiotics like metabolites such as SCFAs, exopolysaccharides, cell wall fragments, enzymes/proteins, and structural postbiotics such as peptides, teichoic acids, and plasmalogens. Some postbiotics produced or secreted by gut microbiota are discussed here.

7.2.1 Short-Chain Fatty Acids (SCFAs)

SCFAs fatty acids synthesized by intestinal microbial strains by anaerobic fermentation of nondigestible dietary fibers or polysaccharides that enter the gut as food. If the consumption of dietary fibers is enhanced, SCFA synthesis is also increased by gut microbes (13). SCFAs are postbiotic organic acids with less than six carbon chains. The most important SCFAs are two-carbon acetate (C2), three-carbon propionate (C3), and four-carbon butyrate (C4). Acetate contributes almost 60%, while propionate and butyrate contribute 20% each of total SCFAs (14). These carbon ions are synthesized during bacterial fermentation in the gut. SCFAs are formed by the digestion of prebiotics by intestinal microbes or when dietary fibers are fermented by intestinal enzymes (5). These postbiotics are secreted during plant polysaccharide fermentation by gut microbes like Bacteroides and Firmicutes. Other strains like *Lactobacillus acidophilus*, *L. fermentum*, *L. paracasei* ATCC 335, and *L. brevis* synthesized SCFAs by cell wall lysis. SCFs have several advantageous effects on health. They improve the colonic function, lower the pH, enhance the blood flow and epithelial cell proliferation in the gut, and significantly decrease the chance of colorectal diseases (7). These acids can increase the synthesis of mucins, interleukin 10 and 18 cytokines, antimicrobial proteins like bacteriocins, and serotonin (neurotransmitter) in the intestine and enhance immune functions (15).

As SCFA has significant antibacterial properties, it can replace antibiotics. Other long-chain fatty acids like eicosapentaenoic acid (EPA), lauric acid, and meristic acid can also inhibit the multiplication of Gram-positive bacteria (16).

7.2.2 Peptides

Postbiotic antimicrobial peptides are produced by microbial strains. These prevent the growth and proliferation of microbes by pleiotropic processes, in which AMP can act in various ways including (1) targeting the microbial membrane and making it porous, (2) targeting the bacterial cytoplasm and disintegrating them, (3) inhibiting the synthesis of microbial macro-molecules, or (4) producing hydrolase enzymes that damage the bacterial cell wall. It was discovered that the growth of *Salmonella enterica* is prevented by *E. coli* antimicrobial peptides that harm the cell wall (16). These are also used as immunomodulatory agents to strengthen innate immunity. AMP produced by genetically modified *Lactococcus lactis* prevents the growth of pathogenic strains like multidrug-resistant *E. faecium* from proliferating in the gut. The most investigated AMP is bacteriocin, which has great antibacterial activity and is synthesized by both Gram-negative and Gram-positive strains (5). For thousands of years, humans have made use of bacteriocins in fermented foods due to their strong antibacterial activities. These also have advantageous properties like resistance to pH and heat. The growth and multiplication of gastrointestinal infections are prevented by bacteriocins, which also decrease the threat of gut infection (11).

7.2.3 Vitamins

Vitamins are important postbiotics produced by probiotic bacteria in the food matrix and the gut of the host. They also have antibacterial activity and can inhibit the growth of harmful and pathogenic microbes in the gut. For example, itamin C blocks the growth of microbes by increasing the acidity of microbial membrane and cell wall, causing the bacterial wall to be lysed. These postbiotic compounds can be utilized in various ways to preserve and enhance the food shelf life (16). Lactic acid bacteria *Bifido* play an important role in the synthesis of de novo vitamins in the human gut, such as vitamin K, which is synthesized by lactobacilli, and citamin B_{12}, which is a vital postbiotic product synthesized by microbiota as it can't be produced by animals, plants, or fungi (5).

7.2.4 Enzymes

Postbiotic enzymes like peptidases, proteases, amylases, antioxidants, ureases, and lipases catalyze multiple biological activities. These enzymes are derived from a breakdown of microbial components like protein, carbohydrates, and fats and assist human digestion and also improve the taste of food (13). The specific activities or functions of different enzymes can vary depending on the composition of microbes, animal species, and food substrate. Microbial enzymes have antioxidant activities, so they can regulate the physiology and biochemical metabolism of animals. Moreover, the enzyme catalase produced from *Lactococcus lactis* can prevent colon cancer. Proteases produced by *Bacillus gibsonii* 6BS15-4 help in the digestion of protein and tissue growth and repair. Other microbial enzymes, like hemicellulases, amylases, or cellulases, can digest the dietary carbohydrates, including hemicellulose, starches, and cellulose, into simpler sugar to aid the host in digestion and absorption (17).

7.2.5 Exopolysaccharides (EPS)

Postbiotic exopolysaccharides are extracellular metabolic products of microbes and are polymers of substances that are either soluble or insoluble and are produced by bacteria, fungi, yeast, and algae. These are high-molecular-weight substances made up of sugar residues of carbohydrates,

DNA, proteins, phospholipids, and noncarbohydrate compounds such as phosphate, acetate, sulfate, glycerol, succinate, pyruvate, and carboxylate. EPSs can be used as bioabsorbents, bioflocculants, and drug delivery agents (18). A great number of researchers have investigated the use of EPS with immune cell macrophages to enhance immune activity. These interact with adaptive immune responses through cytokines production and pathogenic phagocytosis. It is also reported that EPSs also exhibit significant antitumor activities (19). Wang et al. in 2024 studied the in vitro postbiotic fermentative and digestive characteristics of EPS secreted from *Levilactobacillus brevis* M-10. They observed that EPS was not degraded in saliva but very slightly digested in gastric juice and intestinal juice. The EPS could lower the pH of the intestine pH and increase the diversity of the microbiota of the human gut. They conclude that EPS has strong postbiotic characteristics (20). Lactic acid bacteria are an important source of various postbiotic EPS secretions. EPS is produced during the fermentation of live lactic acid bacteria as metabolites, having multiple health-promoting impacts on the host. Lactic acid bacteria encode all the regulatory enzymes that are involved in the biosynthesis of EPS (21).

7.3 MOLECULAR MECHANISM OF POSTBIOTIC ACTION

Postbiotics are a mixture of various bioactive compounds that exert health benefits on the target host (6). These bioactive postbiotics have different mechanisms of action. These mechanisms are sometimes similar to the probiotic mechanism of action (22). Postbiotics exert their beneficial impact through several mechanisms, which are as follows:

- Enhancing the intestinal barrier function
- Modulation of local and systemic immune responses
- Influencing the host systemic metabolic responses

7.3.1 Enhancing the Intestinal Barrier Function

The intestinal epithelial barrier is exposed to various harmful factors that damage the integrity of the barrier and cell survival. Various postbiotics activate the protein kinase B (PKB/Akt), which modulates the apoptotic process, survival of cells, and integrity of the intestine by mucin production. Butyrate produced by *L. plantarum* RG14 enhances the formation of tight junction proteins like claudins-3 and 4 (23). Various postbiotics increase the activity of the mucosal barrier through the alteration of different secreted proteins. When heat-killed *L. rhamnosus* was administered in mice with colitis, an increase in mucosal permeability and repair of barrier function was observed. Moreover, EPS and SCFAs modulate the defense mechanism of the mucosal barrier. It is also reported that the antibacterial activity of EPS postbiotics forms the protective biofilm in host epithelium, which protects them from pathogens or toxins (24). EPS secreted from *Bifidobacterium* sp. (*B. longum*) also prevents or reduces the inflammatory responses that promote the barrier function. SCFAs modulate the activity of the epithelial barrier by protecting against lipopolysaccharide-induced disruption (13). These postbiotics can inhibit the growth of those intestinal pathogens that prevent the formation of epithelial biofilm by its antimicrobial activity (22). Overall, postbiotics exert an effect both locally and systemically in supporting the survival of epithelial cells (25). Postbiotics prevent infection and pathogenic development in the initial phase by binding or adhering to intestinal epithelial cells. These reduce the inflammation and strengthen the epithelial barrier (11).

7.3.2 Modulation of Local and Systemic Immune Responses

Various postbiotic microbial metabolites such as branched chain and SCFA, histamine, and microbial cell wall components like peptidoglycans and muramic acid can modulate the various immune responses by interacting with immune cell receptors like toll-like receptors (TLR) and lectins (22).

Supernatant postbiotics of a *Bifidobacterium breve* can improve the survival and maturation of dendritic cells. Postbiotics modify the immune system and make the body resistant to pathogenic infection. These decrease the inflammatory reactions by synthesizing the anti-inflammatory cytokine IL-10. Postbiotics also enhance mucosal immunity by increasing IgA antibody secretion (11). TLRs are suitable targets for ligand drugs, which make postbiotics possible in reducing inflammatory diseases and autoimmune disorders. Postbiotics also activate immune cell macrophages by producing cytokines and nitric oxide. Teichoic acid (TA) enhances the host immune response to various pathogens and also limits excessive inflammatory responses (24). Postbiotics also regulate immune responses by restoring the imbalance between lymphocytes such as type 1 and 2 T helper cells (Th1 and Th2). Due to their immunomodulatory activities, postbiotics are being utilized in the treatment and inhibition of COVID-19 in patients. Postbiotics also mediate the communication between the host immune system and the mutualistic microbiota to promote health benefits (25).

7.3.3 Influencing the Host Systemic Metabolic Responses

Various postbiotics produced inside the probiotic microbes or present on the surface of probiotics such as different enzymes or metabolites affect the systemic metabolic responses. Postbiotic enzymes such as bile acid hydrolase (BSH) or bile acid-inducible enzymes (BAI) secreted by gut microbiota regulate the metabolism of bile acid by producing unconjugated and secondary bile salts. These enzymes also contribute to the homeostasis of bile acid in the gut, host energy metabolism, and enhance the gut microbiota health (13). Alteration in gut microbiota due to various factors such as drugs, diet, or exercise influence the metabolism of the host. Gut dysbiosis affects the ability of the host to utilize the energy and leads to various metabolic disorders. The use of postbiotics can prevent metabolic diseases without altering the gut microbiota of the host (10). The postbiotics that modulate or cooperate with the host metabolism are glycerol, SCFAs, trimethylamine oxide (TMAO), ethanol, and lactate. These postbiotics are released from both Gram-positive and Gram-negative microorganisms (26).

7.4 MITOCHONDRIA AND MITOCHONDRIAL DISORDERS

Mitochondria are subcellular organelles of eukaryotic cells, originated from endosymbiosis, and have an essential role in energy metabolism and ATP synthesis. They also play a role in cellular metabolism under the pathological and physiological conditions of cells (27). Mitochondria synthesize adenosine triphosphate (ATP) by oxidative phosphorylation (OXPHOS), along with two metabolic pathways: mitochondrial electron transport chain of cellular respiration and fatty acid β-oxidation (FAO). Mitochondria also synthesize 20% ATP by the enzymatic tricarboxylic acid (TCA) cycle. These organelles are also involved in other cellular processes like calcium homeostasis, heme biosynthesis, intracellular signaling, and acute cell death (28).

Due to progress in mitochondrial science, mitochondria are considered the therapeutic target for many different diseases. Mitochondria have their self-replicating genome called mitochondrial DNA (mtDNA), which encodes the essential components of respiratory activities, the so-called powerhouse of the cell. It also plays a role in apoptosis, lipid metabolism, and Ca2+ homeostasis (29). Mitochondrial disorders are defined as decreases in mitochondrial biogenesis, changes in membrane potential, and the drop-off in mitochondrial number and changed activities of oxidative proteins due to the aggregation of reactive oxygen species (ROS) in cells and tissues (30). Human mitochondrial disorders are a group of various diseases that are caused by mutations in both nuclear and mitochondrial DNA. These diseases have adverse effects on many body organs and can occur at any stage of life. Primary mitochondrial diseases (PMD) are encephalopathy, Leigh syndrome, lactic acidosis mitochondrial myopathy, chronic progressive external ophthalmoplegia (CPEO), and stroke-like episodes (MELAS) (31). Mitochondrial disorders lead to a direct fault in the pyruvate pathway of OXPHOS. The clinical diagnosis of mitochondrial disorders causes a great challenge to physicians (32). These disorders also lead to various human diseases like cardiovascular disorders, obesity, cancer, neurodegenerative disorders, and neurometabolic diseases. In the last two decades, studies about mitochondrial DNA (mtDNA) mutation and its pathogenic effect have been rising (33). As mitochondria provide energy essential for smooth cellular activities, all the mutations that destroy normal mitochondrial activities primarily lead to defects in those organs with high energy demand like the brain and muscles (34).

7.4.1 Modulating Mitochondrial Biogenesis and Mitophagy

Mitochondrial biogenesis and mitophagy are two alternative pathways that modulate the mitochondrial number and metabolism in homeostasis. Mitochondrial biogenesis is the proliferation of mitochondria, while mitophagy is the degradation of mitochondria. The balance between these two alternative processes is necessary for cellular survival and cellular metabolic state. An imbalance between these processes can result in various pathological conditions (35).

Mitochondrial biogenesis is the synthesis of new mitochondrial proteins, mtDNA, membrane, and sometimes mitochondrial mass. But new mitochondria are themselves synthesized from preexisting old mitochondria by the process of mitochondrial fission (36). The major driving force for the synthesis of mitochondria is the energetic balance in the cell. This process is very important in neurons to drive nerve impulses to the final position; otherwise, it would take many days to reach the target (37). Postbiotic SCFAs are an important factor in inducing mitochondrial biogenesis by controlling the transcription of co-activators like peroxisome proliferator-activated receptor gamma 1 alpha (PGC-1α), the energy sensor Sirtuin 1 (SIRT1), which is redox-sensitive, and the enzyme activated protein kinase (AMPK). These also suppress the inflammatory response (28). Mitochondrial mitophagy is the process by which old and impaired mitochondria break down and are recycled into new cellular organelles. Postbiotic urolithin A (UA) is the first discovered postbiotic that has a great influence on mitochondrial mitophagy. Abnormality in mitophagy can cause multiple age-related disorders like neurodegenerative disease, cancers, and cardiovascular disease. It is also reported that balanced mitophagy can exert a positive impact on aging and the progress of these harmful chronic diseases (38). UA is secreted from gut probiotics, and it is also a metabolite of ellagitannins and polyphenolic compounds, which are present in food items such as walnuts, berries, and pomegranate. UA can lower the plasma level of C-reactive proteins and acylcarnitines; as a result, it can increase mitochondrial efficiency and also reduce inflammation. Urolithin A administration also increases mitochondrial metabolic processes in skeletal muscles; as a result, muscular performance increases (39). Urolithin A extends the lifespan and reduces the loss of activity as a result of aging. It was also observed that UA also exerts a positive impact on muscular function in mice (40).

7.4.2 Postbiotics in Mitochondrial ROS Management

Oxidative stress is the imbalance between the reactive oxygen species production in the body and the ability of the body to neutralize them. This is a common pathogenic factor that leads to the progression of various pathological diseases like cancer. Higher reactive oxygen species (ROS) disrupt the mitochondrial electron transport chain, which leads to more production of ROS and disrupts the mitochondrial function. All these factors lead to apoptosis of normal cells (41). ROS is both beneficial and harmful to the body. Its beneficial role is in killing pathogenic microbes, bacteria-induced signaling, and tissue repair, while negative impacts appear in the excess level of ROS (42). In normal respiration, reactive ROS is produced by mitochondria as a side product. ROSs have an important role in homeostasis and cell signaling, but higher levels of ROS damage the cell structures or even cause death by triggering the apoptotic activity. As the mitochondrial genome replicates, it accumulates genetic mutations that have the capability of mitochondrial OXPHOS. Gut microbiota reduce the production of ROS by secreting various postbiotics like N-butyrate SCFA. These postbiotics control the oxidative stress of mitochondria and improve metabolism energy production and expenditure (28). Postbiotic vanillic acid secreted by *L. rhamnosus* bacteria has an antioxidant capability that can eliminate the free radicals and can also modulate oxidative stress (43). Postbiotics increase the antioxidant capacity of cells. They exert antioxidant impacts by up-regulating the defense mechanism of endogenous antioxidants or by reducing oxidative stress. The antioxidant activity of postbiotics protects the cell against damage due to oxidative stress (44). Bioactive LGG postbiotics secreted from *Lactobacillus rhamnosus* GG (LGG) also have antioxidant properties that reduce excessive ROS production in diarrhea pathogenesis (45).

7.4.3 Role of Postbiotics in the Treatment of Various Mitochondrial Disorders

In the era of pre-, pro, and postbiotics, studies of new good microbes, their specific products, and their role in the treatment of various disorders have become of great interest. The following disorders have been discussed whose postbiotic treatment becomes possible.

7.4.3.1 Neurogenerative Diseases

The common neurodegenerative diseases (NDs) are Alzheimer's disease (AD) and Parkinson's disease (PD), which are characterized by abnormal activation of glial cells, impaired autophagy, oxidative stress, and neuroinflammation. Furthermore, both mitochondrial dysfunction and increased permeability of the blood–brain barrier have a great contribution to ND pathogenesis (46). A major clinical characteristic of PD is the failure of motor control because of the degradation of dopamine-producing neurons of the substantia nigra, which is caused by the increase of misfolded α-synuclein (46). It generates ROS, which leads to cellular oxidative stress. If oxidative stress prevails, it can cause cell death, especially of neuron cells. Within the cell, a major source of ROS production is considered mitochondria pathogenesis. Different antioxidants can maintain neuronal survival and improve the antioxidant activity in PD patients (47). Postbiotic inosine can increase the antioxidant capacity of plasma in PD patients (43).

AD is a common age-related mitochondrial and neurodegenerative disorder. It is considered a stress-induced disease that damages the neuronal system. Mitochondria are considered the first and most important pathogenetic factor in this disease. Mitochondrial disorders are associated with this disease due to changes in mitochondrial DNA, bioenergetic dysfunctions, damage to mitochondria-associated membrane complexes and calcium channels, and also disruption of mitochondrial control (48). Recently it was observed that mitochondria isolated from the brain of a patient who was affected by AD contain significant amounts of amyloid precursor protein (APP) as compared to controls. Breakdown of APP results in extracellular amyloid plaques called amyloid-β peptide (Aβ). Both peptides APP and Aβ aggregate in the mitochondria of AD patients impair mitochondrial functions and cause cell toxicity. It is suggested that inhibition of the APP can slow down the progress of AD (49).

The postbiotic treatment can decrease the levels of APP and ultimately the level of Aβ both in mitochondria and outside the cell. So postbiotic treatment can diminish mitochondrial dysfunction (50). The administration of postbiotics can also decrease the production of inflammatory cytokines, which can further decrease inflammation and oxidative stress. As a result, the progress of neurodegenerative diseases like Alzheimer's can be slowed down (51).

7.4.3.2 Metabolic Disorders

Metabolic disorders can be defined as a group of various interconnected metabolic, biochemical, and physiological factors that cause obesity, diabetes, glucose intolerance, hyperglycemia, insulin resistance (IR), low high-density lipoprotein (HDL) cholesterol levels, atherogenic dyslipidemia, or hypertension. These increase the chance of type 2 diabetes mellitus, cardiovascular diseases, and neurological impairment. The most common metabolic pathogenesis is obesity and insulin resistance (52).

Obesity has become a major health issue worldwide and also increases the risk of hyperinsulinemia and type 2 diabetes. Mitochondria play an essential role in the metabolism of healthy adipocytes to produce ATP. That's why mitochondrial function becomes diminished in obese people (53). Obesity causes many changes in the body; one is a rise in the production of ROSs, which cause oxidative stress and lead to mitochondrial changes, called mitochondrial disorders. Moreover, an excessive supply of nutrients (in the case of obesity) can overpower the mitochondrial Krebs cycle, which causes higher ROS production and ultimately leads to obesity (54).

Insulin resistance (IR) is a condition that is characterized by a decrease in the capacity of the cells to respond to insulin. Oxidative stress and obesity contribute to insulin resistance. Oxidative stress is further induced by ROS in mitochondria, which ultimately causes insulin resistance. Oxidative stress and mitochondrial dysfunction are also involved in many other metabolic diseases like obesity and diabetes (30). Postbiotics like SCFAs can inhibit excessive fat consumption and also promote the consumption of fats to regulate energy homeostasis. They also activate hormone secretion like glucagon-like peptide-1 (GLP-1) and peptide YY (PYY) in the intestine, which decreases the appetite by giving the feeling of fullness. It also affects the glucose homeostasis. Many postbiotics have antidiabetic roles like propionic and butyric acids can improve energy and glucose homeostasis by inducing gluconeogenesis in the intestine (10).

7.4.3.3 Inflammatory Bowel Diseases (IBD)

IBD is a complex, chronic, and active inflammatory state of the gastrointestinal tract (GIT) that is characterized by various symptoms like diarrhea, abdominal pain, bloody stools, and weight loss. There are two diseases characterized as IBD; one is Crohn's disease, and the other is ulcerative colitis (UC). Crohn's disease is GIT inflammation that can affect any part of the gut, while ulcerative colitis inflammation is limited to the mucosa layer of the rectum and colon. Chronic IBD can also lead to colorectal cancer (55). Recent evidence shows the possible links between mitochondrial dysfunction and the occurrence of IBD. There are strong interactions between the gut microbiota and host mitochondria, which may stimulate intestinal inflammation (56). Mitochondrial dysfunction plays a key role in UC pathogenesis due to energy deficiency, which causes abnormal ATP metabolism in the tricarboxylic acid (TCA) cycle, amino acid biosynthesis, and fatty acid catabolism. It often damages the epithelial barrier function by enhancing the susceptibility to tumor necrosis factor alpha (TNF-α), which induces cell death and reduces the planet cells and goblet cells. That's why the colonic epithelium of UC patients has a high susceptibility to mitochondrial dysfunction (57). Mitochondrial dysfunction in colitis is the reduction in the anion channel-dependent mitochondria voltage, lower level of tissue ATP, and decrease in the mitochondrial network. Mutation in the multidrug resistance 1 gene is also a susceptible factor for ulcerative colitis (58).

The normal function of the intestine disrupted by chronic inflammatory colitis can be inhibited or delayed by several postbiotic factors that are secreted from *:actobacilli* and *Bifidobacteria* strains. These postbiotics have positive health impacts on the host by reducing inflammatory responses. The most attractive postbiotics for the prevention and inhibition of gastrointestinal sickness such as inflammation of the colon, cancer, or metabolic disorders are tryptophan substances such as SCFAs and indole-3-propionic acid (59).

7.5 CONCLUSION

Postbiotics are the components or metabolites secreted by probiotic or gut microbes that exert health benefits or therapeutic advantages to their host. Microbe-derived components such as SCFAs, different proteins and peptides, EPS, various enzymes, antibacterial agents like bacteriocins, and cell wall fragments like teichoic acid and lipoteichoic acid can act as postbiotics. These postbiotics are involved in the treatment of various disorders and malfunctions of the gut. This chapter focused on the beneficial impact of postbiotics on disorders caused by mitochondrial dysfunction. Mitochondria are cellular organelles that are involved in the biosynthesis of energy in the form of ATP. Mitochondrial dysfunction leads to various disorders including NDs like Parkinson's or Alzheimer's diseases, metabolic disorders like obesity or insulin-resistant diabetes, and IBDs like colitis. The use of well-characterized and well-purified postbiotic components can prevent these disorders and also reduce the risk of these inflammatory and metabolic diseases in the host. So postbiotics can be safer and stronger alternatives to therapeutic drugs.

REFERENCES

1. Vinderola G, Sanders ME, Salminen S. The concept of postbiotics. *Foods*. 2022;11(8):1077.
2. Islam F, Azmat F, Imran A, Zippi M, Hong W, Tariq F, et al. Role of postbiotics in food and health: a comprehensive review. *CyTA-Journal of Food*. 2024;22(1):2386412.
3. Zhao X, Liu S, Li S, Jiang W, Wang J, Xiao J, et al. Unlocking the power of postbiotics: a revolutionary approach to nutrition for humans and animals. *Cell Metabolism*. 2024;36(4):725–44.
4. Zółkiewicz J, Marzec A, Ruszczyński M, Feleszko W. Postbiotics—a step beyond pre-and probiotics. *Nutrients*. 2020;12(8):2189.
5. Rafique N, Jan SY, Dar AH, Dash KK, Sarkar A, Shams R, et al. Promising bioactivities of postbiotics: a comprehensive review. *Journal of Agriculture and Food Research*. 2023:100708.
6. Liang B, Xing D. The current and future perspectives of postbiotics. *Probiotics and Antimicrobial Proteins*. 2023;15(6):1626–43.
7. Gurunathan S, Thangaraj P, Kim J-H. Postbiotics: functional food materials and therapeutic agents for cancer, diabetes, and inflammatory diseases. *Foods*. 2023;13(1):89.
8. Nataraj BH, Ali SA, Behare PV, Yadav H. Postbiotics-parabiotics: The new horizons in microbial biotherapy and functional foods. *Microbial Cell Factories*. 2020;19:1–22.
9. Davani-Davari D, Negahdaripour M, Karimzadeh I, Seifan M, Mohkam M, Masoumi SJ, et al. Prebiotics: definition, types, sources, mechanisms, and clinical applications. *Foods*. 2019;8(3):92.
10. Park M, Joung M, Park J-H, Ha SK, Park H-Y. Role of postbiotics in diet-induced metabolic disorders. *Nutrients*. 2022;14(18):3701.
11. Prajapati N, Patel J, Singh S, Yadav VK, Joshi C, Patani A, et al. Postbiotic production: harnessing the power of microbial metabolites for health applications. *Frontiers in Microbiology*. 2023;14:1306192.
12. Wegh CA, Geerlings SY, Knol J, Roeselers G, Belzer C. Postbiotics and their potential applications in early life nutrition and beyond. *International Journal of Molecular Sciences*. 2019;20(19):4673.
13. Hijová E. Postbiotics as metabolites and their biotherapeutic potential. *International Journal of Molecular Sciences*. 2024;25(10):5441.
14. Fusco W, Lorenzo MB, Cintoni M, Porcari S, Rinninella E, Kaitsas F, et al. Short-chain fatty-acid-producing bacteria: key components of the human gut microbiota. *Nutrients*. 2023;15(9):2211.
15. Raghavan ML, Hemalatha S. The functional roles of short-chain fatty acids as postbiotics in the human gut: future perspectives. *Food Science and Biotechnology*. 2024;33(2):275–85.
16. Mishra B, Mishra AK, Mohanta YK, Yadavalli R, Agrawal DC, Reddy HP, et al. Postbiotics: the new horizons of microbial functional bioactive compounds in food preservation and security. *Food Production, Processing and Nutrition*. 2024;6(1):28.
17. Liu C, Ma N, Feng Y, Zhou M, Li H, Zhang X, et al. From probiotics to postbiotics: concepts and applications. *Animal Research and One Health*. 2023;1(1):92–114.
18. Angelin J, Kavitha M. Exopolysaccharides from probiotic bacteria and their health potential. *International Journal of Biological Macromolecules*. 2020;162:853–65.
19. Sheng S, Fu Y, Pan N, Zhang H, Xiu L, Liang Y, et al. Novel exopolysaccharide derived from probiotic *Lactobacillus pantheris* TCP102 strain with immune-enhancing and anticancer activities. *Frontiers in Microbiology*. 2022;13:1015270.
20. Wang Q, Liu P, Peng J, Zhao B, Cai J. Postbiotic properties of exopolysaccharide produced by *Levilactobacillus brevis* M-10 isolated from natural fermented sour porridge through in vitro simulated digestion and fermentation. *Journal of Food Science*. 2024;89(5):3110–28.
21. Erdem TK, Tatar HD, Ayman S, Gezginç Y. Exopolysaccharides from lactic acid bacteria: a review on functions, biosynthesis and applications in the food industry. *Turkish Journal of Agriculture-Food Science and Technology*. 2023;11(2):414–23.
22. Mosca A, Abreu Y Abreu AT, Gwee KA, Ianiro G, Tack J, Nguyen TVH, et al. The clinical evidence for postbiotics as microbial therapeutics. *Gut Microbes*. 2022;14(1):2117508.

23. Jastrząb R, Graczyk D, Siedlecki P. Molecular and cellular mechanisms influenced by postbiotics. *International Journal of Molecular Sciences*. 2021;22(24):13475.
24. Ma L, Tu H, Chen T. Postbiotics in human health: a narrative review. *Nutrients*. 2023;15(2):291.
25. Kumar A, Green KM, Rawat M. A comprehensive overview of postbiotics with a special focus on discovery techniques and clinical applications. *Foods*. 2024;13(18):2937.
26. Fang H, Rodrigues E-Lacerda R, Barra NG, Kukje Zada D, Robin N, Mehra A, et al. Postbiotic impact on host metabolism and immunity provides therapeutic potential in metabolic disease. *Endocrine Reviews*. 2025;46(1):60–79.
27. Javadov S, Kuznetsov AV. Mitochondria: the cell powerhouse and nexus of stress. *Frontiers in Physiology*. 2013;4:207.
28. Clark A, Mach N. The crosstalk between the gut microbiota and mitochondria during exercise. *Frontiers in Physiology*. 2017;8:319.
29. Zong Y, Li H, Liao P, Chen L, Pan Y, Zheng Y, et al. Mitochondrial dysfunction: mechanisms and advances in therapy. *Signal Transduction and Targeted Therapy*. 2024;9(1):124.
30. Bhatti JS, Bhatti GK, Reddy PH. Mitochondrial dysfunction and oxidative stress in metabolic disorders—a step towards mitochondria based therapeutic strategies. *Biochimica et Biophysica Acta (BBA)-Molecular Basis of Disease*. 2017;1863(5):1066–77.
31. Nunnari J, Suomalainen A. Mitochondria: in sickness and in health. *Cell*. 2012;148(6):1145–59.
32. Schlieben LD, Prokisch H. The dimensions of primary mitochondrial disorders. *Frontiers in Cell and Developmental Biology*. 2020;8:600079.
33. Khan NA, Govindaraj P, Meena AK, Thangaraj K. Mitochondrial disorders: challenges in diagnosis & treatment. *Indian Journal of Medical Research*. 2015;141(1):13–26.
34. Dominguez-Martinez A, Molina-Menor E, Blanco-Ramos M, Urpi A, Pereto J, Porcar M, et al. Maltodextrin administration ameliorates brain pathology in a mouse model of mitochondrial disease. *bioRxiv*. 2023:2023–06. 28.546916.
35. Palikaras K, Tavernarakis N. Mitochondrial homeostasis: the interplay between mitophagy and mitochondrial biogenesis. *Experimental Gerontology*. 2014;56:182–8.
36. Selfridge JE, Lezi E, Lu J, Swerdlow RH. Role of mitochondrial homeostasis and dynamics in Alzheimer's disease. *Neurobiology of Disease*. 2013;51:3–12.
37. López-Doménech G, Kittler JT. Mitochondrial regulation of local supply of energy in neurons. *Current Opinion in Neurobiology*. 2023;81:102747.
38. Chen G, Kroemer G, Kepp O. Mitophagy: an emerging role in aging and age-associated diseases. *Frontiers in Cell and Developmental Biology*. 2020;8:200.
39. Singh A, D'Amico D, Andreux PA, Fouassier AM, Blanco-Bose W, Evans M, et al. Urolithin A improves muscle strength, exercise performance, and biomarkers of mitochondrial health in a randomized trial in middle-aged adults. *Cell Reports Medicine*. 2022;3(5):100633.
40. Ryu D, Mouchiroud L, Andreux PA, Katsyuba E, Moullan N, Nicolet-Dit-Félix AA, et al. Urolithin A induces mitophagy prolongs lifespan in *C. elegans* and increases muscle function in rodents. *Nature Medicine*. 2016;22(8):879–88.
41. Yavaş A, Akan E, Aksel M, Erbay Z. *Lactobacillus acidophilus* LA-5 postbiotic induces apoptosis by activating the mitochondrial apoptotic pathway in prostate cancer. *Journal of Food Biochemistry*. 2024;2024(1):7855679.
42. Singh V, Ahlawat S, Mohan H, Gill SS, Sharma KK. Balancing reactive oxygen species generation by rebooting gut microbiota. *Journal of Applied Microbiology*. 2022;132(6):4112–29.
43. López-Almada G, Mejía-León ME, Salazar-López NJ. Probiotic, postbiotic, and paraprobiotic effects of *Lactobacillus rhamnosus* as a modulator of obesity-associated factors. *Foods*. 2024;13(22):3529.
44. Rahimi A, Qaisar SA, Janet T, Karimpour H, Darbandi M, Moludi J. Clinical trial of the effects of postbiotic supplementation on inflammation, oxidative stress, and clinical outcomes in patients with CVA. *Scientific Reports*. 2024;14(1):24021.
45. Poeta M, Cioffi V, Tarallo A, Damiano C, Lo Vecchio A, Bruzzese E, et al. Postbiotic preparation of LACTICASEIBACILLUS RHAMNOSUS GG against diarrhea and oxidative stress induced by spike protein of SARS-CoV-2 in human enterocytes. *Antioxidants*. 2023;12(10):1878.
46. Głowacka P, Oszajca K, Pudlarz A, Szemraj J, Witusik-Perkowska M. Postbiotics as molecules targeting cellular events of aging brain—the role in pathogenesis, prophylaxis and treatment of neurodegenerative diseases. *Nutrients*. 2024;16(14):2244.

47. Duarte-Jurado AP, Gopar-Cuevas Y, Saucedo-Cardenas O, Loera-Arias MdJ, Montes-de-Oca-Luna R, Garcia-Garcia A, et al. Antioxidant therapeutics in Parkinson's disease: current challenges and opportunities. *Antioxidants*. 2021;10(3):453.
48. Sukhorukov VS, Mudzhiri NM, Voronkova AS, Baranich TI, Glinkina VV, Illarioshkin SN. Mitochondrial disorders in Alzheimer's disease. *Biochemistry (Moscow)*. 2021;86(6):667–79.
49. Pavlov PF, Petersen CH, Glaser E, Ankarcrona M. Mitochondrial accumulation of APP and Aβ: significance for Alzheimer disease pathogenesis. *Journal of Cellular and Molecular Medicine*. 2009;13(10):4137–45.
50. Kolonics A, Bori Z, Torma F, Abraham D, Fehér J, Radak Z. Exercise combined with postbiotics treatment results in synergistic improvement of mitochondrial function in the brain of male transgenic mice for Alzheimer's disease. *BMC Neuroscience*. 2023;24(1):68.
51. Bulacios GA, Cataldo PG, Naja JR, de Chaves EP, Taranto MP, Minahk CJ, et al. Improvement of key molecular events linked to Alzheimer's disease pathology using postbiotics. *ACS Omega*. 2023;8(50):48042–9.
52. Bourebaba Y, Marycz K, Mularczyk M, Bourebaba L. Postbiotics as potential new therapeutic agents for metabolic disorders management. *Biomedicine & Pharmacotherapy*. 2022;153:113138.
53. Xia W, Veeragandham P, Cao Y, Xu Y, Rhyne TE, Qian J, et al. Obesity causes mitochondrial fragmentation and dysfunction in white adipocytes due to RalA activation. *Nature Metabolism*. 2024;6(2):273–89.
54. de Mello AH, Costa AB, Engel JDG, Rezin GT. Mitochondrial dysfunction in obesity. *Life Sciences*. 2018;192:26–32.
55. Novak EA, Mollen KP. Mitochondrial dysfunction in inflammatory bowel disease. *Frontiers in Cell and Developmental Biology*. 2015;3:62.
56. Peña-Cearra A, Song D, Castelo J, Palacios A, Lavín JL, Azkargorta M, et al. Mitochondrial dysfunction promotes microbial composition that negatively impacts ulcerative colitis development and progression. *NPJ Biofilms and Microbiomes*. 2023;9(1):74.
57. Zhang Y-f, Fan M-y, Bai Q-r, Zhao R, Song S, Wu L, et al. Precision therapy for ulcerative colitis: insights from mitochondrial dysfunction interacting with the immune microenvironment. *Frontiers in Immunology*. 2024;15:1396221.
58. Mancini NL, Goudie L, Xu W, Sabouny R, Rajeev S, Wang A, et al. Perturbed mitochondrial dynamics is a novel feature of colitis that can be targeted to lessen disease. *Cellular and Molecular Gastroenterology and Hepatology*. 2020;10(2):287–307.
59. Asif A, Afzaal M, Shahid H, Saeed F, Ahmed A, Shah YA, et al. Probing the functional and therapeutic properties of postbiotics about their industrial application. *Food Science & Nutrition*. 2023;11(8):4472–84.

8 Postbiotics
Role in Immune System Modulation

Fernanda Maria Policarpo Tonelli and Flávia Cristina Policarpo Tonelli
Federal University of São João del Rei, Centro-Oeste Campus, Divinópolis-MG, Brazil

8.1 INTRODUCTION

The human gut contains a large variety of bacteria from more than 50 genera [1], most of them present in the large intestine [2].

Among the important functions of these microorganisms are favoring human health by performing various tasks such as participating in immunomodulation and regulating homeostasis; in fact, the gastrointestinal tract is considered by scientists the largest immunological organ in the human body [3].

Prevention is performed by these microorganisms not only against pathogens, but also in reducing the risk of developing diseases like inflammatory bowel disease and type 2 diabetes [4].

The beneficial living beings involved in this scenario are referred to as probiotics, and the substrates they use to perform the mentioned tasks are called prebiotics. However, effects that favor humans' health can also be offered by nonviable microorganisms and most commonly by metabolites produced by probiotics [5]. The terms paraprobiotics [6] and postbiotics [7] are commonly applied to refer to these, but in this chapter the term "postbiotic" will be used.

Postbiotics comprise metabolites secreted and also soluble products or cell wall fragments that contribute in a positive manner to the health of the host [8]. They include proteins and lipopolysaccharides (LPSs) from the cell wall, for example, and polysaccharides and metabolites produced by microorganisms as a consequence of their metabolism [5]. As previously mentioned, even nonviable cells from microorganisms can be considered postbiotics [9].

Exopolysaccharides (EPSs), postbiotics isolated from human *Bifidobacterium longum* and *Bifidobacterium pseudocatenulatum* strain C52, for example, could favor the growth of these bacteria and induce the production of short-chain fatty acids (SCFAs) in vitro. Consequently, the ratio acetic acid/propionic acid decreases, which may indicate inhibition in the synthesis of fatty acids and cholesterol, resulting in a hypolipidemic effect [10].

Although postbiotics present a large array of interesting activities (Figure 8.1) that can be useful to human beings [11], this chapter is dedicated to reviewing the role of postbiotics in immune system modulation, addressing their production, classification, main action mechanisms, and the potential they possess to modulate the immune response. The future perspectives of this field also receive attention.

DOI: 10.1201/9781003607779-8

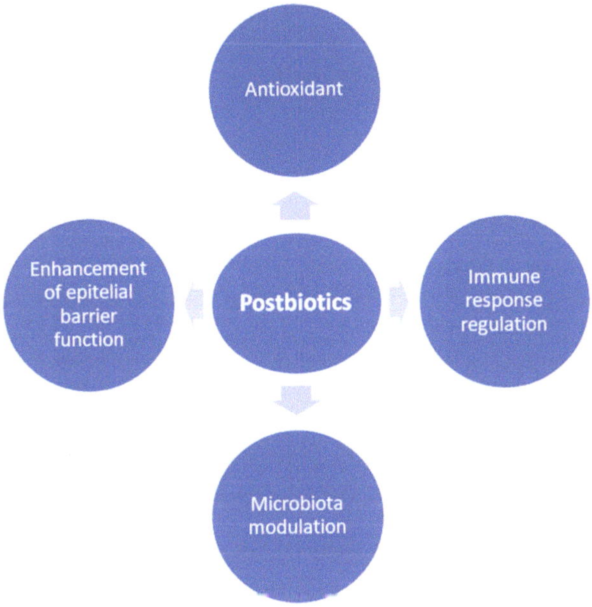

FIGURE 8.1 Main advantageous effects offered by postbiotics.

8.2 POSTBIOTICS: CLASSIFICATION AND IMPORTANT EXAMPLES

Postbiotics, according to their chemical nature, can be classified into groups such as EPSs, enzymes, proteins, peptides, cell wall fragments, and SCFAs. However, sometimes molecules or cell fragments are not isolated before assaying their potential to behave as a probiotic; consequently cell-free supernatants, bacterial lysates, and metabolites produced by gut microbiota are also considered when classifying this type of substance [12, 13].

When it comes to EPSs, they are already known to interfere in the immune response [14]. Besides that, they can also exhibit interesting activities such as antioxidant and antimicrobial ones. The ones produced by *Lactobacillus* strains isolated from tempoyak could present optimized results when compared to ATCC strains. It happened when analyzing *Lactobacillus pentosus* and *Lactobacillus fermentum*. Regarding antimicrobial activity, pathogens' inhibition through coaggregation was exhibited by *Lactobacillus crispatus*, *Lactobacillus reuteri*, *Lactobacillus plantarum*, and *L. pentosus* against, respectively, *Staphylococcus aureus*, methicillin-resistant *Staphylococcus aureus*, *Salmonella typhimurium*, and *Escherichia coli* [15].

Enzymes produced by microorganisms, such as catalase and glutathione peroxidase, that present antioxidant activity, can act to protect these organisms from oxidative stress by fighting reactive oxygen species (ROS). Consequently, they are classified as postbiotics. For example, the superoxide dismutase-like activity presented by *Lactobacillus acidophilus* 900 could efficiently alleviate intestinal inflammation in mice [16].

Cell wall fragments from lactic acid bacteria, for example, can contain lipoteichoic acid (LTA) [11] and teichoic acid (TA), which may influence immune system regulation [17].

SCFA such as butyric acids can originate salts, as sodium butyrate, that can interfere in immune response exerting an anti-inflammatory effect [18] and also exhibiting an anti-atherogenic effect [19].

Cell-free supernatants can exhibit, for example, antioxidant and anti-inflammatory activities such as the ones from *Lactobacillus casei* and *Lactobacillus acidophilus* [20]. Activities associated with antimicrobial effects [21–24] and reducing cancer cell invasion [25] have also already been reported when it comes to this type of postbiotic.

Bacterial lysates in lyophilized versions can induce innate immunity in a manner that depends upon signaling involving toll-like receptors (TLRs) TLR-9 and TLR-2/6. These lysates can reduce children's recurrence in respiratory tract infection and also attenuate chronic obstructive pulmonary disease in adults [26]. The formulation of Broncho-Vaxom®, for example, contains bacterial lysates and has proved to be able to induce immune response, increasing the number of T-lymphocytes and reducing the frequency of respiratory infection in childhood [27]. Children using these lysates could diminish the use of antibiotics, the number and duration of wheezing episodes, and the episodes of asthma exacerbation [28].

Metabolites produced by microorganisms have already been reported to act in stroke prevention (folic acid) [29], improvement of not only cardiovascular but also bone parameters (equol) [30], and also in fighting obesity in mice (urolithin) [31].

Recently, a protein from *Lactobacillus rhamnosus* GG not associated with enzymatic activity also proved to be useful in inhibiting not only adhesion of *E. coli* O157: H7 but also in preventing invasion of HT-29 cells. A pretreatment using the HM0539 exhibited interesting results not only in vitro; in vivo (in mice) it prevented jejunal goblet cells' damage, for example, and also offered other desirable effects such as avoiding body weight loss associated with the infection by the microorganism [32].

Peptides and proteins produced by probiotics and that are considered bacteriocins have already proved to be able to mitigate the inflammatory scenario associated with colorectal cancer by modulating nuclear factor-κB (NF-kB) and NOD-like receptor protein 3 (NLRP3) pathways [33].

8.3 A BIRD'S-EYE VIEW OF POSTBIOTICS' PRODUCTION

Postbiotics can be produced through fermentation, such as acetate, butyrate, and propionate [34]. The latter, for example, can be produced through pathways using lactate and glucose [35].

It is also possible that during microorganisms' growth the postbiotics' synthesis occurs to generate structures capable of performing a structural role (such as EPSs and TAs). EPSs, for example, are commonly generated by lactic acid bacteria such as the ones from *Lactobacillus* and *Streptococcus* genera [13, 36]. TAs are synthesized by bacteria to be part of cell membrane [37]. LTA presents anti-inflammatory properties already verified in mice and that may be also useful to human beings [38, 39].

Antimicrobial structures can also be produced by probiotics. Bacteriocins, for example, can be produced by microorganisms to favor them in competition with other species. However, these molecules can also have beneficial roles to humans. Helveticin-M from *Lactobacillus crispatus*, for example, could exhibit a bacteriostatic effect against *Enterobacter cloacae*, *Staphylococcus saprophyticus*, and *S. aureus*, interfering with the cell wall's structure and permeability [40]. Biosurfactants produced by *Lactobacillus jensenii* P6A, *Lactobacillus gasseri* P6 [41], and *L. casei* ATCC 393 [42], for example, have also been proved to possess antibacterial activity.

The enzymes that are considered postbiotics are produced by microorganisms to perform important roles in their survival and growth. However, these molecules can, commonly, also be useful to human beings such as proteases from *Bacillus pumilus* (that can inhibit *Serratia marcescens*'s biofilm production [43]) or antioxidant enzymes from species such as *L. fermentum* [44].

8.4 MAIN ACTION MECHANISMS OF POSTBIOTICS

The main mechanisms responsible for the advantageous effects of postbiotics, highlighted in Figure 8.1, are still being studied. However, some aspects are already known.

Regarding antioxidant activity, postbiotics can contribute to reducing the amount of ROS by inducing the production of antioxidants, for example [11]. The enzymatic antioxidant system can be favored by postbiotics such as the one from *L. plantarum* RG14; lambs suffered increase in serum glutathione peroxidase' level as well as in the isoforms from the liver when exposed to it; hepatic zinc superoxide dismutase has also undergone an increase in amount resulting in antioxidant protection after exposure [45].

Postbiotics may also offer beneficial effects by modulating microbiota. They can influence the environment's features (such as pH) and impair the growth of undesirable harmful organisms [46]. These postbiotics can consist of substances presenting antimicrobial activity, capable of preventing pathogens from growing and colonizing body environments [47, 48], and/or that can act favoring the establishment of desirable microorganism by assisting their adhesion [49].

When it comes to the epithelial barrier, a postbiotic from *L. rhamnosus* GG (the protein HM0539), for example, could favor this barrier's function by inducing an increase in mucin level. It exerted a protective effect against the infection by *E. coli* K1 as it could avoid damages caused by LPS and tumor necrosis factor α (TNFα) [50]. LTA from lactobacilli strains could favor, in colitis mice models, the regulation of gut permeability preventing colon damages [39].

The immune response regulation (Table 8.1), the focus of this chapter, will be addressed in a more detailed manner in the following section.

8.5 POSTBIOTICS MODULATING IMMUNE RESPONSE

Both adaptive and innate immunities can be affected by postbiotics [11]. Innate immunity response to postbiotics, for example, commonly involves TLRs. Mannan is a component of EPS of some bacteria [51] but also present in yeast's wall. It presents useful potential in modulating immune response in a manner that depends upon TLR4. Broilers that received mannan and live yeast as dietary supplementation, when challenged by *E. coli*, underwent a reduction in signaling involving TLR4, NF-κB, and interleukin (IL)-1β as the expression of their genes was reduced, attenuating inflammation [52]. The supplementation with *S. cerevisiae* and mannan also diminished LPS levels in sheep's plasma [53]. Another important saccharide, β glucan, from *S. cerevisiae*'s cell wall could, in ovine ruminal epithelial cells, enhance the production of β-defensin-1 in a TLR2-MyD88-dependent manner. Mitogen-activated protein kinase (MAPK) and NF-κB pathways seem to be involved in this regulation with the latter being the main one [54]. In fact, β-defensin-1 release could have been enhanced after activation of dendritic-cell-associated C-type lectin 1 receptors once these receptors can trigger signaling involving NF-κB [55].

When studying the effect of the cell-free supernatant from *Lactobacillus lactis*, *L. casei*, and *L. acidophilus* on ex vivo human monocytes differentiated in macrophages, alterations in the level of innate proinflammatory cytokines were noticed. A decrease in TNF-α could be detected as well as a decrease in IL-1β (the latter induced only by *L. acidophilus*'s supernatant). Regarding IL-10 (anti-inflammatory molecule), the supernatants from *L. reuteri* and *L. lactis* induced its secretion. However, when the cells were pre-challenged using LPS, the supernatants from all previously mentioned microorganisms could oppose the inflammatory cell state by increasing IL-10 production [20].

The anti-inflammatory potential previously addressed may also be triggered by LTA together with TA. In experiments using peritoneal macrophages from mice, these substances were involved in a synergistic induction of IL-10 production caused by *L. plantarum* and *L. casei* through a signaling involving TLR2 and extracellular signal-regulated kinase (ERK) [17]. The structure of LTAs influences the signaling triggered by them. *L. plantarum* from different strains (K55-5, K5-5, K88, and K8) produced LTAs with different structures. The ones from K55-5 and K5-5 could associate with CD14 and TLR2 to offer a stronger signaling and consequently an optimized immunomodulatory effect [38].

Butyrate, a SCFA, could also reduce inflammatory signaling associated with colitis resulting in less production of pro-inflammatory molecules such as IL-12, IL-6, IL-8, and TNF-α, and in an increase in anti-inflammatory ones such as IL-10 by macrophages. In vivo, mice NF-κB suffered blockage and their histones, deacetylation [18].

TABLE 8.1
Studies on postbiotics modulation of immune response addressed in this chapter

Postbiotic	Response Offered	Reference
Mannan in the presence of live yeast	Reduction in signaling involving TLR4, NF-κB, and interleukin (IL)-1β in broilers	[52]
Mannan in the presence of *Saccharomyces cerevisiae*	Diminishing in LPSs' level in sheep's plasma	[53]
β-glucan from *S. cerevisiae*'s cell wall	Enhancing the production of β-defensin-1 in a TLR2-myeloid differentiation primary response 88 (MyD88)–dependent manner in ovine ruminal epithelial cells	[54]
β-glucan	Releasing β-defensin-1	[55]
Cell-free supernatant from *Lactobacillus lactis*, *L. casei* and *L. acidophilus*	Decreasing TNF-α on ex vivo human monocytes differentiated in macrophages	[20]
Cell-free supernatant from *L. acidophilus*	Decreasing IL-1β's production by ex vivo human monocytes differentiated in macrophages	[20]
Cell-free supernatant from *L. reuteri* and *L. lactis*	Secreting of IL-10 by ex vivo human monocytes differentiated in macrophages	[20]
LTA/TA	Producing IL-10 by peritoneal macrophages from mice	[17]
LTA	Increasing the production of IL-10 and reducing the one from TNF-α by HT-29 cells	[39]
LTA from *Lactiplantibacillus plantarum* K55-5 and K5-5	Differentiation antigen (CD) number 14 and TLR2-mediated immunomodulatory effect	[38]
Butyrate	Diminishing the production of IL-12, IL-6, IL-8, and TNF-α and increasing the production of IL-10 by macrophages. NF-κB blockage and histone deacetylation in mice	[18]
EPSs from *Lactobacillus delbrueckii* ssp. *bulgaricus* OLL1073R-1	Increasing interferon (IFN) γ production and activating natural killer cells	[14]
EPS0142 from *L. plantarum*	Increasing macrophages' phagocytic activity and nitric oxide production	[57]
EPS from *L. paracasei* L1	Increasing IFN-γ's concentration and decrease IL-10's concentration in mononuclear cells from humans	[58]
Postbiotics produced by *S. cerevisiae* co-fermenting wheat bran with phytase	Decreasing nitric oxide release in chicken peripheral blood mononuclear cells. Reducing IL-1β, NF-κB, and IL-6's levels in broilers	[59]
Protein HM0539 from *L. rhamnosus*	Modulating TLR4/Myd88/NF-κB signaling to an anti-inflammatory effect in RAW 264.7 macrophages	[60]
The supernatants from *Saccharomyces boulardii*, *L. reuteri*, and *L. lactis*	Reducing the production of prostaglandin E2 by HT-29 cells in vitro	[20]
The supernatants from *Saccharomyces boulardii*, *L. reuteri*, *L. lactis*, *L. casei*, and *L. acidophilus*	Reducing the production of IL-8 after the challenge	[20]
Cell-free supernatant from *Lactobacillus plantarum* RG14	Reducing IgA production in lambs	[63]

EPSs from *Lactobacillus delbrueckii* ssp. *bulgaricus* OLL1073R-1 could induce the production of interferon (IFN) γ (through IL-12 and IL-18) that resulted in the activation of natural killer cells [14]. These cells can contribute to fighting, for example, viral infections and cancer as they are cytotoxic lymphoid cells [56]. The EPS0142 from *L. plantarum* could increase macrophages' (RAW 264.7 cells) phagocytic activity and nitric oxide production (favoring a typical behavior for an activated macrophage) [57]. The EPS from *L. paracasei* L1 could affect mononuclear cells from humans inducing an increase in IFN-γ concentration and a decrease in the level of IL-10 [58].

Wheat bran co-fermented by phytase and *S. cerevisiae* could induce in chicken peripheral blood mononuclear cells a decrease in nitric oxide release after challenge. A reduction was observed in IL-1β, NF-κB and IL-6's levels in broilers [59].

The protein HM0539 from *L. rhamnosus* offered a protective effect when it comes to murine colitis, which can be performed by modulating TLR4/Myd88/NF-κB signaling and resulting in an anti-inflammatory effect. In RAW 264.7 macrophages a reduction in TLR4's activation and in MyD88's transduction modulated NF-κB's activation [60].

To modulate immune response postbiotics can also act indirectly. By modulating the production of signaling molecules by human intestinal cells it is also possible to influence the response elaborated by the immune system. LTA from *Lactobacilli* strains could, through its signaling in HT-29 cells (enterocytes), enhance the production of IL-10 and decrease the one from TNF-α [39].

The supernatants from *Saccharomyces boulardii*, *L. reuteri*, and *L. lactis* could reduce the production of prostaglandin E2 (a proinflammatory cytokine) by HT-29 cells in vitro. After a challenge involving LPS, the production of IL-8 (another proinflammatory cytokine) could also be diminished by supernatants obtained from these species and from *L. casei* and *L. acidophilus*. In fact, the material from *L. lactis*'s cell culture could reduce the production of this interleukin previously to stimulation [20]. The mentioned prostaglandin has the ability to stimulate its own synthesis/release and contributes to dendritic cells' activation and the promotion of immune response involving T helper 2, T helper 17, and regulatory T cells; it is an important molecule when it comes to the immunopathology in cancer or chronic infections [61]. IL-8 is crucial to neutrophil recruitment and can contribute to induce monocytes' firm adhesion [62].

Immunoglobulins can have their production modulated by postbiotics. The cell-free supernatant from *Lactobacillus plantarum* RG14, for example, added in lambs' diet contributed to reduce IgA production in jejunum but caused no difference in the levels of platelets, neutrophils, basophils and lymphocytes, as well as in the levels of mRNA coding IL-1β, IL-10, and TNF [63].

8.6 CONCLUSION AND FUTURE PERSPECTIVES

Postbiotics are useful structures, presenting interesting activities that can benefit humans and other animals. However, a large number of gaps still exist in knowledge about them.

There is still a large potential to be revealed when it comes to these structures, and it will be possible to elucidate this potential as more studies address postbiotics mechanisms of action, not only in vitro, but also in vivo.

It is also important to investigate their activities in living beings from different physiological states, age, and sex, also dedicating attention to other variables such as diet intake habits. Microbiome studies can assist in this task and also contribute to elucidating new postbiotics. Regarding understanding their impact in humans' organisms, the contributions will be of special interest to the future of medicine in a personalized manner aiming for more efficiency. Setting the exact dose, and long-term effects of these active substances are important and still need to receive attention from the scientific community. Research on safety aspects associated with their use by humans is also necessary as well as the development of regulations; regulatory aspects still need to advance to guarantee quality and protection to users. Challenging issues are also present when it comes to applying postbiotics not only in the health field but also in the food industry. Solubility and miscibility, for example, are important aspects when it comes to some of these bioactive substances.

So, the future of postbiotics appears to be promising and of special importance to humanity.

REFERENCES

1. Adak, A., & Khan, M.R. (2019). An insight into gut microbiota and its functionalities. *Cell. Mol. Life Sci.*, 76, 473–493.https://doi.org/10.1007/s00018-018-2943-4

2. Fujimura, K.E., Sitarik, A.R., Havstad, S., Lin, D.L., Levan, S., Fadrosh, D. et al. (2016). Neonatal gut microbiota associates with childhood multisensitized atopy and T cell differentiation. *Nat. Med.*, 22, 1187–1191.https://doi.org/10.1038/nm.4176
3. Takiishi, T., Fenero, C.I.M., & Câmara, N.O.S. (2017). Intestinal barrier and gut microbiota: shaping our immune responses throughout life. *Tissue Barriers*, 5, e1373208. https://doi.org/10.1080/21688370.2017.1373208
4. Tsai, Y.L., Lin, T.L., Chang, C.J., Wu, T.R., Lai, W.F., Lu, C.C. et al. (2019). Probiotics, prebiotics and amelioration of diseases. *J. Biomed. Sci.*, 26, 3. https://doi.org/10.1186/s12929-018-0493-6
5. Collado, M.C., Vinderola, G., & Salminen, S. (2019). Postbiotics: Facts and open questions. A position paper on the need for a consensus definition. *Beneficial Microbes*, 10(7), 711–720. https://doi.org/10.3920/BM2019.0015
6. Taverniti, V., & Guglielmetti, S. (2011). The immunomodulatory properties of probiotic microorganisms beyond their viability (ghost probiotics: Proposal of paraprobiotic concept). *Genes Nutr.*, 6, 261–274. https://doi.org/10.1007/s12263-011-0218-x
7. Teame, T., Wang, A., Xie, M., Zhang, Z., Yang, Y., Ding, Q., et al. (2020). Paraprobiotics and postbiotics of probiotic *Lactobacilli*, their positive effects on the host and action mechanisms: A review. *Front. Nutr.*, 7, 570344. https://doi.org/10.3389/fnut.2020.570344
8. Aguilar-Toalá, J.E., Garcia-Varela, R., Garcia, H.S., Mata-Haro, V., González-Córdova, A.F., Vallejo-Cordoba, B. et al. (2018). Postbiotics: An evolving term within the functional foods field. *Trends Food Sci. Technol.* 75, 105–114. https://doi.org/10.1016/j.tifs.2018.03.009
9. Rad, A.H., Samadi Kafil, H., Fathi Zavoshti, H., Shahbazi, N., & Abbasi, A. (2020). Therapeutically effects of functional postbiotic foods. *Clinical Excellence*, 10(2), 33–52. http://ce.mazums.ac.ir/article-1-532-en.html
10. Salazar, N., Gueimonde, M., Hernandez-Barranco, A.M., Ruas-Madiedo, P., & de los Reyes-Gavilan, C.G. (2008). Exopolysaccharides produced by intestinal bifidobacterium strains act as fermentable substrates for human intestinal bacteria. *App.l Environ. Microbiol.* 74, 4737–4745. https://doi.org/10.1128/AEM.00325-08
11. Zhao, X., Liu, S., Li, S., Jiang, W., Wang, J., Xiao, J., et al. (2024). Unlocking the power of postbiotics: A revolutionary approach to nutrition for humans and animals. *Cell Metabolism*, 36(4), 725–744. https://www.cell.com/cell-metabolism/fulltext/S1550-4131(24)00083-4
12. Zółkiewicz, J., Marzec, A., Ruszczyński, M., & Feleszko, W. (2020) Postbiotics: A step beyond pre- and probiotics. *Nutrients*, 12, 2189. https://doi.org/10.3390/nu12082189
13. Homayouni-Rad, A., Pouragha, B., Houshyar, J. et al. Postbiotic application: A review on extraction, purification, and characterization methods. *Food Bioprocess Technol.* (2024). https://doi.org/10.1007/s11947-024-03701-9
14. Makino, S., Sato, A., Goto, A., Nakamura, M., Ogawa, M., Chiba, Y. et al. (2016) Enhanced natural killer cell activation by exopolysaccharides derived from yogurt fermented with *Lactobacillus delbrueckii* ssp. bulgaricus OLL1073R-1. *J. Dairy Sci.*, 99, 915–923. https://doi.org/10.3168/jds.2015-10376
15. Khalil, E.S., Abd Manap, M.Y., Mustafa, S., Alhelli, A.M., & Shokryazdan, P. (2018) Probiotic properties of exopolysaccharide-producing *Lactobacillus* strains isolated from tempoyak. *Molecules*, 23, 398. https://doi.org/10.3390/molecules23020398
16. Tomusiak-Plebanek, A., Heczko, P., Skowron, B., Baranowska, A., Okoń, K., Thor, P.J., & Strus, M. (2018). *Lactobacilli* with superoxide dismutase-like or catalase activity are more effective in alleviating inflammation in an inflammatory bowel disease mouse model. *Drug Design, Develop Ther*, 12, 3221–3233. https://doi.org/10.2147/DDDT.S164559
17. Kaji, R., Kiyoshima-Shibata, J., Nagaoka, M., Nanno, M., & Shida, K. (2010) Bacterial teichoic acids reverse predominant IL-12 production induced by certain *Lactobacillus* strains into predominant IL-10 production via TLR2-dependent ERK activation in macrophages. *J. Immunol.*, 184, 3505–3513. https://doi.org/10.4049/jimmunol.0901569
18. Lee, C., Kim, B.G., Kim, J.H., Chun, J., Im, J.P., & Kim, J.S. (2017) Sodium butyrate inhibits the NF-kappa B signaling pathway and histone deacetylation, and attenuates experimental colitis in an IL-10 independent manner. *Int. Immunopharmacol.*, 51, 47–56. https://doi.org/10.1016/j.intimp.2017.07.023
19. Kasahara, K., Krautkramer, K.A., Org, E., Romano, K.A., Kerby, R.L., Vivas, E.I., et al. (2018) Interactions between *Roseburia intestinalis* and diet modulate atherogenesis in a murine model. *Nat. Microbiol.*, 3, 1461–1471. https://doi.org/10.1038/s41564-018-0272-x

20. De Marco, S., Sichetti, M., Muradyan, D., Piccioni, M., Traina, G., Pagiotti, R. et al. (2018) Probiotic cell-free supernatants exhibited anti-inflammatory and antioxidant activity on human gut epithelial cells and macrophages stimulated with LPS. *Evid. Based Complementary Altern. Med.* 2018, 1756308. https://doi.org/10.1155/2018/1756308
21. Khodaii, Z., Ghaderian, S.M.H., & Natanzi, M.M. (2017) Probiotic bacteria and their supernatants protect enterocyte cell lines from enteroinvasive *Escherichia coli* (EIEC) invasion. *Int. J. Mol. Cell. Med.* 6, 183–189. https://doi.org/10.22088/acadpub.BUMS.6.3.183
22. Abdelhamid, A.G., Esaam, A., & Hazaa, M.M. (2018) *Cell free preparations of probiotics exerted anti-bacterial and antibiofilm activities against multidrug resistant E. coli. Saudi Pharma. J.*, 26, 603–607. https://doi.org/10.1016/j.jsps.2018.03.004
23. Khusro, A., Aarti, C., Barbabosa-Pilego, A., & Rojas Hernández, S. (2019) Anti-pathogenic, antibiofilm, and technological properties of fermented food associated *Staphylococcus succinus* strain AAS2. *Prep. Biochem. Biotechnol.* 49, 176–183. https://doi.org/10.1080/10826068.2019.1566149
24. Ebrahimi, M., & Sadeghi, A. (2021) Postbiotic and anti-aflatoxigenic capabilities of *Lactobacillus kunkeei* as the potential probiotic LAB isolated from the natural honey. *Probiotics Antimicrob. Proteins*, 13, 343–355. https://doi.org/10.1007/s12602-020-09697-w
25. Amaretti, A., Di Nunzio, M., Pompei, A., Raimondi, S., Rossi, M., & Bordoni, A. (2013) Antioxidant properties of potentially probiotic bacteria: In vitro and in vivo activities. *Appl. Microbiol. Biotechnol.* 97, 809–817. https://doi.org/10.22088/acadpub.BUMS.6.3.183
26. Kearney, S.C., Dziekiewicz, M., & Feleszko, W. (2015) Immunoregulatory and immunostimulatory responses of bacterial lysates in respiratory infections and asthma. *Ann. Allergy Asthma Immunol.*, 114, 364–369. https://doi.org/10.1016/j.anai.2015.02.008
27. Yin, J., Xu, B., Zeng, X., & Shen, K. (2018) Broncho-Vaxom in pediatric recurrent respiratory tract infections: A systematic review and meta-analysis. *Int. Immunopharmacol.*, 54, 198–209. https://doi.org/10.1016/j.intimp.2017.10.032
28. De Boer, G., Żółkiewicz, J., Strzelec, K., Ruszczyński, M., Hendriks, R., Braunstahl, G., Feleszko, W., & Tramper-Stranders, G. (2020) Bacterial lysate add-on therapy for the prevention of wheezing episodes and asthma exacerbations: A systematic review and meta-analysis. *Eur. Respir. Rev.*, 29(158), 190175. https://doi.org/10.1183/16000617.0175-2019
29. Hsu, C.Y., Chiu, S.W., Hong, K.S., Saver, J.L., Wu, Y.L., Lee, J.D., Lee, M., & Ovbiagele, B. (2018) Folic acid in stroke prevention in countries without mandatory folic acid food fortification: A meta-analysis of randomized controlled trials. *J. Stroke.*, 20, 99–109. https://doi.org/10.5853/jos.2017.01522
30. Yoshikata, R., Myint, K.Z.Y., & Ohta, H. (2018) Effects of Equol supplement on bone and cardiovascular parameters in middle-aged Japanese women: A prospective observational study. *J. Altern. Complement. Med.*, 24, 701–708. https://doi.org/10.1089/acm.2018.0050
31. Xia, B., Shi, X.C., Xie, B.C., Zhu, M.Q., Chen, Y., Chu, X.Y., Cai, G.H., Liu, M., Yang, S.Z., Mitchell, G.A., Pang, W.J., & Wu, J.W. (2020) Urolithin A exerts antiobesity effects through enhancing adipose tissue thermogenesis in mice. *PLoS Biol.*, 18, e3000688. https://doi.org/10.1371/journal.pbio.3000688
32. Zhang, H., Gao, J., He, X., Gong, Z., Wan, Y., Hu, T., et al. (2020) The postbiotic HM0539 from Lactobacillus rhamnosus GG prevents intestinal infection by enterohemorrhagic *E. coli* O157: H7 in mice. *J. South. Med. Univ.*, 40, 211–218. https://doi.org/10.12122/j.issn.1673-4254.2020.02.12
33. Patra, S., Sahu, N., Saxena, S., Pradhan, B., Nayak, S.K., & Roychowdhury, A. (2022). Effects of probiotics at the interface of metabolism and immunity to prevent colorectal cancer-associated gut inflammation: A systematic network and meta-analysis with molecular docking studies. *Front. Microbiol.*, 13, 878297. https://doi.org/10.3389/fmicb.2022.878297
34. Vera-Santander, V.E., Hernández-Figueroa, R.H., Jiménez-Munguía, M.T., Mani-López, E., & López-Malo, A. (2023). Health benefits of consuming foods with bacterial probiotics, postbiotics, and their metabolites: A review. *Molecules*, 28(3), 1230. https://doi.org/10.3390/molecules28031230
35. Markowiak-Kopeć, P., & Śliżewska, K. (2020). The effect of probiotics on the production of short-chain fatty acids by human intestinal microbiome. *Nutrients*, 12(4), 1107. https://doi.org/10.3390/nu12041107
36. Gezginç, Y., Karabekmez-Erdem, T., Tatar, H.D., Ayman, S., Ganiyusufoğlu, E., & Dayisoylu, K.S. (2022). Health promoting benefits of postbiotics produced by lactic acid bacteria: Exopolysaccharide. *Biotech Studies*, 31(2), 61–70. https://doi.org/10.38042/biotechstudies.1159166
37. Nataraj, B.H., Ali, S.A., Behare, P.V., & Yadav, H. (2020). Postbiotics-parabiotics: The new horizons in microbial biotherapy and functional foods. *Microbial Cell Factories*, 19(1), 1–22. https://doi.org/10.1186/s12934-020-01426-w

38. Jung, B.-J., Kim, H., & Chung, D.-K. (2022). Differential immunostimulatory effects of lipoteichoic acids isolated from four strains of *Lactiplantibacillus plantarum. Applied Sciences*, 12(3), 954. https://doi.org/10.3390/app12030954
39. Pradhan, D., Gulati, G., Avadhani, R., Rashmi, H.M., Soumya, K., Kumari, A., & Grover, S. (2023). Postbiotic lipoteichoic acid of probiotic *Lactobacillus* origin ameliorates inflammation in HT-29 cells and colitis mice. *Int. J. Biol. Macromol.*, 236, 123962. https://doi.org/10.1016/j.ijbiomac.2023.123962
40. Sun, Z., Wang, X., Zhang, X., Wu, H., Zou, Y., Li, P., Sun, C., Xu, W., Liu, F., & Wang, D. (2018). Class III bacteriocin Helveticin-M causes sublethal damage on target cells through impairment of cell wall and membrane. *J. Indust. Microbiol. Biotechnol.*, 45(3), 213–227. https://doi.org/10.1007/s10295-018-2008-6
41. Morais, I., Cordeiro, A., Teixeira, G., Domingues, V., Nardi, R., Monteiro, A., Alves, R., Siqueira, E., & Santos, V. (2017). Biological and physicochemical properties of biosurfactants produced by *Lactobacillus jensenii* P 6A and *Lactobacillus* gasseri P 65. *Microbial Cell Factories*, 16, 1–15. https://doi.org/10.1186/s12934-017-0769-7
42. Merghni, A., Dallel, I., Noumi, E., Kadmi, Y., Hentati, H., Tobji, S., Amor, A.B., & Mastouri, M. (2017). Antioxidant and antiproliferative potential of biosurfactants isolated from *Lactobacillus casei* and their anti-biofilm effect in oral *Staphylococcus aureus* strains. *Microbial Pathogenesis*, 104, 84–89. https://doi.org/10.1016/j.micpath.2017.01.017
43. Mitrofanova, O., Mardanova, A., Evtugyn, V., Bogomolnaya, L., & Sharipova, M. (2017). Effects of *Bacillus* serine proteases on the bacterial biofilms. *Biomed. Res. Int.*, 2017(1), 8525912. https://doi.org/10.1155/2017/8525912
44. LeBlanc, J.G., del Carmen, S., Miyoshi, A., Azevedo, V., Sesma, F., Langella, P., Bermúdez-Humarán, L.G., Watterlot, L., Perdigon, G., & de LeBlanc, A.D. M. (2011). Use of superoxide dismutase and catalase producing lactic acid bacteria in TNBS induced Crohn's disease in mice. *J. Biotechnol.*, 151(3), 287–293. https://doi.org/10.1016/j.jbiotec.2010.11.008
45. Izuddin, W.I., Humam, A.M., Loh, T.C., Foo, H.L., & Samsudin, A.A. (2020). Dietary postbiotic *Lactobacillus plantarum* improves serum and ruminal antioxidant activity and upregulates hepatic antioxidant enzymes and ruminal barrier function in post-weaning lambs. *Antioxidants.*, 9(3), 250. https://doi.org/10.3390/antiox9030250
46. Thu, T.V., Loh, T.C., Foo, H.L., Yaakub, H., & Bejo, M.H. (2011). Effects of liquid metabolite combinations produced by *Lactobacillus plantarum* on growth performance, faeces characteristics, intestinal morphology and diarrhoea incidence in postweaning piglets. *Trop. Anim. Health Product.*, 43, 69–75. https://doi.org/10.1007/s11250-010-9655-6
47. Martinez, F.A.C., Balciunas, E.M., Converti, A., Cotter, P.D., & de Souza Oliveira, R.P. (2013). Bacteriocin production by *Bifidobacterium* spp.: A review. *Biotechnol. Adv.*, 31(4), 482–488. https://doi.org/10.1016/j.biotechadv.2013.01.010
48. Xavier-Santos, D., Padilha, M., Fabiano, G.A., Vinderola, G., Cruz, A.G., Sivieri, K., & Antunes, A.E.C. (2022). Evidences and perspectives of the use of probiotics, prebiotics, synbiotics, and postbiotics as adjuvants for prevention and treatment of COVID-19: A bibliometric analysis and systematic review. *Trends Food Sci. Technol.*, 120, 174–192. https://doi.org/10.1016/j.tifs.2021.12.033
49. Lebeer, S., Claes, I., Tytgat, H.L., Verhoeven, T.L., Marien, E., von Ossowski, I., & Vanderleyden, J. (2012). Functional analysis of *Lactobacillus rhamnosus* GG pili in relation to adhesion and immunomodulatory interactions with intestinal epithelial cells. *Appl. Environ. Microbiol.*, 78(1), 185–193. https://doi.org/10.1128/aem.06192-11
50. Gao, J., Li, Y., Wan, Y., Hu, T., Liu, L., Yang, S., & Cao, H. (2019). A novel postbiotic from *Lactobacillus rhamnosus* GG with a beneficial effect on intestinal barrier function. *Front. Microbiol., 10*, 477. https://doi.org/10.3389/fmicb.2019.00477
51. Corsaro, M.M., Evidente, A., Lanzetta, R., Lavermicocca, P., & Molinaro, A. (2001). Structural determination of the phytotoxic mannan exopolysaccharide from *Pseudomonas syringae* pv. ciccaronei. *Carbohydrate Res.*, 330(2): 271–277. https://doi.org/10.1016/S0008-6215(00)00279-2
52. Wang, W., Li, Z., Han, Q., Guo, Y., Zhang, B., & D'inca, R. (2016). Dietary live yeast and mannan-oligosaccharide supplementation attenuate intestinal inflammation and barrier dysfunction induced by *Escherichia coli* in broilers. *Br. J. Nutr.*, 116(11): 1878–1888. https://doi.org/10.1017/S0007114516004116
53. Garcia Diaz, T., Ferriani Branco, A., Jacovaci, F.A., Cabreira Jobim, C., Bolson, D.C., & Pratti Daniel, J.L. (2018). Inclusion of live yeast and mannan-oligosaccharides in high grain-based diets for sheep: Ruminal parameters, inflammatory response and rumen morphology. *PLoS ONE.*, 13(2): e0193313. https://doi.org/10.1371/journal.pone.0193313

54. Jin, X., Zhang, M., & Yang, Y.F. (2019). Saccharomyces cerevisiae β-glucan-induced SBD-1 expression in ovine ruminal epithelial cells is mediated through the TLR-2-MyD88-NF-κB/MAPK pathway. *Vet. Res. Commun.*, 43: 77–89. https://doi.org/10.1007/s11259-019-09747-x
55. Zhang, M., Jin, X., & Yang, Y.F. (2019). β-Glucan from *Saccharomyces cerevisiae* induces SBD-1 production in ovine ruminal epithelial cells via the Dectin-1–Syk–NF-κB signaling pathway. *Cell. Signal.*, 53, 304–315. https://doi.org/10.1016/j.cellsig.2018.10.018
56. Wolf, N.K., Kissiov, D.U., & Raulet, D.H. (2023). Roles of natural killer cells in immunity to cancer, and applications to immunotherapy. *Nat. Rev. Immunol.* 23, 90–105. https://doi.org/10.1038/s41577-022-00732-1
57. Wang, J., Wu, T., Fang, X., Min, W., & Yang, Z. (2018). Characterization and immunomodulatory activity of an exopolysaccharide produced by *Lactobacillus plantarum* JLK0142 isolated from fermented dairy tofu. *Int. J. Biol. Macromol.*, 115, 985–993. https://doi.org/10.1016/j.ijbiomac.2018.04.099
58. Kıray, E., & Raheel, N.M. (2023). A new approach to exopolysaccharides of post probiotic *Lactobacillus paracasei* L1 strain: Anti-quarum sensing activity. *Balkan Med. J.*, 40(5), 351–357. https://doi.org/10.4274/balkanmedj.galenos.2023.2023-1-96
59. Chuang, W.Y., Lin, L.J., Hsieh, Y.C., Chang, S.C., & Lee, T.T. (2021). Effects of saccharomyces cerevisiae and phytase co-fermentation of wheat bran on growth, antioxidation, immunity and intestinal morphology in broilers. *Animal Bioscience*., 34(7), 1157–1168. https://doi.org/10.5713/ajas.20.0399
60. Li, Y., Yang, S., Lun, J., Gao, J., Gao, X., Gong, Z., Wan, Y., He, X., & Cao, H. (2020). Inhibitory effects of the *Lactobacillus rhamnosus* GG effector protein HM0539 on inflammatory response through the TLR4/MyD88/NF-κB axis. *Front. Immunol.*, 11, 551449. https://doi.org/10.3389/fimmu.2020.551449
61. Kalinski, P. (2012). Regulation of immune responses by prostaglandin E2. *J. Immunol.*, 188(1), 21–28. https://doi.org/10.4049/jimmunol.1101029
62. Griffith, J.W., Sokol, C.L., & Luster, A.D. (2014). Chemokines and chemokine receptors: positioning cells for host defense and immunity, *Ann. Rev. Immunol.*, 32, 659–702. https://doi.org/10.1146/annurev-immunol-032713-120145
63. Izuddin, W.I., Loh, T.C., Foo, H.L., Samsudin, A.A., & Humam, A.M. (2019). Postbiotic *L. plantarum* RG14 improves ruminal epithelium growth, immune status and upregulates the intestinal barrier function in post-weaning lambs. *Sci. Rep.*, 9, 9938. https://doi.org/10.1038/s41598-019-46076-0

9 Postbiotics and Their Role in Modulating Inflammation-Driven Metabolic Disorders

Syed Sohail Ahmad
Rehman College of Allied Health Sciences, Rehman Medical Institute, Peshawar, Pakistan

Maria Nazir
Cholistan University of Veterinary and Animal Sciences, Bahawalpur, Pakistan

Aqel Albutti
College of Applied Medical Sciences, Qassim University, Saudi Arabia

9.1 INTRODUCTION

The global population is now affected by metabolic disorders such as obesity, type 2 diabetes, and cardiovascular diseases. These conditions are rising in prevalence and exert a profound strain on healthcare systems worldwide. Today, people of all ages experience new types of pathology that are associated with obesity, type 2 diabetes, and cardiovascular diseases as illustrated in Table 9.1. They are increasingly experienced and have an extensive impact on healthcare systems globally. The common denominator behind these chronic diseases is a hidden culprit: Low-grade inflammation most of the time (Sinha, n.d.). This low-grade chronic inflammation interferes with every metabolic process and speeds up the progression of these diseases. This inflammation is not elicited by conventional drugs but by stimulating the normal end products of the gut microbiota. The human body, and in particular the gastrointestinal tract, harbors diverse and complex bacterial communities, together with fungi and viral particles referred to as gut microbiota. This microbial community including bacteria is very important for controlling almost all body processes ranging from digestion to immunity (Ohland & Jobin, 2015).

Until recently, most research has focused on prebiotics, which are the live bacteria, in your gut that offer health benefits when ingested. However, a new class of compounds is gaining attention for its potential to manage chronic diseases: postbiotics. These nonliving byproducts are some of those formed when healthy gut bacteria digest prebiotics and other such fibers in foods. Unlike probiotics, postbiotics don't require live organisms to exert their therapeutic effects, offering a significant advantage: stability and precision. These postbiotics, the short-chain fatty acids (SCFAs), peptides, and lipopolysaccharides, are effective in modulating systemic inflammation and metabolism (Fang et al., 2024). The works clearly show that they have the potential for use in the cure of major diseases such

 DOI: 10.1201/9781003607779-9

TABLE 9.1
Metabolic Disorders and Their Association with Chronic Inflammation (Navab et al., 2008)

Metabolic Disorder	Chronic Inflammation Mechanism	Impact on Metabolism	Associated Risk Factors
Obesity	Increased adiposity, secretion of inflammatory cytokines	Insulin resistance, altered lipid metabolism	Poor diet, sedentary lifestyle
Type 2 diabetes	Elevated pro-inflammatory cytokines (TNF-α, IL-6)	Impaired insulin sensitivity	Obesity, genetic predisposition
Cardiovascular disease	Inflammatory markers leading to endothelial dysfunction	Dysregulated lipid profiles, arterial plaque formation	High blood pressure, high cholesterol, smoking

as obesity and type II diabetes, among others, which are worrisome global ailments further characterized by chronic inflammation. Thus, the presented postbiotics, being the natural byproducts of the bacteria, present a possibility to act therapeutically to modulate the dysbiosis and inflammation state in the gut and to influence the essential metabolic processes including insulin sensitivity, fat utilization, and energy management (Park et al., 2023). It is also well-described that chronic low-grade inflammation can play a critical role in the development of metabolic diseases. In obesity, type 2 diabetes, and cardiovascular diseases, inflammation is not an acute process characteristic of injury or infection but rather a chronic, low-level process that dysregulated metabolism. In obesity, for instance, excess fat stimulates the immune system to release pro-inflammatory agents, and thus inflammation becomes systemic. This inflammation causes insulin resistance, meaning the body's cells do not respond as they should to insulin, a feature of type 2 diabetes (Perseghin et al., 2003). Consequently, inflammation in the cardiovascular system leads to atherosclerosis and the arteries thicken and narrow, which poses a threat to heart attacks and stroke. A major culprit in these diseases is constant low-grade inflammation, which needs to be treated to prevent or even cure these diseases. These bioactive compounds can alter the composition of the gut microbiota, enhance the permeability of the intestinal tract, and, more importantly, decrease the inflammation that gives rise to metabolic disorders. Therefore, postbiotics are a promising new concept for managing some of the most prevalent and difficult health problems of the 21st century (Brizita & Nevena, 2022).

9.2 UNDERSTANDING POSTBIOTICS

Postbiotics are bioactive compounds as a result of the action by metabolically active microbial strains on prebiotics or dietary fibers. Unlike probiotics, which are live microorganisms, and prebiotics, which are nondigestible fibers, postbiotics offer a unique advantage: they can still impart health benefits for those who cannot use living bacteria. Some of these compounds include SCFAs, peptides, enzymes, and cell wall structural polymers including lipopolysaccharides (Ali et al., 2022). The place where the benefits of postbiotics come from is in the acting of inflammation, metabolism regulation, and immune function. B-culture postbiotics include butyrate and acetate propionate, which is derived from fibers by the gut bacteria. Some of these metabolites are involved in the regulation of gut function, inflammation levels, and immune system function. Probiotics, prebiotics, and postbiotics each play a distinct but complementary role in gut health. Probiotics are products containing viable beneficial microorganisms in sufficient quantities that, upon ingestion, provide health benefits to the host by affecting the gut flora or acting on the contemplated or proposed mechanisms (Kumar et al., 2022). Prebiotics are dietary substances that can neither be broken down by human digestion enzymes nor absorbed in the upper gastrointestinal tract for metabolism. Examples are fructooligosaccharides, inulin, and fiber, which act as food sources for selected friendly bacteria in the large intestine, thereby improving the effects of the probiotics (Kozłowska et al., 2016).

9.2.1 Key Bioactive Compounds in Postbiotics

In particular, postbiotics include several bioactive compounds that may possess rather important therapeutic properties. These include the following.

1. **Short-Chain Fatty Acids (SCFAs):** Butyrate, acetate, and propionate are the SCFAs that are formed when the gut bacteria overrun the dietary fibers in the colon (Yüksel et al., 2024). Increases in SCFAs supply energy to the intestinal cells, accelerate gut motility, improve the integrity of the barrier, and regulate inflammation and metabolic disorders. However, it is particularly well-documented that butyrate has anti-inflammatory activity and stabilizes insulin levels.
2. **Lipopolysaccharides (LPS):** LPSs are molecules located in the outer membrane of Gram-negative bacteria. LPS found in moderate amounts from friendly gut bacteria has a positive effect on the immune system and helps to ward off invaders (Chen et al., 2023). However, an imbalance can cause something like chronic inflammation, which is associated with metabolic diseases.
3. **Peptides and Proteins:** Several peptides generated from intestinal bacteria are anti-inflammatory and can regulate immune activity. Some of these peptides can affect cytokine release in inflammation and also be used to control appetite and alimentary balance (Mashaqi & Badr, 2019).
4. **Bile Acids:** Bile acids produced after the administration of probiotics exert an impact on lipid metabolism and control fat absorption and cholesterol content (Wu et al., 2022). It also plays a role in the regulation of bacterial balance in the gut which is normally disturbed by certain conditions.
5. **Exopolysaccharides:** These complex sugars synthesized by bacteria are currently understood to help regulate the immune system and enhance the mucosal barrier. They help maintain the integrity of the gut and decrease the permeability of the gut wall and therefore minimize leakage of toxic substances into the blood with all its attendant pro-inflammatory effects (Eilam et al., 2023). The postbiotics contribute to anti-inflammatory properties, metabolic function, and integrity of the gastrointestinal tract, underscoring the therapeutic values of postbiotics.

9.3 INFLAMMATION AND METABOLIC DISORDERS

9.3.1 Mechanisms of Chronic Low-Grade Inflammation

The primary mechanisms behind chronic low-grade inflammation include the following.

1. **Immune Cell Activation:** In obesity, elevated visceral obesity leads to a stress response in macrophages and other immune central nodes, which are constantly active. These immune cells secrete pro-inflammatory cytokines including TNF-α and IL-6, thereby maintaining an inflammatory condition (Wang & He, 2020). This immune activation in response to elevated levels of leaky gut and oxidative stress contributes to the creation of systemic inflammation.
2. **Obesity:** An excess of fat, especially intra-abdominal, is an endocrine organ producing adipokines as signaling molecules that regulate inflammation. While obesity is characterized by elevated levels of pro-inflammatory adipokines such as leptin and decreased levels of the anti-inflammatory adipokine adiponectin, local and systemic inflammation damages insulin sensitivity and leads to metabolic imbalance (Singh & Benencia, 2019).
3. **Endothelial Dysfunction:** Endothelial cells, and in particular the vessels that exert the heart, can become damaged from chronic inflammation. This damage reduces the synthesis

of the important molecule nitric oxide, which keeps blood vessels healthy and increases the permeability of blood vessels, letting in toxic substances such as oxidized low-density lipoprotein (LDL) cholesterol into the arteries (Sheikh et al., 2024). It advances cardiovascular diseases as well as extends inflammation, thereby heightening further escalation.

9.4 MECHANISMS OF POSTBIOTICS IN INFLAMMATION MODULATION

It is strongly suggested that postbiotics, the prescriptive metabolites formed from the fermentation of dietary fibers and prebiotics by gut microbiota, have powerful immunomodulatory effects on metabolism. These bioactive compounds such as SCFAs, peptides, and LPS presented in Figure 9.1 modify immune response, regulate intestinal permeability, and modulate gut microbiota homeostasis (Mousavi Ghahfarrokhi et al., 2024).

9.4.1 Postbiotics and Their Role in Modulating Inflammation and Metabolic Health

Postbiotics including nonviable molecules generated during the fermentation of prebiotics and dietary fibers by the host's symbiotic bacteria have recently emerged as potential therapies for addressing chronic low-grade inflammation with a potential for improving metabolism. In addition to maintaining gut health, these bioactive compounds, including SCFAs, peptides, and LPS, are endowed with other functions (Thoda & Touraki, 2023). They shape immune cells, promote intestinal integrity, and control systemic inflammation—important factors in fighting metabolic disorders such as obesity, diabetes, and cardiovascular diseases.

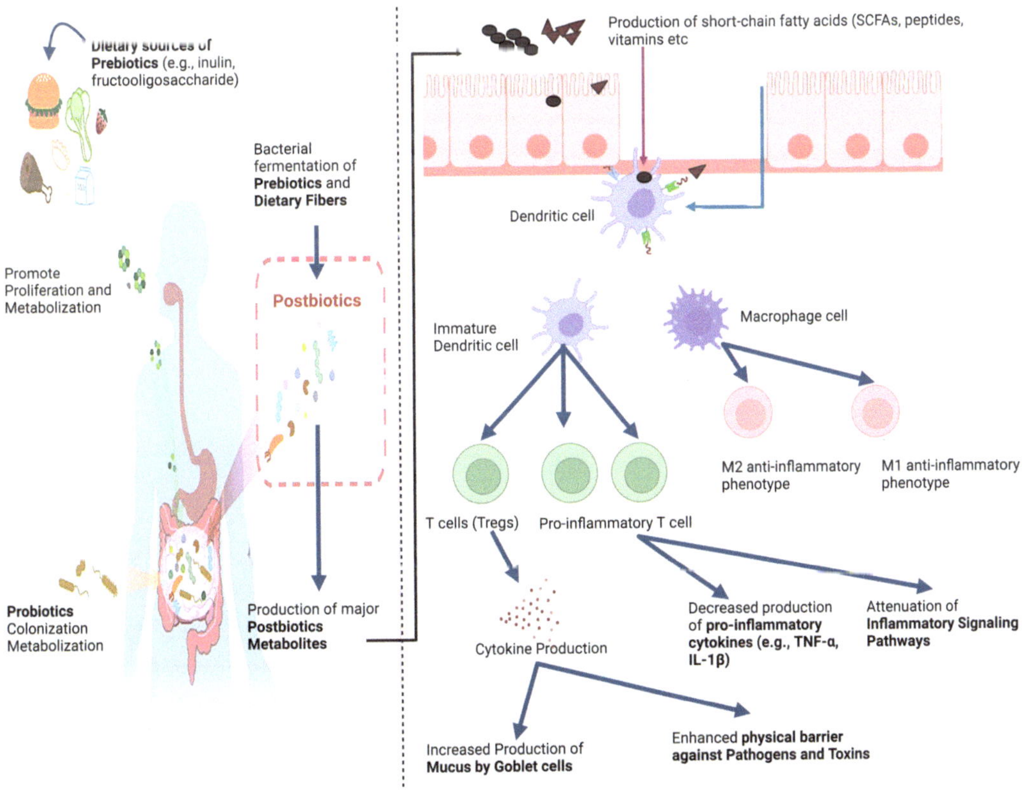

FIGURE 9.1 Mechanisms of postbiotics in modulating inflammation and metabolic health.

9.4.2 Modulation of Gut Microbiota by Postbiotics

It is vital to notice how the gut microbiota contributes to the organism's immune response and overall health. Dysbiosis, or the state of microbial ecology imbalance, outweighs the onset of chronic inflammations and metabolic diseases. Dysbiosis, or microbial ecology imbalance, causes the development of chronic inflammations and metabolic disorders. Postbiotics offer the ability to balance microbiota in favour of beneficial bacteria like *Bifidobacteria* and *Lactobacilli* (Al-Habsi et al., 2024). They, in turn, generate further metabolites that provide anti-inflammatory properties that enhance the gut microbiome. Since postbiotics, especially SCFAs, remain the preferred energy substrate for such favorable microbes, incubation results in the growth and enhanced function of these microbes.

9.4.3 Improvement of Intestinal Barrier Function

The lining of the intestine represents a vital defense mechanism that seals off the gastrointestinal tract lumen from the blood. When this barrier is damaged, known as leaky gut, any pathogen, toxin, or pro-inflammatory molecule is free to enter the bloodstream and impact the body, causing metabolic syndrome through inflammation and leading to obesity, insulin resistance, and heart diseases. Further, they are involved with butyrate, which has given its name to the weak link between intestinal epithelial cells as it maintains tight junctions to prevent toxic agents from leaking into the body. Further, postbiotics have a positive effect on mucus synthesis and contribute to a layer adhesion that barr the pathogens and bad bacteria from coming closer, thus eliminating inflammation agents (Yan et al., 2024).

9.4.4 Regulation of Immune Responses to Reduce Inflammation

Postbiotics also control immune system reactions, influence the overall lean toward inflammation, and reduce inflammation levels. The butyrate, acetate, and propionate SCFAs interfere with immune cells in the gut-associated lymphoid tissue (GALT) linking with Tregs to encourage their differentiation. These Tregs have vital functions in the immune tolerance system and contain chronic inflammatory processes. Furthermore, postbiotics are capable of modulating macrophages from Macrophage 1 (M1), which is inflammation, to Macrophage 2 (M2), which is anti-inflammatory efficiency (S. Wang et al., 2024b). This shift is necessary to diminish inflammation and prevent the disorders of immune regulation, which participates in the development of metabolic disorders.

9.4.5 Harnessing the Power of Postbiotics in Combating Chronic Inflammation

Postbiotics have increased the possibility of managing inflammation and consequently preventing metabolic disorders. Practical supplementation of postbiotics has the potential to fend off chronic low-grade inflammation, which is responsible for obesity, type 2 diabetes, and cardiovascular pathology (Luzzi et al., 2024). The following dietary approaches should be taken: increase the consumption of postbiotics, that is, fiber, and increase use of prebiotics. Furthermore, it is possible that the administration of probiotics or indeed synbiotics could provide another avenue by which postbiotic production could be increased and overall metabolism enhanced.

9.5 POSTBIOTICS AND METABOLIC PATHWAYS

Specifically, postbiotics are involved in the regulation of metabolic signaling, participating in the regulation of energy production, utilization of insulin, and lipid metabolism. Postbiotics interact with different metabolic pathways in the enhancement of metabolic health and avoiding ailments like obesity and type 2 diabetes.

9.5.1 The Role of Postbiotics in Modulating Metabolic Pathways and Promoting Health

Postbiotics are the compounds that are generated from fermenting prebiotics by the positive modulating gut bacteria, and these compounds have attracted considerable attention for modulating diverse metabolic processes. These compounds, including the SCFAs butyrate, acetate, and propionate, are integral to modulating some primary metabolism pathways including insulin signaling, glucose utilization, lipid handling, and energy homeostasis (Rekha et al., 2024).

9.5.2 Impact on Insulin Sensitivity and Glucose Metabolism

In other words, the capacity of target tissues to recognize insulin and act consequently is a fundamental part of metabolic health. Type 2 diabetes results in impaired glucose metabolism whereby insulin resistance causes high blood sugar levels and devastating metabolic complications (Hayden, 2002). Some of the postbiotics, mainly SCFAs, have been demonstrated to have a strong positive effect on insulin sensitization and glucose homeostasis.

9.5.3 Role in Lipid Metabolism and Fat Storage Regulation

The maintenance of lipid homeostasis has a major influence on obesity and cardiovascular diseases. Normal postbiotics, specifically butyrate and propionate, have an impact on lipids as they concern both fat deposition and mobilization. The butyrate down-regulates the sterol regulatory element-binding proteins, which are important transcription factors for fatty acid and cholesterol biosynthesis (Mu et al., 2021). This inhibition aids in reducing the deposition of fats in organs like hepatic and adipose tissues. Further, postbiotics aid in fat breakdown, known as fat oxidation, and lipolysis, the breakdown of stored fat for energy. This dual process not only prevents dyslipidemia, which is an inherent characteristic of obesity and metabolic syndrome, but also assists in the reduction of lipid storage.

9.5.4 Effects on Appetite Regulation and Energy Expenditure

Postbiotics are involved in regulating appetite and energy intake through their endocrine actions on hormones in the gastrointestinal tract and the gut–brain sign language. Both SCFAs including butyrate and propionate have been found to stimulate the release of the satiety hormones PYY and GLP-1 that help control the intake of food (Alhabeeb et al., 2021) (Table 9.2).

TABLE 9.2
Impact of Postbiotics on Metabolic Pathways and Disorders (Park et al., 2022)

Postbiotic Type	Impact on Obesity	Impact on Type 2 Diabetes	Impact on Cardiovascular Disease
Short-chain fatty acids (SCFAs)	Reduces fat accumulation, and improves gut health	Enhances insulin sensitivity, reduces glucose levels	Decreases inflammatory markers, improves lipid profiles
Lipopolysaccharides (LPS)	Modulates gut microbiota to reduce adiposity	Reduces inflammation, and improves insulin function	Reduces vascular inflammation, improves endothelial function
Bile acids	Reduces visceral fat, modulates gut microbiota	Enhances glucose homeostasis, improves insulin secretion	Regulates cholesterol levels, reduces plaque buildup

9.6 THERAPEUTIC POTENTIAL OF POSTBIOTICS

9.6.1 The Therapeutic Potential of Postbiotics in Metabolic Health

Postbiotics as bioactive compounds manufactured by gut bacteria during the process of fermenting prebiotics have recently attracted more attention due to their possible therapeutic function by regulating main metabolic pathways. The concerned pathways include inflammation, glucose homeostasis, lipid regulation, and energy expenditure. The potential replacement of postbiotics for traditional therapies for the management of metabolic disorders that include obesity, diabetes, and cardiovascular diseases, and they are considered to create a promising future therapeutic model (Zhong et al., 2024).

9.6.2 Summary of Preclinical and Clinical Studies

Numerous in vivo trials conducted during the past decades have shown that postbiotics can enhance metabolic health by lowering inflammation, increasing insulin sensitivity, decreasing fat deposition, and improving lipid profiles (Park et al., 2023). These are more postbiotics studies, and most of them have been done on animal models before the research on postbiotics in humans. Butyrate, surely, is the most researched postbiotic that possesses the potential to enhance insulin sensitivity, prevent inflammation of the liver, and strengthen the gut barrier in animal models. Another SCFA, acetate, has shown evidence of the ability to decrease appetite and enhance lipid concentrations, especially in obese people. Li et al. (2024), a case-control clinical trial, concluded that the intake of acetate lowered food intake and improved lipid metabolism, therefore effectively reducing obesity and its related diseases (Table 9.3).

9.6.3 Specific Postbiotics and Their Therapeutic Effects

Several specific postbiotics, particularly butyrate, acetate, and propionate, have demonstrated substantial therapeutic effects in metabolic disorders:

1. **Butyrate:** Butyrate, which is famous for its impact on the gut, has benefits such as increasing insulin sensitivity, decreasing fat storage, increasing fat-burning rate, and reducing inflammation levels. It has a significant role in obesity and type 2 diabetes treatment as well as for influencing the gut–brain axis and appetite (Richards et al., 2021).
2. **Acetate:** Apart from controling appetite through decreasing food consumption, acetate also adjusts lipid profile. it also possesses anti-inflammatory effects, which make it a potential therapeutic molecule in the management of metabolic diseases that are characterized by systemic inflammation (Bagheri et al., 2022).

TABLE 9.3
Summary of Key Clinical and Preclinical Studies on Postbiotics (Eslami et al., 2024)

Study Type	Study Focus	Postbiotic Used	Key Findings	Limitations
Preclinical	Obesity and inflammation	SCFAs (acetate, butyrate)	Reduced fat mass, improved insulin sensitivity, reduced cytokines	Limited human applicability, animal models only
Clinical	Type 2 diabetes	LPS, SCFAs	Enhanced insulin sensitivity, reduced HbA1c levels	Small sample size, short duration of follow-up
Preclinical	Cardiovascular disease	Bile acids	Decreased vascular inflammation, reduced cholesterol levels	Lack of diversity in animal models, high variability in results

3. **Propionate:** Propionate increases insulin responsiveness, decreases fat content, and increases fatty acid oxidation. It can also regulate cholesterol levels and the gut hormones linked with metabolic regulation, possibly making the agent suitable for managing obesity and type 2 diabetes. Concisely, postbiotics appear to hold great potential as a safe, natural strategy for targeting metabolic conditions. Because they directly modulate metabolic pathways and are highly stable and biologically active compounds, these products may be promising in terms of therapeutic applications, especially as add-ons or even a replacement for conventional treatment of metabolic diseases (Rafique et al., 2023).

9.7 CHALLENGES IN THE CLINICAL APPLICATION OF POSTBIOTICS

The products of postbiotics, particularly the postbiotics like SCFAs, peptides, and LPS, have immense therapeutic potential in the management of inflammation-mediated metabolic disorders like obesity, diabetes, and cardiovascular diseases; however, before they can be used in clinics, certain challenges have to be addressed (Table 9.4).

9.7.1 Formulation and Standardization of Postbiotic Products

One of the major developing concerns in the clinical application of postbiotics is the stability and therapeutic effects of these products. Postbiotics are then products resulting from the action of the fermentation of prebiotics by gut bacteria and therefore may deteriorate due to environmental factors such as temperature, light, pH, and even oxygen (Balthazar et al., 2022). These conditions may alter the bioactive content of compounds such as butyrate, acetate, and propionate and their effectiveness in treating illnesses.

9.7.2 Variability in Patient Responses

Another problem is the ability of patients to manage postbiotic therapy, which very often differs in individuals. There are some things like the composition of the human gut microbiome, one's genetic makeup, diet, and underlying health conditions that determine the efficacy of postbiotics. For example, it would be possible for people having dysbiosis to have a completely different reaction to postbiotics than people with healthy microbiota (Yeşilyurt et al., 2021). Also, outcomes among many patients with obesity, type 2 diabetes, or cardiovascular disease may differ as their given metabolic and immune systems may also differ. This variability makes it difficult for the definition of standard treatments to establish the diagnosis of the mental disorder. Personalized medicine may be an effective solution that delivers interventions regarding the microbiome and health of a particular recipient.

TABLE 9.4
Challenges and Future Directions in Clinical Use of Postbiotics (Suthar et al., 2023)

Challenge/Issue	Impact on Clinical Use	Proposed Solution	Future Research Focus
Formulation inconsistency	Variable effectiveness, poor reproducibility	Development of standardized postbiotic formulations	Clinical trials testing uniform postbiotic products
Dose variability	Uncertain therapeutic window, difficulty in dose optimization	Research on optimal dosing and long-term safety	Trials determining optimal postbiotic doses for specific disorders
Limited long-term clinical data	Lack of understanding of long-term benefits and risks	Extended clinical trials, focusing on safety profiles	Large-scale longitudinal studies

It is, therefore, conceivable that progress in microbiome characterization and genomic biomarker identification could improve the treatment outcomes associated with postbiotic therapy across population subgroups (Motei et al., 2023).

9.7.3 Need for Rigorous Clinical Trials

While many benefits of postbiotics have been observed in preclinical studies and phase 1 and phase 2 clinical trials, it's essential to carry out more substantial, high-quality phase 3 trials to evaluate postbiotics' efficacy entirely. The most compelling need in the area of postbiotic research is the conduct of definite large-scale, long-term RCTs for deriving efficacy and safety data and to determine the all-human postbiotic dosing strategies (Ceapa, 2016). It must also take into account the age groups, the sex, the patient's pre-existing conditions as well as the microbial richness of human gut microbiota to check the generalizability of such trials. Further, these trials should examine if postbiotics are safe alone or together with other therapies used for managing metabolic disorders, for example, insulin or statin.

9.7.4 Regulatory Considerations

Postbiotics have some issues, most of which stem from the fact that there are no specific standards on how to approve them as therapeutic substances. All these regulatory agencies, including the Food and Drug Administration and European Medicines Agency, have not recognized postbiotics as a separate group of therapeutic products, which leads to confusion in regulation (Waheed et al., 2024). Essentially, postbiotics are usually in a limbo between food supplements, and drugs, which makes their regulation challenging. As there remains no consistent set of indicators across the globe, there are concerns about the potential for postbiotics to be sold with misleading promises to consumers as well as to clinicians (Chugh et al., 2023). Standard specifications of postbiotics' utilization as well as safety, efficacy, and dosage parameters that accompany it have to be developed and unambiguous to allow postbiotics' safe and effective implementation into the practices of clinicians and other users.

9.8 FUTURE DIRECTIONS

9.8.1 The Future of Postbiotics in Managing Metabolic Disorders

Nonliving products of microbial fermentation, postbiotics, receive a lot of focus in the biology sector because of their ability to self-administer inflammation and metabolic health. As work on the gut microbiota progresses, postbiotics may be an important avenue for targeting inflammation-associated metabolic diseases including obesity, type 2 diabetes, and cardiovascular disease.

9.8.2 Emerging Trends in Postbiotic Research

Recent advancements in omics technologies such as metagenomics, metabolomics, and transcriptomics provide deeper insights into the gut microbiome's role in metabolic health (Mathuria et al., 2024). Researchers are increasingly focusing on isolating specific bioactive compounds within postbiotics, such as butyrate, acetate, and propionate, to target inflammation, metabolic regulation, and immune responses effectively.

Moreover, artificial intelligence (AI) is playing a growing role in identifying novel postbiotics and their mechanisms of action. Combining postbiotics with probiotics, prebiotics, and other treatments could enhance therapeutic outcomes by addressing multiple biological pathways, such as immune modulation, gut health, and metabolic regulation (Al-Habsi et al., 2024).

Contemporary developments related to metagenomics, metabolomics, and transcriptomics enable us to explore the gut microbiome's role in metabolism in depth. Many scientists are now noting the possibilities of deploying individual molecules in postbiotics, especially butyrate, acetate, and propionate, to fight inflammations and regulate metabolism and immune response. Furthermore, postbiotics, by influencing the human gut microbiome and its composition, are increasingly being identified by AI with their corresponding action mechanisms (G. Wang et al., 2024a). About combining postbiotics with probiotics, prebiotics, and other treatments: Postbiotic treatment could be more effective when used together with probiotics, prebiotics, and other treatment methods as a result of the impact upon several biological processes, including immunomodulation, gut function, and metabolism.

9.8.3 Personalized Medicine and Postbiotics

Postbiotics are particularly well-suited for personalized medicine, where treatments are tailored to an individual's unique microbiome, genetics, and health conditions. Using microbiome profiling, healthcare providers can select the most effective postbiotics for reducing inflammation or improving metabolic functions like insulin sensitivity and fat metabolism. For instance, patients with microbiomes rich in specific bacterial taxa may respond better to certain SCFAs (Al-Qadami et al., 2022). Personalized approaches could also incorporate microbiome-based pharmacology, in which specific gut bacteria are targeted to modulate metabolic health, enhancing the precision and efficacy of postbiotic therapies.

9.8.4 Therapeutic Value in Metabolic Disorder Management

From their ability to modulate inflammation levels and enhance cardinal metabolic pathways, postbiotics are particularly relevant to the treatment of metabolic diseases. Specifically, butyrate has been found to improve insulin response and control fat deposition and appetite-regulating mechanisms (Yu et al., 2024). These effects make postbiotics a useful supplement to traditional remedies with lifelong medication, such as obesity and type 2 diabetes. Furthermore, postbiotics contain preventative potential, especially in targeted groups with high risk of disease, targeting inflammation and metabolic disarray before a clinical setting can be diagnosed. The biggest advantages of these treatments would include long-term effects which may include halting the development of diseases and lowering the use of drugs (Hajek et al., 2014).

9.9 CONCLUSION

Postbiotics are a promising area for the elimination of inflammation-mediated metabolic disturbances. As microbiome and super personalized medicine advance postbiotics may be at the forefront in treating chronic diseases and people's metabolic health Postbiotics are used as a patient-specific therapeutic for chronic diseases. However, formidable barriers to postbiotic formulation standardization, interpatient variability, and regulatory approval will be crucial for optimal postbiotic-based therapy. The advancement in research and technology will enhance the outcome of postbiotics toward global metabolic health.

REFERENCES

Al-Habsi, N., Al-Khalili, M., Haque, S. A., Elias, M., Olqi, N. A., & Al Uraimi, T. (2024). Health benefits of prebiotics, probiotics, synbiotics, and postbiotics. *Nutrients*, *16*(22), 3955.

Al-Qadami, G. H., Secombe, K. R., Subramaniam, C. B., Wardill, H. R., & Bowen, J. M. (2022). Gut microbiota-derived short-chain fatty acids: Impact on cancer treatment response and toxicities. *Microorganisms*, *10*(10), 2048.

Alhabeeb, H., AlFaiz, A., Kutbi, E., AlShahrani, D., Alsuhail, A., AlRajhi, S., Alotaibi, N., Alotaibi, K., AlAmri, S., & Alghamdi, S. (2021). Gut hormones in health and obesity: The upcoming role of short chain fatty acids. *Nutrients*, *13*(2), 481.

Ali, Q., Ma, S., La, S., Guo, Z., Liu, B., Gao, Z., Farooq, U., Wang, Z., Zhu, X., & Cui, Y. (2022). Microbial short-chain fatty acids: A bridge between dietary fibers and poultry gut health—a review. *Animal Bioscience*, *35*(10), 1461.

Bagheri, S., Zolghadri, S., & Stanek, A. (2022). Beneficial effects of anti-inflammatory diet in modulating gut microbiota and controlling obesity. *Nutrients*, *14*(19), 3985.

Balthazar, C. F., Guimarães, J. F., Coutinho, N. M., Pimentel, T. C., Ranadheera, C. S., Santillo, A., Albenzio, M., Cruz, A. G., & Sant'Ana, A. S. (2022). The future of functional food: Emerging technologies application on prebiotics, probiotics and postbiotics. *Comprehensive Reviews in Food Science and Food Safety*, *21*(3), 2560–2586.

Brizita, D. I., & Nevena, I. D. (2022). Precise nutrition and metabolic syndrome: Remodeling the microbiome with polyphenols, probiotics, and postbiotics. In *Advances in Precision Nutrition, Personalization and Healthy Aging* (pp. 145–178). Springer.

Ceapa, C. D. (2016). *Phenotypic and genetic diversity of the species Lactobacillus rhamnosus*. Wageningen University and Research.

Chen, S., Lei, Q., Zou, X., & Ma, D. (2023). The role and mechanisms of Gram-negative bacterial outer membrane vesicles in inflammatory diseases. *Frontiers in Immunology*, *14*, 1157813.

Chugh, P., Misra, S., Dhar, M. S., & Raghuwanshi, S. (2023). Regulatory aspects relevant to probiotic products. In *Probiotics, prebiotics, synbiotics, and postbiotics: Human microbiome and human health* (pp. 513–534). Springer.

Eilam, Y., Khattib, H., Pintel, N., & Avni, D. (2023). Microalgae—Sustainable source for alternative proteins and functional ingredients promoting gut and liver health. *Global Challenges*, *7*(5), 2200177.

Eslami, M., Pakmehr, A., Pourghazi, F., Kami, A., Ejtahed, H.-S., Mohajeri-Tehrani, M., Hasani-Ranjbar, S., & Larijani, B. (2024). *The anti-obesity effects of postbiotics: A systematic review of pre-clinical and clinical studies*. Clinical Nutrition ESPEN.

Fang, H., Rodrigues, E., Lacerda, R., Barra, N. G., Kukje Zada, D., Robin, N., Mehra, A., & Schertzer, J. D. (2024). Postbiotic impact on host metabolism and immunity provides therapeutic potential in metabolic disease. *Endocrine Reviews*, *46*(1), bnae025.

Hajek, P., Etter, J. F., Benowitz, N., Eissenberg, T., & McRobbie, H. (2014). Electronic cigarettes: Review of use, content, safety, effects on smokers and potential for harm and benefit. *Addiction*, *109*(11), 1801–1810.

Hayden, M. R. (2002). Islet amyloid, metabolic syndrome, and the natural progressive history of type 2 diabetes mellitus. *Jop*, *3*(5), 126–138.

Kozłowska, I., Marć-Pieńkowska, J., & Bednarczyk, M. (2016). 2. Beneficial aspects of inulin supplementation as a fructooligosaccharide prebiotic in monogastric animal nutrition–A review. *Annals of Animal Science*, *16*(2), 315–331.

Kumar, A., Singh, R., Sehrawat, R., Kumar, N., & Pratibha. (2022). Probiotics. In *Advances in food chemistry: Food components, processing and preservation* (pp. 497–518). Springer.

Li, J., Li, Y., Zhang, S., Wang, C., Mao, Z., Huo, W., Yang, T., Li, Y., Xing, W., & Li, L. (2024). Association of the short-chain fatty acid levels and dietary quality with type 2 diabetes: A case-control study based on Henan rural cohort. *British Journal of Nutrition*, *131*(10), 1668–1677.

Luzzi, A., Briata, I. M., Di Napoli, I., Giugliano, S., Di Sabatino, A., Rescigno, M., & Cena, H. (2024). Prebiotics, probiotics, synbiotics and postbiotics to adolescents in metabolic syndrome. *Clinical Nutrition*, *46*(3), 1433–1446.

Mashaqi, S., & Badr, M. S. (2019). The impact of obstructive sleep apnea and positive airway pressure therapy on metabolic peptides regulating appetite, food intake, energy homeostasis, and systemic inflammation: A literature review. *Journal of Clinical Sleep Medicine*, *15*(7), 1037–1050.

Mathuria, A., Chaudhary, A., Sharma, H., & Mani, I. (2024). Multi-omics in gut microbiome. In *Multi-omics analysis of the human microbiome: From technology to clinical applications* (pp. 181–213). Springer.

Motei, D. E., Beteri, B., Hepsomali, P., Tzortzis, G., Vulevic, J., & Costabile, A. (2023). Supplementation with postbiotic from *Bifidobacterium Breve* BB091109 improves inflammatory status and endocrine function in healthy females: A randomized, double-blind, placebo-controlled, parallel-groups study. *Frontiers in Microbiology*, *14*, 1273861.

Mousavi Ghahfarrokhi, S. S., Mohamadzadeh, M., Samadi, N., Fazeli, M. R., Khaki, S., Khameneh, B., & Khameneh Bagheri, R. (2024). Management of cardiovascular diseases by short-chain fatty acid postbiotics. *Current Nutrition Reports*, *13*, 294–313.

Mu, T., Hu, H., Ma, Y., Feng, X., Zhang, J., & Gu, Y. (2021). Regulation of key genes for milk fat synthesis in ruminants. *Frontiers in Nutrition*, *8*, 765147.

Navab, M., Gharavi, N., & Watson, A. D. (2008). Inflammation and metabolic disorders. *Current Opinion in Clinical Nutrition & Metabolic Care*, *11*(4), 459–464.

Ohland, C. L., & Jobin, C. (2015). Microbial activities and intestinal homeostasis: A delicate balance between health and disease. *Cellular and Molecular Gastroenterology and Hepatology*, *1*(1), 28–40.

Park, M., Joung, M., Park, J.-H., Ha, S. K., & Park, H.-Y. (2022). Role of postbiotics in diet-induced metabolic disorders. *Nutrients*, *14*(18), 3701.

Park, S.-J., Sharma, A., & Lee, H.-J. (2023). Postbiotics against obesity: Perception and overview based on pre-clinical and clinical studies. *International Journal of Molecular Sciences*, *24*(7), 6414.

Perseghin, G., Petersen, K., & Shulman, G. (2003). Cellular mechanism of insulin resistance: Potential links with inflammation. *International Journal of Obesity*, *27*(3), S6–S11.

Rafique, N., Jan, S. Y., Dar, A. H., Dash, K. K., Sarkar, A., Shams, R., Pandey, V. K., Khan, S. A., Amin, Q. A., & Hussain, S. Z. (2023). Promising bioactivities of postbiotics: A comprehensive review. *Journal of Agriculture and Food Research*, *14*, 100708.

Rekha, K., Venkidasamy, B., Samynathan, R., Nagella, P., Rebezov, M., Khayrullin, M., Ponomarev, E., Bouyahya, A., Sarkar, T., & Shariati, M. A. (2024). Short-chain fatty acid: An updated review on signaling, metabolism, and therapeutic effects. *Critical Reviews in Food Science and Nutrition*, *64*(9), 2461–2489.

Richards, P., Thornberry, N. A., & Pinto, S. (2021). The gut–brain axis: Identifying new therapeutic approaches for type 2 diabetes, obesity, and related disorders. *Molecular Metabolism*, *46*, 101175.

Sheikh, A. M., Yano, S., Tabassum, S., & Nagai, A. (2024). The role of the vascular system in degenerative diseases: Mechanisms and implications. *International Journal of Molecular Sciences*, *25*(4), 2169.

Singh, M., & Benencia, F. (2019). Inflammatory processes in obesity: Focus on endothelial dysfunction and the role of adipokines as inflammatory mediators: We reviewed obesity-induced metabolic and immunological changes at the level of vasculature and emphasize on the importance of adipokines. *International Reviews of Immunology*, *38*(4), 157–171.

Sinha, G. (n.d.). Infectious microbes: The real culprits behind today's most debilitating chronic degenerative diseases.

Suthar, P., Kumar, S., Kumar, V., Sharma, V., & Dhiman, A. (2023). Postbiotics: An exposition on next generation functional food compounds—opportunities and challenges. *Critical Reviews in Food Science and Nutrition*, *65*(6), 1–20.

Thoda, C., & Touraki, M. (2023). Immunomodulatory properties of probiotics and their derived bioactive compounds. *Applied Sciences*, *13*(8), 4726.

Waheed, R., Farooq, A. Z., & Asma, L. (2024). Regulatory considerations for microbiome-based therapeutics. In *Human microbiome: Techniques, strategies, and therapeutic potential* (pp. 657–689). Springer.

Wang, G., Ding, T., & Ai, L. (2024a). Effects and mechanisms of probiotics, prebiotics, synbiotics and postbiotics on intestinal health and disease. *Frontiers in Cellular and Infection Microbiology*, *14*, 1430312.

Wang, S., Wang, P., Wang, D., Shen, S., Wang, S., Li, Y., & Chen, H. (2024b). Postbiotics in inflammatory bowel disease: Efficacy, mechanism, and therapeutic implications. *Journal of the Science of Food and Agriculture*, *105*(2), 721–734.

Wang, T., & He, C. (2020). TNF-α and IL-6: The link between immune and bone system. *Current Drug Targets*, *21*(3), 213–227.

Wu, T., Wang, G., Xiong, Z., Xia, Y., Song, X., Zhang, H., Wu, Y., & Ai, L. (2022). Probiotics interact with lipids metabolism and affect gut health. *Frontiers in Nutrition*, *9*, 917043.

Yan, R., Zeng, X., Shen, J., Wu, Z., Guo, Y., Du, Q., Tu, M., & Pan, D. (2024). New clues for postbiotics to improve host health: A review from the perspective of function and mechanisms. *Journal of the Science of Food and Agriculture*, *104*(11), 6376–6387.

Yeşilyurt, N., Yılmaz, B., Ağagündüz, D., & Capasso, R. (2021). Involvement of probiotics and postbiotics in the immune system modulation. *Biologics*, *1*(2), 89–110.

Yu, M., Yu, B., & Chen, D. (2024). The effects of gut microbiota on appetite regulation and the underlying mechanisms. *Gut Microbes*, *16*(1), 2414796.

Yüksel, E., Voragen, A. G., & Kort, R. (2024). The pectin metabolizing capacity of the human gut microbiota. *Critical Reviews in Food Science and Nutrition*, 1–23.

Zhong, Y., Wang, T., Luo, R., Liu, J., Jin, R., & Peng, X. (2024). Recent advances and potentiality of postbiotics in the food industry: Composition, inactivation methods, current applications in metabolic syndrome, and future trends. *Critical Reviews in Food Science and Nutrition*, *64*(17), 5768–5792.

10 Postbiotics and Gut Microbiota

Gut Health and Oxidative Stress Modulation

Sui Sien Leong
Department of Animal Sciences and Fishery, Faculty of Agricultural and Forestry Sciences, Universiti Putra Malaysia Bintulu Sarawak Campus, 97008 Bintulu, Sarawak, Malaysia.
Lactic Acid Bacteria Biota Technology Research Program, Research Laboratory of Probiotics and Cancer Therapeutics, UPM-MAKNA Cancer Research Laboratory (CANRES), Institute of Bioscience, Universiti Putra Malaysia, 43400 Serdang, Selangor, Malaysia.

Teck Chwen Loh
Lactic Acid Bacteria Biota Technology Research Program, Research Laboratory of Probiotics and Cancer Therapeutics, UPM-MAKNA Cancer Research Laboratory (CANRES), Institute of Bioscience, Universiti Putra Malaysia, 43400 Serdang, Selangor, Malaysia.
Department of Animal Science, Faculty of Agriculture, Universiti Putra Malaysia, 43400 Serdang, Selangor, Malaysia.

Hooi Ling Foo
Lactic Acid Bacteria Biota Technology Research Program, Research Laboratory of Probiotics and Cancer Therapeutics, UPM-MAKNA Cancer Research Laboratory (CANRES), Institute of Bioscience, Universiti Putra Malaysia, 43400 Serdang, Selangor, Malaysia.
Department of Bioprocess Technology, Faculty of Biotechnology and Biomolecular Sciences, Universiti Putra Malaysia, 43400 Serdang, Selangor, Malaysia.

10.1 INTRODUCTION

The gut is a key essential organ in the gastrointestinal tract, comprising small and large intestines. It hosts a complex gut microbiome, which has an important function in nutrient uptake by breaking down complex food molecules and enhancing nutrient utilisation efficiency (Leong et al., 2023). The intestinal mucosa is vital as it protects, supports nutrient absorption, and creates an optimal environment for microbial colonisation (Rinttilä & Apajalahti, 2013). The microbiota shapes gut structure and function, fortifies the mucosal barrier, supports digestion, and assists nutrient absorption.

DOI: 10.1201/9781003607779-10

A healthy microbial balance prevents infectious intrusion and modulates immune responses (Sugiharto, 2016). The development of a balanced intestinal mucosal barrier relies on the interactions between the mucus membrane, epithelial tissues, intestinal microbiome, and immune system cells (Takiishi et al., 2017).

However, the gut remains vulnerable to extrinsic pathogens and stressors such as diet changes, nutrition, and infectious agents, which can disrupt mucosal dynamics. These disruptions negatively affect the health and productivity of livestock, particularly cattle, swine, and poultry (Lovera et al., 2017). For the past common practices, farmers used feed additives and antibiotics to enhance livestock health and growth. Yet, due to adverse effects and concerns over antibiotic resistance, regulations have limited antibiotic use in animal feed. This shift has sparked interest in safe alternatives, such as probiotics, to achieve similar benefits without the risks. With antibiotic restrictions, probiotics have gained popularity as feed additives to enhance nutrient bioavailability, prevent pathogen infections, and reduce environmental impacts associated with livestock production (Lingoh et al., 2020). Lactic acid bacteria (LAB) have proven beneficial in promoting gut health by supporting microbiome balance, exhibiting antimicrobial properties, and even reducing carcinogenic risks (Foo et al., 2019; Yang et al., 2019). However, concerns regarding live probiotics have emerged, particularly under inflammation or dysbiosis, which may increase vulnerability to pathogenic infections (Tsilingiri et al., 2013; Khairul et al., 2024).

To address the negative effects of viable probiotic supplementation, alternative products known as postbiotic metabolites have been developed, the soluble bioactive compounds developed from viable probiotics via a fermentation process using a defined growth medium. The postbiotic metabolites also defined as metabiotics or culture supernatants have been described as metabolic byproducts or cell-free extracts generated or excreted by living bacteria or during bacterial breakdown (Aguilar-Toalá et al., 2018). Postbiotics, including organic acids, enzymes, peptides, bacteriocins, polysaccharides, and plasmalogens, exert health benefits without introducing live microorganisms. They help restore microbial balance (dysbiosis) in the gut and induce beneficial physiological responses (Cicenia et al., 2014; Loh et al., 2021). *Lactiplantibacillus plantarum*, a widely studied and commercially important species of LAB (Chen et al. 2017) isolated from Malaysian fermented food (Foo et al., 2019) shows considerable promise as a bioresource for the production of effective postbiotics (Rajoka et al., 2017). *L. plantarum* produces beneficial compounds like short-chain fatty acids (SCFA), pyrrole compounds, and bacteriocins, which demonstrate antimicrobial and antioxidant properties (Kareem et al., 2016; Humam et al., 2019; Izuddin et al., 2020). Postbiotic mixtures of several *L. plantarum* strains exhibited superior antimicrobial effects compared to individual strains, attributed to different lactic acid, acetic acid, and bacteriocin compositions (Muhammad et al., 2019). Oxidative stress caused by an imbalance of pro-oxidants and antioxidants commonly affects livestock productivity, which may further lead to cell damage and tissue injury in animals. Postbiotics offer excellent antioxidant advantages by neutralising free radicals, lowering oxidative stress, and boosting antioxidant mechanisms (Rahimi et al., 2024). Many studies reported their capacity to affect gut microorganisms and metabolic routes associated with physiological, immune and hormonal processes, leading to improved health (Yang et al., 2019; Foo et al., 2019).

Postbiotics are gaining prominence attention in Malaysian livestock management as sustainable alternatives to antibiotic growth promoters, addressing concerns on antibiotic resistance. The soluble bioactive compounds of postbiotics, such as SCFA and bacteriocins, enhance gut health, digestion, and immune responses while improving growth performance and reducing disease risks. The adoption of postbiotic alternatives aligns with national goals for sustainable farming by reducing antibiotic use and fostering healthier and more resilient livestock populations. This chapter explores the functionalities of *Lactiplantibacillus plantarum*–derived postbiotics, focusing on their roles in gut microbiota modulation, oxidative stress mitigation, and redox balance in livestock, supporting sustainable agriculture and food security.

10.2 MECHANISMS AND FUNCTIONALITIES OF POSTBIOTIC BIOACTIVE COMPOUNDS

The term "postbiotics" was initially proposed by Tsilingiri and Rescigno in 2013, and it originates from the Greek prefix "post," meaning "after," and "biotic," meaning "living things." There have been numerous definitions of postbiotics proposed in the decade after the suggestion of Tsilingiri and Rescigno in 2013. Recently, postbiotics have been defined as nonviable microorganism components derived from microbial cells or metabolites produced by microbes that offer health advantages to the host (Vinderola et al., 2024), similar to the probiotic paradox. The production of currently defined postbiotic products involves the separation or inactivation of live microorganisms, resulting in the production of bioactive substances present in supernatants without cells, exopolysaccharide compounds, antioxidant enzymes, fragments of the cell wall, bacterial lipoteichoic acids, SCFA, lysates from bacteria, peptidoglycan muropeptides, and pili structures (Sánchez et al., 2017; Zółkiewicz et al., 2020).

The efficacy of postbiotics relies on interactions between the host and microbial-derived soluble metabolites (Loh et al., 2013). Postbiotics contain soluble bioactive compounds exhibiting diverse functionalities, although the exact roles or mechanisms of postbiotics in promoting health are not fully understood (Salminen et al., 2022). Recently, Hijová (2024) described four primary postbiotic mechanisms: alteration of host microbiota composition, strengthening of gut barrier integrity, systemic metabolic pathways management, and communication via the gut-brain connection. First, postbiotics influence the host microbiota directly through their antimicrobial ability mediated by bioactive compounds such as lactic acid, SCFA, and bacteriocins. Subsequently, postbiotics exert indirect effects by promoting metabolic interactions within the microbiota. For instance, the lactic acid composition of postbiotics consumed by specific microbiota will produce butyrate SCFA, which is crucial for intestinal health (Salminen et al., 2022). The antimicrobial activity of postbiotics can inhibit intestinal pathogens mediated by soluble bioactive metabolites and bacteriocins, disrupt biofilm formation, and inhibit the growth of harmful microorganisms (Hernández-Granados & Franco-Robles, 2020). Second, postbiotics improve gut barrier function by secreting proteins like Msp1/p75 and Msp1/p40 from *Lactobacillus rhamnosus GG* (Bäuerl et al., 2019). SCFA and exopolysaccharides enhance the integrity of tight junctions and reduce inflammation (Feng et al., 2018; Engevik et al., 2019).

Third, postbiotics display a vital role in modulating systemic immune responses through microbial metabolites, bile acid regulation, and SCFA production, which collectively influence host metabolism and help prevent metabolic disorders. Immunomodulation initiates when molecular patterns associated with microorganisms (MAMPs) interact with pattern recognition receptors (PRRs) present in immune cells, activating pathways that produce cytokines and immune modulators (Thangaraju et al., 2009). The key metabolites from inactivated postbiotic microorganisms include lactic acid (Morita et al., 2019), tryptophan-derived indole compounds (Barragan et al., 2018), histamine (Thomas et al., 2012) and SCFA influence bile acid metabolism. The enzymes like both bile salt hydrolase (BSH) and also bile acid-inducible enzymes (BAI) play critical roles in maintaining gut microbiome health and host metabolic functions, such as glucose and lipid metabolism (Khairul et al., 2024). SCFA, extensively study postbiotic metabolites, regulate host metabolism and show potential in managing cardiovascular diseases, type 2 diabetes, and obesity. Additionally, muramyl dipeptide (MDP) in bacterial cell walls enhances insulin sensitivity, regulates glucose tolerance, and modulates GLP-1 secretion (Williams et al., 2020).

Finally, postbiotics may offer enhanced clinical benefits by regulating the immune system and enhancing gut barrier function, achieving through the enhancement of tight junctions or the encouragement of mucus secretion for the intestinal lining (Hijová, 2024). Postbiotics impact the gut-brain axis, showing potential as novel therapies for depression and neurodegenerative diseases by targeting gut microbiome changes (Flux & Lowry, 2020). The exact causal relationship within the microbiome–gut-brain axis remains unclear. Nevertheless, the gut microbiome holds potential as a source of indicators and a goal for depression treatment. Moreover, postbiotics might contribute to addressing the pathogenesis of multiple sclerosis (MS), a condition also linked to gut microbiota alterations. Specifically, the reduction in SCFA-producing bacteria observed in MS patients presents a promising target for postbiotic interventions to restore microbial balance and mitigate disease progression.

10.3 GUT MICROBIOTA MODULATION BY POSTBIOTICS

The gut microbiota can be broadly categorised into symbionts and pathobionts, exhibiting distinct immunomodulatory functions. Symbionts primarily promote beneficial and tolerogenic effects, contributing to the host's overall health, while pathobionts are associated with pro-inflammatory responses and the development of inflammatory immune cells (Liu et al., 2024). Pathobionts can dominate their ecological niche during episodes of severe damage or stress, potentially exacerbating disease. Maintaining a balance between these two microbial groups is critical for proper immune development, tolerance, and the upkeep of intestinal equilibrium (Yang et al., 2023). Tremendous research has emphasised the vital role of gut health and microbiota balance in the overall wellness of animals. The digestive system is not just crucial for breaking down and absorbing nutrients but also serves as a defence barrier against external pathogens through the secretion of immune-related substances (Lingoh et al., 2020). Additionally, it serves as a reservoir for a diverse microbial community essential for immunological balance. Alterations in the composition or reduced diversity of gut commensals have been implicated in numerous pathological conditions. These conditions often feature an overactive immune response targeting self-antigens or commensal-derived antigens, underscoring the essential function of gut microbiota in preserving health. The application of probiotics in livestock gained attention in the late 1990s and early 2000s as a sustainable alternative to antibiotics in animal agriculture practice. This transition was primarily driven by global concerns about antibiotic resistance and the introduction of regulations to minimise antibiotic use as growth promoters (Leong et al., 2023). In Malaysia, the incorporation of probiotics reflects efforts to enhance livestock productivity and promote animal health to ensure food safety and security. During the early 2000s, government initiatives and academic research prioritized evaluating the effectiveness of probiotics in enhancing the wellness and productivity of poultry, aquaculture, and ruminants. More recently, growing recognition of the benefits of postbiotics, such as enhancing gut health, optimising feed efficiency, and mitigating disease risk, has further supported their integration into Malaysian livestock production practices (Loh et al., 2021).

Supplementation of postbiotics in poultry offers multiple benefits, primarily through enhancing gut health and immunity. Postbiotics promote a healthy gut microbiota that safeguards the host by generating and secreting antimicrobial substances (SCFA, bacteriocins, and lipopolysaccharides) that will prevent pathogenic infiltration in the intestine (Chang et al., 2021). Over time, the gut matures along with strengthened mucosal integrity, supported by tight junction proteins, mucins like acidic and sulfated mucins, and immune-related compounds such as cytokines and immunoglobulin A (Chang et al., 2021). The engagement of postbiotics with the host microbiome, intestinal epithelial cells (IEC), and the immune system is dynamic, influencing immune responses and gut barrier function. IECs serve as the primary barrier, governing immune reactions through cytokines and chemokines that condition dendritic cells for either tolerogenic or immunogenic phenotypes (Goto, 2019). Postbiotics strengthen the epithelial barrier, preventing "leaky gut" conditions and shielding against pathogenic invasion. Postbiotics can shield the IEC barrier and prevent them from the apoptosis process.

Postbiotics produced by *L. plantarum* exhibit antimicrobial, anti-inflammatory, and antioxidant properties. They lower gut pH to inhibit pathogen growth, stimulate mucin production, and enhance the regeneration of epithelial cells (Kareem et al., 2014). Moreover, *L. plantarum* has been shown to prevent the proliferation of dangerous bacteria, such as *Listeria monocytogenes*, *Salmonella typhimurium*, and *Escherichia coli* (Kareem et al., 2014). The efficiency of *L. plantarum* was demonstrated in direct rivalry among pathogens for adhesion locations and the enhancement of the speed of epithelial repair (Aguilar-Toalá et al., 2018). Postbiotics *L. plantarum* display both bactericidal and bacteriostatic properties, successfully inhibiting the growth and spread of harmful bacteria in the gut tested *in vivo* (Loh et al., 2014; Kareem et al., 2016). Supplementing poultry feed with postbiotics has improved growth performance, feed conversion ratios (FCR), and immune responses in broilers, laying hens, pigs, and lambs (Choe et al., 2012; Izuddin et al., 2020). Notably, postbiotic-treated broilers displayed increased beneficial LAB populations and reduced pathogenic *E. coli* counts, contributing to better intestinal health (Kareem et al., 2016).

Postbiotic supplementations have been shown to positively impact gut microbiota composition, fostering the expansion of advantageous bacteria such as *Bifidobacterium* and LAB, while minimising pathogenic microorganisms. The benefits of postbiotic supplementation extend beyond poultry, as evidenced by improved fermentation in the rumen and the development progress of lambs (Izuddin et al., 2019a). The diverse functions of postbiotics in supporting gut health, boosting immunity, and improving productivity underscore their potential as sustainable feed additives in livestock management. This aligns with the growing need to reduce antibiotic reliance while ensuring animal welfare and food safety throughout the world, especially in Malaysia's agricultural landscape (Yang et al., 2019). Kareem et al. (2016) have proven that supplementing *Lactiplantibacillus plantarum* RG14 in combination with inulin significantly promoted the growth of advantageous *Bifidobacterium* while decreasing the number of pathogens in comparison to untreated groups. Likewise, broilers and layer poultry consuming *L. plantarum*–based postbiotics showed higher faecal LAB levels (Choe et al., 2012). Humam et al. (2019) reported that broilers kept in heat stress conditions with postbiotics supplementation had increased cecal number of total bacteria, *Lactiplantibacillus*, and *Bifidobacterium*, while demonstrating markedly lower levels of pathogenic *Enterobacteriaceae*, *E. coli*, and *Salmonella*. The changes in the gut microbiome are closely associated with the cecal pH, which is lowered by the organic acids in postbiotics. A lower cecal pH inhibits the growth of acid-intolerant pathogens while fostering the increase of acid-resistant beneficial bacteria of *Lactiplantibacillus* and *Bifidobacterium*. Furthermore, in lambs, adding *L. plantarum* postbiotics to the diet increased the presence of fiber-degrading bacteria in the rumen while lowering the numbers of total protozoa and methanogens, which promote better ruminal health and efficiency (Izuddin et al., 2019b). Preserving the equilibrium of the gastrointestinal system is vital for the ideal physiological outcomes, as it sustains a balanced microbial community and offers defence against harmful agents (Adedokun & Olojede, 2019). The interaction between postbiotics and gut morphology structure is often evaluated through changes in intestinal villi height and crypt depth, which serve as key portals for nutrient uptake (Choe et al., 2012). Assessments of gut health commonly use parameters of villi height and crypt depth acting as indicators of the gut mucosal condition (Heak et al., 2017). Supplementation with postbiotics in laying hens has been shown to improve nutrient digestibility by maintaining healthy villus elevation and crypt depth (Choe et al., 2012). Optimal gut function is associated with taller villi and shallower crypts, which enhance the efficiency of nutrient absorption into the blood circulation (Marković et al., 2009).

Studies have shown that combinations of postbiotics from *L. plantarum* enhance villus height within the small intestines of broiler chickens (Kareem et al., 2016) and pigs (Loh et al., 2013). Similarly, the supplementation of 6% (v/v) liquid metabolites derived from *L. plantarum* improved villus height in laying hens (Choe et al., 2012). Postbiotic supplementation in post-weaning lamb feed was also reported to increase villus height, with corresponding improvements in the small intestinal morphology, such as taller villi and shallower crypts (Izuddin et al., 2019a). These morphological changes have been linked to enhanced nutrient absorption and overall gut health (Humam et al., 2019). According to Marković et al. (2009), the observed improvements in villi height and crypt depth may explain the better growth and health outcomes in broiler chickens receiving postbiotic supplements. These structural improvements contribute significantly to overall gut health and nutrient utilisation, demonstrating the benefits of postbiotic interventions in poultry. Additionally, *L. plantarum*–derived postbiotics have been shown to promote the expression of intestinal mucin, which is essential for forming a protective barrier and activating immune responses against pathogens. Extensive reports have consistently described the beneficial impacts of postbiotics on enhanced immune responses in meat chicken (Kareem et al., 2016) and even lambs (Izuddin et al., 2019a), while improving overall gut health in animals. The benefits extend further to the reduction of pathogenic bacterial populations, including *E. coli*, *Enterobacteria*, and *Salmonella*, in the caecum of broiler chickens. This effect is particularly valuable under heat stress conditions, where postbiotics help mitigate negative impacts and support intestinal integrity (Humam et al., 2019). These results highlight the role of *L. plantarum*–based postbiotics as a vital alternative in enhancing livestock productivity and health through improved gut morphology and immune modulation.

10.4 OXIDATIVE STRESS AND REDOX BALANCE BY POSTBIOTICS

Oxidative stress occurs when the rise of reactive oxygen species (ROS) surpasses the capacity of antioxidant defence mechanisms, resulting in the overoxidation of biomolecules and the disturbance of signalling pathways (Sies & Jones, 2020). ROS comprise minor, diffusible substances containing one or several unpaired electrons, including peroxides, superoxide, and hydroxyl radicals, along with nonradical compounds such as hydrogen peroxide and peroxynitrite. ROS substances are waste products generated from oxygen metabolism in animal cells (Navarro-Yepes et al., 2014). In moderate amounts, ROS play beneficial roles in several biological processes, including stem cell specialization, lineage determination, self-replication, and cellular equilibrium (Ren et al., 2015). They act as signalling molecules to regulate gene expression associated with cell development and growth. Nonetheless, an overproduction of ROS can result in oxidative harm to proteins, lipids, and DNA, leading to the decline of cellular function, which can ultimately cause cell death. This oxidative damage has been associated with the initiation and advancement of different illnesses and can have detrimental effects on critical cellular components like proteins, lipids, and nucleic acids (Wu et al., 2013). Farm animals are particularly vulnerable to oxidative stress due to various environmental challenges and physiological constraints throughout their productive lives. These stressors can disrupt their physiological balance, resulting in reduced performance, compromised immunity, and overall health deterioration (Abuelo et al., 2019). Persistent oxidative stress can induce temporary or permanent cellular and systemic perturbations, leading to adverse physiological consequences at both cellular and organismal levels (Pignatelli et al., 2018). Understanding oxidative stress and its impact is crucial for developing strategies to enhance the health and productivity of farm animals under challenging conditions.

Probiotics like LAB, *Bifidobacterium*, and yeast display antioxidative properties via various mechanisms, including metal ion chelation, the production of antioxidants, enhancement of host antioxidant functions and gene activity, alteration of signalling, decrease of ROS enzyme activities, and the adjustment of gut microflora composition (Wang et al., 2017). Postbiotics, derived from probiotics, display comparable antioxidant characteristics, promoting animal health and improving production performance. Specific metabolites produced by LAB, such as lipoteichoic acid, exopolysaccharides, and even ell-surface proteins, have been identified as key contributors to the antioxidant potential of postbiotics (Zhao et al., 2024). Interestingly, the composition of the fermentation medium serves a crucial effect in shaping the antioxidant activities of postbiotics, influencing the types and concentrations of volatile compounds and organic acids produced during the fermentation process of probiotics in the production of postbiotics (Chang et al., 2021). The postbiotic components are instrumental in bolstering the antioxidant protection capacity by actively scavenging ROS, including hydroxyl radicals, and inhibiting their production.

The testing of *L. plantarum* postbiotics has reported significant reductions in malondialdehyde (MDA) levels in the serum of post-weaned lambs, a marker of lipid peroxidation resulting from weakened antioxidant defences and the build-up of free radicals (Yang et al., 2008). Weaning stress, a significant challenge in young mammals, adversely affects the epithelial, immune, and enteric nervous systems, eventually diminishing overall health and efficiency (Moeser et al., 2017). Postbiotics have demonstrated their potential in alleviating the adverse effects of ROS. For example, supplementation with *L. plantarum* RG14 postbiotics in lambs enhanced serum and ruminal glutathione peroxidase (GPX) enzymatic activity, although superoxide dismutase (SOD) concentration was reported unchanged (Izuddin et al., 2020). Additionally, supplementation with *L. plantarum* RG14 postbiotics enhanced the gene expression activity of liver antioxidants, including GPX1, GPX4, and Cu/Zn SOD (Izuddin et al., 2020). Increased SCFA production, a known result of postbiotic activity, has also been noted in broiler chickens supplemented with *L. plantarum* postbiotic metabolites and prebiotics (Kareem et al., 2016) and in heat-stressed broilers (Humam et al., 2019). Heat stress in animals leads to excessive ROS production, including hydroxyl free radicals and superoxide anions, which can cause tissue damage (Akbarian et al., 2016). In poultry, the production of antioxidants acts as a defence mechanism to counteract ROSs to reduce their detrimental effects. (Surai, 2020). Studies have shown that supplementation with postbiotic RI11 in chickens during heat stress conditions enhanced overall antioxidant capacity, catalase, and glutathione functions while decreasing concentrations of alpha-1-acid glycoprotein and ceruloplasmin in blood plasma (Humam et al., 2020). Nonetheless, there were no notable alterations in the levels of plasma corticosterone and heat shock protein. The study conducted by Izuddin et al. (2020) reported that postbiotic supplementation influenced the activities of glutathione peroxidase and thiobarbituric acid reactive substances in serum. Our present knowledge of the antioxidant properties of postbiotics mainly relies on investigating particular bioactive elements generated by microorganisms or found in their cell walls, including SCFA and various other bioactive compounds. The effects of postbiotic metabolites have been demonstrated both in vitro and in vivo, but their precise roles and mechanisms within the complex postbiotic matrix remain challenging to elucidate. The potential interactions or synergistic effects among the bioactive compounds further complicate the understanding of postbiotic beneficial mechanisms. Therefore, more in vivo studies are necessary to explore the pivotal factors influencing antioxidant responses and elucidate the intricate mechanisms underlying the antioxidative effects of postbiotics.

10.5 CONCLUSION

This chapter highlights the benefits of postbiotics, emphasising their antioxidative and gut health–enhancing properties. Postbiotics supplementations have improved animal health by reducing oxidative stress, enhancing gut morphology structure, and modulating gut microbiota balance. Key metabolites like SCFA boost immunity and nutrient absorption while managing stress-related challenges such as heat and weaning stress in livestock. Despite promising findings, further in vivo studies are required to understand the pathway or mechanism underlying the synergistic effects in advancing animal welfare and productivity.

REFERENCES

Adedokun, S.A., & Olojede, O.C. (2019). Optimizing gastrointestinal integrity in poultry: The role of nutrients and feed additives. *Frontiers in Veterinary Science*, 5, 348.

Aguilar-Toalá, J.E., Garcia-Varela, R., Garcia, H.S., Mata-Haro, V., González-Córdova, A. F., Vallejo-Cordoba, B., & Hernández-Mendoza, A. (2018). Postbiotics: An evolving term within the functional foods field. *Trends in Food Science and Technology*, 75, 105–114.

Abuelo, A., Hernández, J., Benedito, J.L., & Castillo, C. (2019). Redox biology in transition periods of dairy cattle: Role in the health of periparturient and neonatal animals. *Antioxidants*, 8(1), 20.

Akbarian, A., Michiels, J., Degroote, J., Majdeddin, M., Golian, A., & De Smet, S. (2016). Association between heat stress and oxidative stress in poultry; mitochondrial dysfunction and dietary interventions with phytochemicals. *Journal of Animal Science and Biotechnology*, 7(1).

Barragan, L.C., Chai, J., Tianero, M.D., Di Luccia, B., Ahern, P., Merriman, J., Cortez, V.S., Caparon, M.G., Donia, M.S., Gilfillan, S., Cella, M., Gordon, J., Hsieh, C.S., & Colonna, M. (2018). *Lactobacillus reuteri* induces gut intraepithelial CD4+CD8αα+ T cells. *The Journal of Immunology*, 200(1_Supplement), 119.2–119.2.

Bäuerl, C., Abitayeva, G., Sosa-Carrillo, S., Mencher-Beltrán, A., Navarro-Lleó, N., Coll-Marqués, J.M., Zúñiga-Cabrera, M., Shaikhin, S., & Pérez-Martinez, G.. (2019). P40 and P75 are singular functional muramidases present in the *Lactobacillus Casei/paracasei/rhamnosus* taxon. *Frontiers in Microbiology*, 10, 1420.

Cicenia, A., Scirocco, A., Carabotti, M., Pallotta, L., Marignani, M., & Severi, C. (2014). Postbiotic activities of lactobacilli-derived factors. *Journal of Clinical Gastroenterology*, 48(1), 18.

Chang, H.M., Foo, H.L., Loh, T.C., Lim, E.T.C., & Abdul Mutalib, N.E. (2021). Comparative studies of inhibitory and antioxidant activities, and organic acids compositions of postbiotics produced by probiotic *Lactiplantibacillus plantarum* strains isolated from Malaysian foods. *Frontiers in Veterinary Science*, 7, 602280.

Chen, Z.Y., Hsieh, Y.M., Huang, C.C., & Tsai, C.C. (2017). Inhibitory effects of probiotic *Lactobacillus* on the growth of human colonic carcinoma cell line ht-29. *Molecules*, 22, 107.

Choe, D.W., Loh, T.C., Foo, H.L., Hair-Bejo, M., & Awis, Q.S. (2012). Egg production, faecal pH and microbial population, small intestine morphology and plasma and yolk cholesterol in laying hens given liquid metabolites produced by *Lactobacillus plantarum* strains. *British Poultry Science*, 53(1), 106.

Engevik, M.A., Luk, B., Chang-Graham, A.L., Hall, A., Herrmann, B., Ruan, W., Endres, B.T., Shi, Z., Garey, K.W., Hyser, J.M., & Versalovic, J. (2019). *Bifidobacterium dentium* fortifies the intestinal mucus layer via autophagy and calcium signaling pathways. *mBio* 10, e01087–19.

Feng, Y., Wang, Y., Wang, P., Huang, Y., & Wang, F. (2018). Short chain fatty acids manifest stimulative and protective effects on intestinal barrier function through the inhibition of nlrp3 inflammasome and autophagy. *Cellular Physiology and Biochemistry*, 49, 190–205.

Flux, M.C., & Lowry, C.A. (2020). Finding intestinal fortitude: Integrating the microbiome into a holistic view of depression mechanisms, treatment, and resilience. *Neurobiology of Disease*, 135, 104578.

Foo, H.L., Loh, T.C., Abdul Mutalib, N.E., & Rahim, R.A. (2019). The myth and therapeutic potentials of postbiotics (Chapter 21). In *Microbiome and metabolome in diagnosis, therapy, and other strategic applications* (pp. 201–211). Academic Press: New York.

Goto, Y. (2019). Epithelial cells as a transmitter of signals from commensal bacteria and host immune cells. *Frontiers in Immunology*, 10, 2057.

Hijová, E. (2024). Postbiotics as metabolites and their biotherapeutic potential. *International Journal of Molecular Sciences*, 25(10), 5441.

Hernández-Granados, M.J., & Franco-Robles, E. (2020). Postbiotics in human health: Possible new functional ingredients? *Food Research International*, 137, 109660.

Izuddin, W.I., Loh, T.C., Foo, H.L., Samsudin, A.A., & Humam, A.M. (2019a). Postbiotic *L. plantarum* RG14 improves ruminal epithelium growth, immune status and up-regulates the intestinal barrier function in postweaning lambs. *Scientific Reports*, 9(1), 9938.

Izuddin, W.I., Loh, T.C., Samsudin, A.A., Foo, H.L., Humam, A.M., & Shazali, N. (2019b). Effects of postbiotic supplementation on growth performance, ruminal fermentation and microbial profile, blood metabolite and G.H.R., IGF-1 and MCT-1 gene expression in postweaning lambs. *BMC Veterinary Research*, 15(1), 1.

Izuddin, W.I., Humam, A.M., Loh, T.C., Foo, H.L., & Samsudin, A.A. (2020). Dietary postbiotic *Lactobacillus plantarum* improves serum and ruminal antioxidant activity and up-regulates hepatic antioxidant enzymes and ruminal barrier function in postweaning lambs. *Antioxidants* 9(3), 250.

Kareem, K.Y., Ling, F.H., Chwen, L.T., Foong, O.M., & Asmara, S.A. (2014). Inhibitory activity of postbiotic produced by strains of *Lactobacillus plantarum* using reconstituted media supplemented with inulin. *Gut Pathogen*, 6(1), 23.

Kareem, K.Y., Loh, T.C., Foo, H.L., Asmara, S.A., & Akit, H. (2016). Influence of postbiotic RG14 and inulin combination on cecal microbiota, organic acid concentration, and cytokine expression in broiler chickens. *Poultry Science*, 96(4), 966.

Khairul, S.R., Leong, S.S., Korel, F., Lingoh, A.D., Toh, S.C. (2024). Systematic review of emerging trends in soil-based probiotic. *Malaysian Journal of Soil Science*, 28, 369–381.

Heak, C., Sukon, P., Kongpechr, S., Tengjaroen, B., & Chuachan, K. (2017). Effect of direct-fed microbials on intestinal villus height in broiler chickens: A systematic review and meta-analysis of controlled trials. *International Journal of Poultry Science*, 16(10), 403–414.

Humam, A.M., Loh, T.C., Foo, H.L., Samsudin, A.A., Mustapha, N.M., Zulkifli, I., & Izuddin, W.I. (2019). Effects of feeding different postbiotics produced by *Lactobacillus plantarum* on growth performance, carcass yield, intestinal morphology, gut microbiota composition, immune status, and growth gene expression in broilers under heat stress. *Animals*, 9(9), 644.

Humam, A.M., Loh, T.C., Foo, H.L., Izuddin, W.I., Zulkifli, I., Samsudin, A.A., & Mustapha, N.M. (2020). Supplementation of postbiotic RI11 improves antioxidant enzyme activity, upregulated gut barrier genes, and reduced cytokine, acute phase proteins, and HSP70 gene expression levels in heat-stressed broilers. *Poultry Science*, 100(3). https://doi.org/10.1016/j.psj.2020.12.011

Leong, S.S., Korel, F., Lingoh, A.D., Sarbini, S.R., Toh, S.C., Abit, L.Y., & Wong, S.C. (2023). Current probiotics application for aquaculture feed: A review. *Borneo Science Journal*, 44(2), 1–13.

Lingoh, A.D., Leong, S.S., & Sarbini, S.R. (2020). Detection of lactic acid bacteria (LAB) from local chicken breed gut as probiotic agent in livestock. *Pakistan Journal of Nutrition*, 19(4), 197–203.

Liu, Y., Luo, R., Bai, S., Lemaitre, B., Zhang, H., & Li, X. (2024). Pathobiont and symbiont contribute to microbiota homeostasis through Malpighian tubules–gut countercurrent flow in *Bactrocera dorsalis*. *The ISME Journal*. https://doi.org/10.1093/ismejo/wrae221

Loh, T.C., Choe, D.W., Foo, H.L., Sazili, A.Q., & Bejo, M.H. (2014). Effects of feeding different postbiotic metabolite combinations produced by *Lactobacillus plantarum* strains on egg quality and production performance, faecal parameters and plasma cholesterol in laying hens. *BMC Veterinary Research*, 10(1), 149.

Loh, T.C., Foo, H.L., & Chang, H.M. (2021). Postbiotic metabolites of probiotics in animal feeding. In N. Mojgani & M. Dadar (eds.), *Probiotic bacteria and postbiotic metabolites: Role in animal and human health. Microorganisms for Sustainability*, vol 2. Springer.

Loh, T.C., Thu, T.V., Foo, H.L., & Bejo, M.H. (2013). Effects of different levels of metabolite combination produced by *Lactobacillus plantarum* on growth performance, diarrhoea, gut environment and digestibility of postweaning piglets. *Journal of Applied Animal Research*, 41(2), 200–207.

Lovera, R., Fernández, M.S., Jacob, J., Lucero, N., Morici, G., Brihuega, B., Farace, M.I., Caracostantogolo, J., & Cavia, R. (2017). Intrinsic and extrinsic factors related to pathogen infection in wild small mammals in intensive milk cattle and swine production systems. *PLOS Neglected Tropical Diseases*, 11(6), e0005722.

Markovi, R., Sefer, D., Krsti, M., & Petrujki, B. (2009). Effect of different growth promoters on broiler performance and gut morphology. *Archivos De Medicina Veterinaria*, 41(2), 163–169.

Moeser, A. J., Pohl, C. S., & Rajput, M. (2017). Weaning stress and gastrointestinal barrier development: Implications for lifelong gut health in pigs. *Animal Nutrition*, 3(4), 313–321.

Morita, N., Umemoto, E., Fujita, S., Hayashi, A., Kikuta, J., Kimura, I., Haneda, T., Imai, T., Inoue, A., Minuro, H., Maeda, Y., Kayama, H., Okumura, R., Aoki, J., Okada, N., Kida, T., Ishii, M., Nabeshima, R., & Takeda, K. (2019). GPR31-dependent dendrite protrusion of intestinal CX3CR1+ cells by bacterial metabolites. *Nature*, 566, 110–114.

Muhammad, Z., Ramzan, R., Abdelazez, A., Amjad, A., Afzaal, M., Zhang, S., & Pan, S. (2019). Assessment of the antimicrobial potentiality and functionality of *Lactobacillus plantarum* strains isolated from the conventional inner Mongolian fermented cheese against foodborne pathogens. *Pathogens*, 8(2), 71.

Navarro-Yepes, J., Zavala-Flores, L., Anandhan, A., Wang, F., Skotak, M., & Chandra, N. (2014). Antioxidant gene therapy against neuronal cell death. *Pharmacology & Therapeutics*, 142, 206–230.

Pignatelli, P., Menichelli, D., Pastori, D., & Violi, F. (2018). Oxidative stress and cardiovascular disease: New insights. *Kardiologia Polska*, 76(4), 713–722.

Rahimi, A., Qaisar, S.A., Janeh, T., Karimpour, H., Darbandi, M., & Moludi, J. (2024). Clinical trial of the effects of postbiotic supplementation on inflammation, oxidative stress, and clinical outcomes in patients with CVA. *Scientific Reports*, 14, 24021.

Rajoka, M.S.R., Shi, J., Zhu, J., Shao, D., Huang, Q., Yang, H., & Jin, M. (2017). Capacity of lactic acid bacteria in immunity enhancement and cancer prevention. *Applied Microbiology and Biotechnology*, 101, 35–45.

Ren, F., Wang, K., Zhang, T., Jiang, J., Nice, E. C., & Huang, C. (2015). New insights into redox regulation of stem cell self-renewal and differentiation. *Biochimica et Biophysica Acta (BBA) - General Subjects*, 1850(8), 1518–1526.

Rinttilä, T., & Apajalahti, J. (2013). Intestinal microbiota and metabolites: Implications for broiler chicken health and performance. *Journal of Applied Poultry Research*, 22(3), 647–658.

Salminen, S., Collado, M.C., Endo, A., Hill, C., Lebeer, S., Quigley, E.M.M., Sanders, M.E., Shamir, R., Swann, J. R., Szajewska, H., & Vinderola, G. (2022). Author correction: The International Scientific Association of Probiotics and Prebiotics (ISAPP) consensus statement on the definition and scope of postbiotics. *Nature Reviews Gastroenterology & Hepatology*, 19(8), 551–551.

Sánchez, B., Delgado, S., Blanco-Míguez, A., Lourenço, A., Gueimonde, M., & Margolles, A. (2017). Probiotics, gut microbiota, and their influence on host health and disease. *Molecular Nutrition & Food Research*, 61, 1600240.

Sies, H., & Jones, D. P. (2020). Reactive oxygen species (ROS) as pleiotropic physiological signalling agents. *Nature Reviews Molecular Cell Biology*, 21(7), 363–383.

Sugiharto, S. (2016). Role of nutraceuticals in gut health and growth performance of poultry. *Journal of the Saudi Society of Agricultural Sciences*, 15(2), 99–111.

Surai, P.F. (2020). Chapter 2: Antioxidant systems in animal body. In *Vitagenes in Avian Biology and Poultry Health*, vol. 1, 53–92.

Takiishi, T., Fenero, C.I.M., & Câmara, N.O.S. (2017). Intestinal barrier and gut microbiota: Shaping our immune responses throughout life. *Tissue Barriers*, 5(4), e1373208.

Thangaraju, M., Cresci, G.A., Liu, K., Ananth, S., Gnanaprakasam, J.P., Browning, D.D., Mellinger, J.D., Smith, S.B., Digby, G.J., Lambert, N.A., Prasad, P.D., & Ganapathy, V. (2009). GPR109A is a G-protein–coupled receptor for the bacterial fermentation product butyrate and functions as a tumor suppressor in colon. *Cancer Research*, 69(7), 2826–2832.

Thomas, C.M., Hong, T., van Pijkeren, J.P., Hemarajata, P., Trinh, D.V., Hu, W., Britton, R.A., Kalkum, M., & Verasalovic, J. (2012). Histamine derived from probiotic *Lactobacillus reuteri* suppresses TNF via modulation of PKA and ERK signaling. *PLoS ONE* 2012, 7, e31951.

Tsilingiri, K., & Rescigno, M. (2013). Postbiotics: What else? *Beneficial Microbes*, 4(1), 101–107.

Vinderola, G., Sanders, M.E., Cunningham, M., & Hill, C. (2024). Frequently asked questions about the ISAPP postbiotic definition. *Frontiers in Microbiology*, 14, 1324565.

Wang, Y., Wu, Y., Wang, Y., Xu, H., Mei, X., Yu, D., Wang, Y., & Li, W. (2017). Antioxidant properties of probiotic bacteria. *Nutrients*, 9(5), 521.

Williams, L., Alshehri, A., Robichaud, B., Cudmore, A., Gagnon, J. (2020). The role of the bacterial muramyl dipeptide in the regulation of glp-1 and glycemia. *International Journal of Molecular Sciences*, 21, 5252.

Wu, J.Q., Kosten, T.R., & Zhang X.Y. (2013). Free radicals, antioxidant defense system, and schizophrenia. *Progress in Neuro-Psychopharmacology & Biological Psychiatry*, 46, 200–206.

Yang, R., Li, W., Shi, Y.-H., & Le, G.-W. (2008). Lipoic acid prevents high-fat diet–induced dyslipidemia and oxidative stress: A microarray analysis. *Nutrition*, 24(6), 582–588.

Yang, J., Wang, C., Liu, L., & Zhang, M. (2019) *Lactobacillus reuteri* KT260178 supplementation reduced morbidity of piglets through its targeted colonization, improvement of cecal microbiota profile, and immune functions. *Probiotics and Antimicrobial Proteins*, 1, 10.

Yang, X., Wang, Z., Niu, J., Zhai, R., Xue, X., Wu, G., Fang, Y., Meng, G., Yuan, H., Zhao, L., & Zhang, C. (2023). Pathobionts from chemically disrupted gut microbiota induce insulin-dependent diabetes in mice. *Microbiome*, 11(1), 62.

Zhao, X., Liu, S., Li, S., Jiang, W., Wang, J., Xiao, J., Chen, T., Ma, J., Khan, M.Z., Wang, W., Li, M., Li, S., & Cao, Z. (2024). Unlocking the power of postbiotics: A revolutionary approach to nutrition for humans and animals. *Cell Metabolism*, 36(4), 725–744.

Zółkiewicz, J., Marzec, A., Ruszczyński, M., & Feleszko, W. (2020). Postbiotics: A step beyond pre- and probiotics. *Nutrients*, 12, 2189.

11 Postbiotics
Current and Future Perspectives in Human Health

Mahdieh Iranmanesh
Islamic Azad University, Tehran, Iran

Masoumeh Bagheri, Azadeh Zahmatkesh and Naheed Mojgani
Razi Vaccine and Serum research Institute, Agriculture Research, Education and Extension Organization, Karaj, Iran

11.1 INTRODUCTION

Postbiotics are compounds that provide direct or indirect benefits to the host's body. Derived from the Greek terms "post" (meaning after) and "bios" (meaning life), the term "biotic" refers to a strategy for promoting gut health through nutrition. This approach focuses on monitoring dietary effects that encourage a beneficial state in the host. The term "biotic" itself comes from the Greek word "biotikos," meaning "pertaining to life," and emphasizes the study of living organisms and their ecological environments (Durack et al., 2018).

Postbiotics are defined as preparations of inactivated microorganisms and/or their components that confer health benefits on the host (Zulkhairi Amin et al., 2020). Here, "inactivated" refers to microorganisms that are no longer alive, though their functional metabolites and structures remain intact, thereby continuing to deliver health benefits (Salminen et al., 2021). Like the live probiotics found in sources such as human milk (Mojgani et al., 2024), postbiotics are known to provide comparable, significant health benefits (Martín & Langella, 2019; Nataraj et al., 2020). Important postbiotic metabolites include bacteriocins, hydrogen peroxide, enzymes, amino acids, vitamins, short-chain fatty acids, organic acids, exopolysaccharides, and surfactants (Anderson et al., 2013).

Antimicrobial compounds produced by lactic acid bacteria (LAB) can be classified as low molecular weight (such as carbon dioxide, diacetyl, and hydrogen peroxide) and high molecular weight (like bacteriocins) compounds, commonly known as postbiotics (Vinderola et al., 2022). Bacteriocins, which are antimicrobial substances modified after expression, are produced by Gram-positive bacteria and have notable antibacterial properties (Khoso et al., 2024). Additionally, various Bacillus species produce enzymes and antibiotic-like compounds that support carbohydrate digestion and inhibit harmful organisms; attributes particularly beneficial in honeybees (Mojgani et al., 2023).

The antimicrobial effects of these metabolites on honeybee diseases have been widely studied. Mojgani et al. (2023) have investigated how metabolites from LAB species act against bacterial agents responsible for American and European foulbrood (AFB and EFB). They found that metabolites produced by probiotic bacteria demonstrated strong antibacterial activity against *Melissococcus. plutonius*. Similarly, postbiotics derived from LAB species displayed comparable antibacterial effects, inhibiting the growth of *Melissococcus. plutonius*. This highlights a potential research area

DOI: 10.1201/9781003607779-11

focused on using postbiotics as safe, cost-effective antimicrobials to control pathogens, with potential applications as biocontrol agents.

Postbiotics present an alternative to probiotics, especially in food industry. Although probiotics are valued for their health benefits, incorporating live bacteria into foods can pose challenges due to their limited survival during processing, storage, and interaction with other food components. Postbiotics, however, are bioactive compounds generated by microorganisms during their growth and fermentation in culture media, which are either separated as cell free supernatant fluid (CFS) or metabolites with cellular lysates. These bioactive ingredients may not only be advantageous to the food products but also provides health benefits to the consumers (Moradi et al., 2020; Wu et al., 2024). Typically, postbiotics are derived from Lactic acid bacteria (LAB) and contain a variety of bioactive substances such as organic acids, antimicrobial peptides, short-chain fatty acids, vitamins, cofactors, carbohydrates, enzymes, flavonoids, immune-signaling compounds, phenolics, etc. (Aguilar-Toalá et al., 2018).

Compared to probiotics, postbiotics offer several distinct benefits. Their nonliving nature results in extended shelf life, offers stability over a broader range of pH and temperatures, and eliminates concerns regarding possible antibiotic resistance or production of biogenic amine. Furthermore, their chemical structures allow for uniform production and enhanced antimicrobial capabilities, rendering them an attractive alternative or complement to viable probiotics (Barros et al., 2020).

In food products, the effectiveness of a postbiotic depends on factors like the type of LAB strain and target microorganism, concentration of the postbiotic and its application form, and the properties of the food material (Moradi et al., 2020). Research has shown that postbiotics from *Lactobacillus rhamnosus* and *Limosilactobacillus reuteri* possess prominent antimicrobial and antioxidant qualities, suggesting they could serve as natural and safe alternatives to chemical preservatives, particularly in decontaminating red meat surfaces in the meat industry (Jalali et al., 2024).

Presently, postbiotics have gained attention for their preservative role and decontamination properties in the food industry (Letizia et al., 2022). These metabolites offer promising effects in food preservation and packaging, having the potential to control or limit the growth of a wide number of pathogen, prevent microbial biofilm formation, and deteriorate contaminants such as pesticides, mycotoxins, and heavy metals (Yilmaz et al., 2022). On the other hand, postbiotics are shown to offer health benefits, potentially providing protective effects similar to the therapeutic actions linked to probiotics (Bhat & Bajaj, 2019).

11.2 CLASSIFICATION OF POSTBIOTICS

Postbiotics are categorized based on factors such as the type of microorganism, their structural composition, and their physiological functions. Due to their diverse physical, chemical, and functional properties, postbiotics can be divided into several types, including exopolysaccharides, peptidoglycans, cell wall fragments, short-chain fatty acids, inactivated and dead probiotics, cell-free supernatants, enzymes, vitamins, extracellular vehicles (EVs), neurotransmitters, and bacteriocins.

11.2.1 Exopolysaccharides

These compounds have protective properties and help defend against phages, immune cells, and toxins. They also have influence on host immunity, physiological functions, lipid metabolism, and the establishment of pathogens (Dinić et al., 2018).

11.2.2 Short-Chain Fatty Acids (SCFAs)

These molecules inhibit the production of pro-inflammatory cytokines, thus preventing the activation of nuclear factor-kappa B (NF-κB) (Kasahara et al., 2019).

11.2.3 Inactivated and Dead Probiotics

Inactivated or dead probiotics (paraprobiotics) are commonly produced by heat treatment. Research on experimental models has shown that nonviable cells retain many beneficial properties of their live counterparts, such as antioxidant activity, anti-inflammatory effects, and the ability to influence host physiology (Sugahara et al. 2017).

11.2.4 Cell-Free Supernatants (CFSs)

Cell-free supernatants (CFSs) extracted from a probiotic microbe are important active metabolites that include organic acids, diacetylene, carbon dioxide, and bacteriocin-like compounds. These metabolites, typically secreted by LAB, may help maintain homeostasis (Siedler et al. 2019) by potentially enhancing the production of anti-inflammatory cytokines like IL-10 while reducing pro-inflammatory cytokines such as TNF-α and IL-1β (Mani-López et al., 2022).

11.2.5 Enzymes

A limited number of bacterial strains, particularly *Bacillus subtilis* and *Bacillus licheniformis*, serve as primary enzyme sources involved in numerous physiological, metabolic, and regulatory functions. For example, two strains of *Lactobacillus fermentum* were found to contain high levels of glutathione peroxidase, which exhibited strong antioxidant activity in laboratory studies. Additionally, the catalase enzyme derived from a genetically modified strain of *Lactococcus lactis* has shown promising results in protecting mice from experimental colon cancer (Contesini et al., 2018).

11.2.6 Vitamins

They are vital for various physiological functions, such as brain function, blood clotting, and bone health. Riboflavin, for example, acts as a hydrogen carrier in redox reactions, and vitamin K functions as a cofactor in blood clotting via gamma carboxylase activity. Some essential vitamins, such as vitamin K and several B-group vitamins (like thymine, riboflavin, niacin, pyridoxine, folate, and cobalamin), are synthesized by LAB and *Bifidobacterium* species (LeBlanc et al., 2013).

11.2.7 Extracellular Vesicles

These structures help control gut barrier permeability and homeostasis, modulate signaling pathways, improve lipid profiles, and facilitate gut–brain communication. They play roles in bacterial survival, competition, pathogenesis, and immune control (de Jong et al., 2020).

11.2.8 Cell Wall Fragments

The bacterial cell wall contains different components, such as teichoic and lipoteichoic acids, which play immunogenic roles, and can induce immune responses. Different forms of lipoteichoic acid in various *Lactobacillus plantarum* strains cause specific immune responses in immune cells. Teichoic acids are involved in functions related to immune regulation, antioxidant effects, and anticancer activities and also participate in the development of antibiotic resistance (Lebeer et al. 2012).

11.2.9 Neurotransmitters

These molecules significantly impact brain function through the gut–brain axis by regulation of signaling in the enteric nervous system (Liu et al., 2020).

11.2.10 Bacteriocins

There are four primary classes of bacteriocins. The first class includes heat-resistant, low-molecular-weight peptides modified by enzymes during transcription, such as sactipeptides, glycokines, and lantibiotics. The second class of bacteriocins is subdivided into pediocin-like circular and non-pediocin-like linear forms; they are characterized by proper pH stability, low melting points, and minimal posttranscriptional processing. The third class includes large, thermolabile proteins over 10 kDa, while the fourth class bacteriocins are complex proteins bound to carbohydrates or lipids (Hols et al., 2019).

11.3 MECHANISM OF ACTION OF POSTBIOTICS

Postbiotic metabolites have shown promising results as an alternative to in-feed antimicrobials in animals (Mantziari et al., 2020). Additionally, postbiotics offer safe and effective alternatives for vulnerable groups such as preterm infants, in whom viable probiotics might pose risks due to immature gut barriers or immune defenses (Rad et al., 2022). Despite the health benefits of postbiotics, their exact mechanisms of action remain largely unknown. Although these mechanisms may vary, postbiotics may exert protective effects similar to those of probiotics. Similar to their viable parent cells (probiotics), postbiotics also exhibit health effects by modulation of the defense barriers and preventing the entry of pathogens (Ma et al., 2023). In addition to their antimicrobial mode of actions, the postbiotics could exert effects by the production of pro-inflammatory and anti-inflammatory cytokines, modulating the epithelial barriers and enhancing the immune responses (Peluzio et al., 2021; Mosca et al., 2022).

A major activity of postbiotics is their antimicrobial action, which has been detected against a number of Gram-positive and Gram-negative bacterial pathogens. In a study, the postbiotics extracted from *Lactobacillus rhamnosus* and *Limosilactobacillus reuteri* have shown strong inhibitory effects on common foodborne pathogens such as *Staphylococcus aureus* and *Escherichia coli*. The bioactive compounds of postbiotics, like enzymes, organic acids, and antimicrobial peptides, inhibit pathogens by decreasing pH and disrupting the cell membranes (Stoyanova et al., 2012). It has been reported that the strong reduction in *Escherichia coli* numbers may result from its higher susceptibility to acidic conditions and antimicrobial agents produced by probiotics. In contrast, *Staphylococcus aureus*, as a Gram positive bacterium with a thicker cell wall, shows greater resistance to postbiotic effects, with a less noticeable reduction. Postbiotics derived from LAB have also exhibited antibacterial activity against other foodborne pathogens, including *Listeria monocytogenes* and *Klebsiella pneumoniae* (Dejene et al., 2021).

In the gastrointestinal tract, microbial metabolism of indigestible carbohydrates produces significant amounts of butyrate and SCFAs (Fong et al., 2020). These metabolites, along with protective enzymes, immunostimulant teichoic acids, antimicrobial peptides, and plasmalogens, contribute to cellular defense against oxidative stress, inflammation, and cancer. The high levels of total phenolic content (TPC) and total flavonoid contents (TFC) in the postbiotics are related to their potent antioxidant activity. In a study, the postbiotics from *L. rhamnosus* and *L. reuteri* reduced the oxidative decay in meat products and increased their shelf life (Jalali et al., 2024). These researchers related the anti-oxidative role of postbiotics to their TPC and TFC. Some other major effects of postbiotics include their ability to activate the peroxisome proliferator-activated receptor, which enhances fatty acid oxidation and reduces triglyceride levels (Nakamura et al., 2016). Additionally, postbiotics increase antimicrobial peptide (AMP)–activated protein kinase activity in both muscle tissue and the liver while reducing the activity of hepatic enzymes hydroxylmethylglutaryl-CoA reductase and hydroxylmethylglutaryl-CoA synthase, which leads to autonomous cell lipolysis (Sabahi et al., 2023). Muramyl dipeptide-based postbiotics, for example, activate the transcription factor IRF4 and a nucleotide-binding oligomerization domain-containing protein, which have been shown to reduce adipose inflammation and improve glucose tolerance in obese mice (Cavallari et al., 2017).

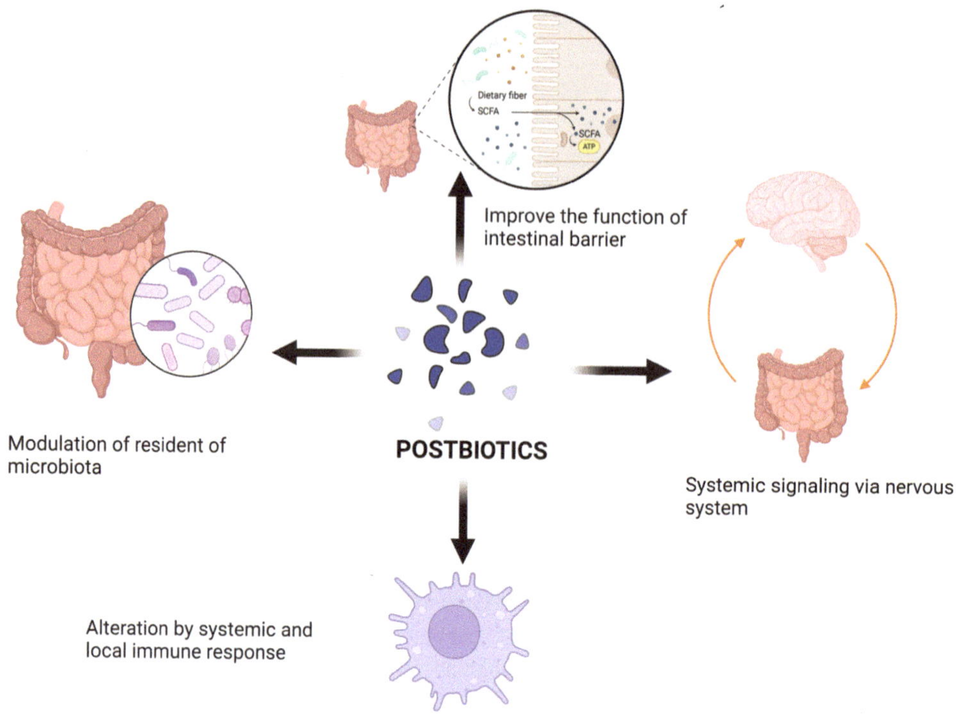

FIGURE 11.1 Major mechanistic actions of postbiotics.

This postbiotic improved insulin resistance in the liver of humans with low-grade endotoxemia. Postbiotics may also have an anticancer effect on colon cells by promoting immune pathways that trigger pro-apoptotic cell death (Tiptiri-Kourpeti et al., 2016). Postbiotics from *Lactobacillus* strains were especially able to inhibit colon cancer cell invasion by reducing metalloproteinase-9 activity (Rad et al., 2021). Several different mechanisms of actions by which postbiotics exert their health effects are indicated in Figure 11.1.

11.4 FUNCTIONAL PROPERTIES OF POSTBIOTIC

Postbiotics are recognized as bioactive substances that can have potential effects on our physiological system if used appropriately. The main effects bestowed by these bioactive compounds include their antimicrobial, anti-inflammatory and immune modulatory effects.

11.4.1 Antimicrobial Activity

Postbiotics can be inhibited both pathogenic and food spoilage bacteria. The antibacterial activity of postbiotics relied on factors like category of probiotic bacteria which the postbiotic is organized, as well as the type of target bacterium (Gram-positive or gram-negative), and the concentration of postbiotics (Moradi et al., 2020). The antibacterial activity of postbiotics from *Pediococcus acidilactici*, *Latilactobacillus sakei*, and *Staphylococcus xylosus* on chicken drumsticks indicated that 10% postbiotics+1% LA was the best combination, which could lower the count of mesophilic bacteria (İncili et al., 2022). In addition, the later combination lowers the *L. monocytogenes* counts to be 1.1 $\log_{10}$ compared to the control group in chicken drumsticks. In this regard, the antimicrobial activity of postbiotics from *Saccharomyces cerevisiae* (PTCC 5269) in two various conditions include in vitro and food models (whole milk and ground meat) was investigated (Hosseini et al., 2024). This bacterium produced phenolics and flavonoids, which have antibacterial activity against *L. monocytogenes*, *Streptococcus mutans*, *Salmonella typhi*, and *Escherichia coli* in both conditions. In another research, the minimum inhibitory concentration (MIC) values of combination postbiotic of *L. rhamnosus* and *L. reuteri* samples against *E. coli* were lower than control samples (Jalali et al., 2024). They result also showed that the minimum bactericidal concentration (MBC) values for *E. coli* were significantly lower than *S. aureus*, so they indicated that postbiotics had higher effect on Gram-negative like *E. coli* in comparison to Gram-positive bacteria like *S. aureus*. The capacity of postbiotics of *Latilactobacillus curvatus* BGMK2-41 in vitro and ex vivo (on infected keratinocytes and infected human skin, respectively) was studied by Dinić et al. (2024). They found that both dilutions of the postbiotic, including 10× and 100×, have equal effect on infection of keratinocytes with *S. aureus*. They also suggested that the antimicrobial activity was associated to AMPs.

11.4.2 Antioxidant Activity

Oxidative metabolism of cells results in peroxidant production, followed by producing the reactive oxygens species (ROS) and free radicals, and finally leads to oxidative damage (Schogor et al., 2013). In the body, antioxidants as well as intracellular enzymes such as superoxide dismutase (SOD), glutathione peroxidase (GPX), and catalase (CAT) can protect cells by eliminating peroxides and superoxides (Miller et al., 1993). Since these antioxidants may not be enough, attention is given to other source like postbiotics that exhibit antioxidant potential. Generally, the antioxidative activity of probiotics related to various factors like the generation of antioxidant metabolites, chelating the metal ion, producing the own antioxidases, upregulation of antioxidase action as well as improvement of antioxidant metabolites of the host, decrease of the enzyme activities of ROS, and regulation of microbiota of the gastrointestinal tract (Wang et al., 2017). In this regard, the antioxidant activity of various lactic acid bacteria such as 7 *Bifidobacterium*, 11 *Lactobacillus*, 6 *Lactococcus*, and 10 *Streptococcus thermophilus* against ascorbic and linolenic acid oxidation (TAAAA and TAALA) in vitro was studied by Amaretti et al. (2013). The postbiotics of *Lactobacillus* and *Lactococcus* strains presented higher TAA_{LA} compared to *Bifidobacteria*, whereas the TAA_{AA} was not significantly different between postbiotics of tested strains. Additionally, the SOD activity of postbiotics from *Lactococcus* was higher than *S. thermophilus*. The antioxidant activity of postbiotics from *Lactiplantibacillus plantarum* was affected by media and was higher in formulated media compared to the control de Man, Rogosa, and Sharpe (MRS) medium (Aguilar-Toalá et al., 2020). According to Izuddin et al. (2020), postbiotic *L. plantarum*, the RG14 strain had the highest antioxidant activity in 2,2′-azino-bis (3-ethylbenzothiazoline-6-sulfonic acid) (ABTS) assay, while there was no difference between postbiotics from *L. plantarum* RG14, RG11, and TL1 in 2,2-diphenyl-1-picrylhydrazyl (DPPH) assay. Furthermore, they indicated that the dietary postbiotic from *L. plantarum* RG14 couldn't affect SOD in both rumen fluid and lambs' serum, whereas the GPX in serum and ruminal was higher in the postbiotic group. Similarly, the postbiotics of *Lacticaseibacillus paracasei* were shown to exhibit DPPH radical scavenging activity with the rate of 38.9–92.4% scavenging at 2.5–50% and also hydroxyl radical scavenging activity with 4.66–10.38% scavenging degree at 0.5–4% concentrations (Dong et al., 2024). According to these reports, the antioxidant activity was related to EPS as well as other compounds like organic matter, proteins, peptides, and amino acids.

11.4.3 Immunomodulatory Activity

The body's defense mechanism provides protection against a number of unwanted invaders and pathogens by setting various natural barriers like skin, mucosa, tear, earwax, and stomach acids, as well as by the humoral barriers such as production of antibodies, complement proteins, antimicrobial peptides, and so on (Mowat & Viney, 1997). The first line of defense is skin and mucous membranes, which act as physical obstructions and inhibit the entry of microbes into the host. When pathogens pass through first line of defense, the second one—the adaptive immune system—will be activated. In this situation, the inflammatory response is created, which include eosinophils, neutrophils, dendritic cells, basophils, Kupffer cells, tissue macrophages such as alveolar macrophages, and phagocytic cells called monocytes in the blood, as well as the IgG and IgA that can bind to the pathogens with their special receptors (Lubbers et al., 2017). In addition, gut microbiota or intestinal microbiota have a vital role in human health by improving the immune system and regulating the metabolic functions (Whitman et al., 1998). Postbiotics can modulate immune system through various mechanisms like reducing pro-inflammatory cytokine production as well as motivating the production of inflammatory cytokine, regulating the Th1 and Th2 in immune response, preventing the entry of pathogens throughout epithelial cells, stimulating the expression of TLR, and so on (Yeşilyurt et al., 2021).

According to the studies of Tarique and his colleagues (2022), the postbiotic of *Lactobacillus delbrueckii* and *Streptococcus thermophilus* isolated from traditional yogurt-like products could reduce the IFN-γ release. They furthermore stated the induction of IFN-γ was much higher in C57BL/6 cells than in BALB/c (Tarique et al., 2022). In another research, the immunomodulating activity of postbiotics from a number of *L. plantarum* strains of plant origin were investigated on human macrophages in vitro. The findings of this study showed that the cell-free supernatant of these strains could modulate the cytokine patterns in human macrophages and also induced the pro-inflammatory cytokine TNF-α (Rocchetti et al., 2024). Furthermore, the tested postbiotic showed increased IL-10 transcription, whereas the IL-8 secreted significantly decreased. Magryś and Pawlik (2023) investigated the immune modulating effects of *L. plantarum* and *L. rhamnosus* in vitro system comprised of co-cultures of intestinal epithelial cells with macrophage cells. These researchers reported that the heat killed portions of the mentioned probiotic strains and their isolated proteins could suppress inflammation, activate the anti-inflammatory cytokine IL-10, and modulate the IL-18 related response. The immunostimulatory effects of different concentrations of postbiotics extracted from a commercially obtainable active multispecies probiotic on RAW264.7 murine macrophages has also been studied (Al-Najjar et al., 2024). During studies, these researchers observed that the postbiotics enhanced the immunostimulatory factors in the studied murine macrophages as evident in cell viability assay and in the production of pro-inflammatory cytokine tumor necrosis factor alpha (TNF-α). According to present research, the bioactive compounds in the postbiotics, such as phenols and flavonoids, might be responsible for their anti-inflammatory and antioxidant properties that consequently are responsible for modulating the immune responses (Badhani et al., 2015; Davarzani et al., 2024; Hamad et al., 2020).

11.5 FUTURE PERSPECTIVES

In last decade, the use of bioactive micro and macromolecules termed postbiotics, which are the cellular lysate and metabolic mixture of the beneficial microorganisms, has been widely expanded owing to their physiological health benefits that have placed them among the most advantageous natural chemical compounds with therapeutic potentials. A vast array of studies have proven the superiority of these bioactive components of bacterial origin to their viable parent cells (probiotics). According to reports, postbiotics have health effects comparable to probiotics while having decreased risk of antibiotic resistance, better stability, and improved safety. All these have placed postbiotic compounds as an alternative advantageous approach instead of probiotics, particularly for use in preterm infants and individuals with risk of exposured immune systems. Some of the essential functional properties of postbiotics including their antibacterial, antioxidant, anti-inflammatory, and immune regulatory capacity have been proven. Available evidence recommends that postbiotics have a more pronounced influence in improving allergic reactions, acute/chronic diarrhea, neurodegenerative diseases, and immune function. Current investigations have further shown the role of postbiotics in the management of diseases such as Alzheimer's and multiple sclerosis, for which effective treatments remain elusive (Goswami et al., 2018).

Although considerable research has been conducted on postbiotics, there remains a significant gap in understanding their mechanism of action in order to prove their efficiency either in functional food products or for substantiating their health claims. It is of utmost importance to study the array of bioactive components present in a specific species of the probiotic bacteria and to assess the health potential of individual bioactive components in order to better understand their mechanism of action and safety parameters (Sharifi-Rad et al., 2020). In recent years, new approaches are being investigated to better understand and identify the specific components of postbiotic formulations that could lead to development and production of "precision postbiotics" that could target specific physiological function in the body and be tailored according to individual patients need (Goswami et al., 2018; Veiga et al., 2020). In spite of all these gaps in research regarding postbiotics, we are witnessing the outpouring of postbiotic products into the world market that are being approved by regulatory agencies in various countries and released into the market.

11.6 CONCLUSIONS

The present review presents an overview of main classes of postbiotic components including the cell lysates, their metabolites, microbial cell components, CFS, and EPS, highlighting their role in modulating the antimicrobial, antioxidant, and immunomodulatory activities that might have significant health impacts. The current research in the field indicates that although the wide variety of postbiotics has proved to be beneficial for the host health, yet the precise mechanisms remain unclear and require further investigations, including in vitro, in vivo, and clinical trial studies. It is anticipated that further extensive research in the field could shed light on the existing gaps regarding the biological and functional properties of these metabolites that would allow novel application of postbiotics in medicine and beyond.

REFERENCES

Aguilar-Toalá, J. E., Garcia-Varela, R., Garcia, H. S., Mata-Haro, V., González-Córdova, A. F., Vallejo-Cordoba, B., & Hernández-Mendoza, A. (2018). Postbiotics: An evolving term within the functional foods field. *Trends in Food Science & Technology*, *75*, 105–114. https://doi.org/10.1016/j.tifs.2018.03.009

Aguilar-Toalá, J. E., Hall, F. G., Urbizo-Reyes, U. C., Garcia, H. S., Vallejo-Cordoba, B., González-Córdova, A. F., … Liceaga, A. M. (2020). In silico prediction and in vitro assessment of multifunctional properties of postbiotics obtained from two probiotic bacteria. *Probiotics and Antimicrobial Proteins*, *12*, 608–622. https://doi.org/10.1007/s12602-019-09568-z

Al-Najjar, M. A. A., Abdulrazzaq, S. B., Alzaghari, L. F. et al. (2024). Evaluation of immunomodulatory potential of probiotic conditioned medium on murine macrophages. *Scientific Reports 14*, 7126. https://doi.org/10.1038/s41598-024-56622-0

Amaretti, A., Di Nunzio, M., Pompei, A., Raimondi, S., Rossi, M., & Bordoni, A. (2013). Antioxidant properties of potentially probiotic bacteria: In vitro and in vivo activities. *Applied Microbiology and Biotechnology*, *97*, 809–817. https://doi.org/10.1007/s00253-012-4241-7

Anderson, K. E., Johansson, A., Sheehan, T. H., Mott, B. M., Corby-Harris, V., Johnstone, L., & Fitz, W. (2013). Draft genome sequences of two *Bifidobacterium* sp. from the honey bee (*Apis mellifera*). *Gut Pathogens*, *5*, 1–3. https://doi.org/10.1186/1757-4749-5-42

Badhani, B., Sharma, N., & Kakkar, R. (2015). Gallic acid: A versatile antioxidant with promising therapeutic and industrial applications. *RSC Advances* **5**, 27540–27557.

Barros, C. P., Guimarães, J. T., Esmerino, E. A., Duarte, M. C. K., Silva, M. C., Silva, R., … Cruz, A. G. (2020). Paraprobiotics and postbiotics: Concepts and potential applications in dairy products. *Current Opinion in Food Science*, *32*, 1–8. https://doi.org/10.1016/j.cofs.2019.12.003

Bhat, B., & Bajaj, B. K. (2019). Hypocholesterolemic potential of probiotics: Concept and mechanistic insights. http://nopr.niscpr.res.in/handle/123456789/45758

Cavallari, J. F., Fullerton, M. D., Duggan, B. M., Foley, K. P., Denou, E., Smith, B. K., … Schertzer, J. D. (2017). Muramyl dipeptide-based postbiotics mitigate obesity-induced insulin resistance via IRF4. *Cell Metabolism*, *25*(5), 1063–1074. https://doi.org/10.1016/j.cmet.2017.03.021

Contesini, F. J., Melo, R. R. D., & Sato, H. H. (2018). An overview of *Bacillus* proteases: From production to application. *Critical Reviews in Biotechnology*, *38*(3), 321–334. https://doi.org/10.1080/07388551.2017.1354354

Davarzani, S., Sanjabi, M. R., Mojgani, N., Mirdamadi, S., & Soltani, M. (2024). Investigating the antibacterial, antioxidant, and cholesterol-lowering properties of yogurt fortified with postbiotic of *Lactobacillus acidophilus* and *Lactiplantibacillus plantarum* in the Wistar rat model. *Journal of Food Protection*, *87*(12), 100408. https://doi.org/10.1016/j.jfp.2024.100408

Dejene, F., Regasa Dadi, B., & Tadesse, D. (2021). In vitro antagonistic effect of lactic acid bacteria isolated from fermented beverage and finfish on pathogenic and foodborne pathogenic microorganism in Ethiopia. *International Journal of Microbiology*, *2021*(1), 5370556. https://doi.org/10.1155/2021/5370556

de Jong, B., Barros, E. R., Hoenderop, J. G., & Rigalli, J. P. (2020). Recent advances in extracellular vesicles as drug delivery systems and their potential in precision medicine. *Pharmaceutics*, *12*(11), 1006. https://doi.org/10.3390/pharmaceutics12111006

Dinić, M., Pecikoza, U., Djokić, J., Stepanović-Petrović, R., Milenković, M., Stevanović, M., … Lukić, J. (2018). Exopolysaccharide produced by probiotic strain *Lactobacillus* paraplantarum BGCG11 reduces inflammatory hyperalgesia in rats. *Frontiers in Pharmacology*, *9*, 1. https://doi.org/10.3389/fphar.2018.00001

Dinić, M., Burgess, J. L., Lukić, J., Catanuto, P., Radojević, D., Marjanović, J., … Pastar, I. (2024). Postbiotic lactobacilli induce cutaneous antimicrobial response and restore the barrier to inhibit the intracellular invasion of *Staphylococcus aureus* in vitro and ex vivo. *The FASEB Journal*, *38*(14), e23801. https://doi.org/10.1096/fj.202400054RR

Dong, H., Ren, X., Song, Y., Zhang, J., Zhuang, H., Peng, C., … Li, Z. (2024). Assessment of multifunctional activity of a postbiotic preparation derived from *Lacticaseibacillus paracasei* postbiotic-P6. *Foods*, *13*(15), 2326. https://doi.org/10.3390/foods13152326

Durack, J., Kimes, N. E., Lin, D. L., Rauch, M., McKean, M., McCauley, K., … Lynch, S. V. (2018). Delayed gut microbiota development in high-risk for asthma infants is temporarily modifiable by *Lactobacillus* supplementation. *Nature Communications*, *9*(1), 707. https://doi.org/10.1038/s41467-018-03157-4

Fong, W., Li, Q., & Yu, J. (2020). Gut microbiota modulation: A novel strategy for prevention and treatment of colorectal cancer. *Oncogene*, *39*(26), 4925–4943. https://doi.org/10.1038/s41388-020-1341-1

Goswami, C., Iwasaki, Y., & Yada, T. (2018). Short-chain fatty acids suppress food intake by activating vagal afferent neurons. *The Journal of Nutritional Biochemistry*, *57*, 130–135. https://doi.org/10.1016/j.jnutbio.2018.03.009

Hamad, G. M., Abdelmotilib, N. M., Darwish, A. M. & Zeitoun, A. M. (2020). Commercial probiotic cell-free supernatants for inhibition of Clostridium perfringens poultry meat infection in Egypt. *Anaerobe*, *62*, 102181

Hols, P., Ledesma-García, L., Gabant, P., & Mignolet, J. (2019). Mobilization of microbiota commensals and their bacteriocins for therapeutics. *Trends in Microbiology*, *27*(8), 690–702. https://doi.org/10.1016/j.tim.2019.03.007

Hosseini, H., Abbasi, A., Sabahi, S., Akrami, S., & Yousefi-Avarvand, A. (2024). Assessing the potential biological activities of postbiotics derived from *Saccharomyces cerevisiae*: An in vitro study. *Probiotics and Antimicrobial Proteins*, *16*(4), 1348–1364. https://doi.org/10.1007/s12602-023-10117-y

İncili, G. K., Karatepe, P., Akgöl, M., Güngören, A., Koluman, A., İlhak, O. İ., … Hayaloğlu, A. A. (2022). Characterization of lactic acid bacteria postbiotics, evaluation in-vitro antibacterial effect, microbial and chemical quality on chicken drumsticks. *Food Microbiology*, *104*, 104001. https://doi.org/10.1016/j.fm.2022.104001

Izuddin, W. I., Humam, A. M., Loh, T. C., Foo, H. L., & Samsudin, A. A. (2020). Dietary postbiotic Lactobacillus plantarum improves serum and ruminal antioxidant activity and upregulates hepatic antioxidant enzymes and ruminal barrier function in post-weaning lambs. *Antioxidants*, *9*(3), 250. https://doi.org/10.3390/antiox9030250

Jalali, S., Mojgani, N., Haghighat, S., Sanjabi, M. R., & Sarem-Nezhad, S. (2024). Investigation of antimicrobial and antioxidant properties of postbiotics produced by *Lactobacillus rhamnosus* and *Limosilactobacillus reuteri* and their potential application in surface decontamination of red meat. *LWT*, *209*, 116758. https://doi.org/10.1016/j.lwt.2024.116758

Kasahara, Y., Sakurai, T., Matsuda, R., Narukawa, M., Yasuoka, A., Mori, N., … Asakura, T. (2019). Novel indole and benzothiophene ring derivatives showing differential modulatory activity against human epithelial sodium channel subunits, ENaC β and γ. *Bioscience, Biotechnology, and Biochemistry*, *83*(2), 243–250. https://doi.org/10.1080/09168451.2018.1533802

Khoso, A., Hussain, A., Rehman, M., Akram, S., Ahmad, D., Bin-Asif, H., … Ali, S. A. (2024). Molecular assessments of antimicrobial protein enterocins and quorum sensing genes and their role in virulence of the genus *Enterococcus*. *Probiotics and Antimicrobial Proteins*, 1–12. https://doi.org/10.1007/s12602-024-10278-4

Lebeer, S., Claes, I. J., & Vanderleyden, J. (2012). Anti-inflammatory potential of probiotics: Lipoteichoic acid makes a difference. *Trends in Microbiology*, *20*(1), 5–10. https://doi.org/10.1016/j.tim.2011.09.004

LeBlanc, J. G., Milani, C., De Giori, G. S., Sesma, F., Van Sinderen, D., & Ventura, M. (2013). Bacteria as vitamin suppliers to their host: A gut microbiota perspective. *Current Opinion in Biotechnology*, *24*(2), 160–168. https://doi.org/10.1016/j.copbio.2012.08.005

Letizia, F., Albanese, G., Testa, B., Vergalito, F., Bagnoli, D., Di Martino, C., … Iorizzo, M. (2022). In vitro assessment of bio-functional properties from *Lactiplantibacillus plantarum* strains. *Current Issues in Molecular Biology*, *44*(5), 2321–2334. https://doi.org/10.3390/cimb44050158

Liu, Y., Hou, Y., Wang, G., Zheng, X., & Hao, H. (2020). Gut microbial metabolites of aromatic amino acids as signals in host–microbe interplay. *Trends in Endocrinology & Metabolism*, *31*(11), 818–834. https://doi.org/10.1016/j.tem.2020.02.012

Lubbers, R., Van Essen, M. F., Van Kooten, C., & Trouw, L. A. (2017). Production of complement components by cells of the immune system. *Clinical & Experimental Immunology*, *188*(2), 183–194. https://doi.org/10.1111/cei.12952

Ma, L., Tu, H., & Chen, T. (2023). Postbiotics in human health: A narrative review. *Nutrients*, *15*(2), 291. https://doi.org/10.3390/nu15020291

Magryś A, & Pawlik M. (2023). Postbiotic fractions of probiotics *Lactobacillus plantarum* 299v and *Lactobacillus rhamnosus* GG show immune-modulating effects. *Cells*, *12*(21), 2538. https://doi.org/10.3390/cells12212538

Mani-López, E., Arrioja-Bretón, D., & López-Malo, A. (2022). The impacts of antimicrobial and antifungal activity of cell-free supernatants from lactic acid bacteria in vitro and foods. *Comprehensive Reviews in Food Science and Food Safety*, *21*(1), 604–641. https://doi.org/10.1111/1541-4337.12872

Mantziari, A., Salminen, S., Szajewska, H., & Malagón-Rojas, J. N. (2020). Postbiotics against pathogens commonly involved in pediatric infectious diseases. *Microorganisms*, *8*(10), 1510. https://doi.org/10.3390/microorganisms8101510

Martín, R., & Langella, P. (2019). Emerging health concepts in the probiotics field: Streamlining the definitions. *Frontiers in Microbiology*, *10*, 1047. https://doi.org/10.3389/fmicb.2019.01047

Miller, J. K., Brzezinska-Slebodzinska, E., & Madsen, F. C. (1993). Oxidative stress, antioxidants, and animal function. *Journal of Dairy Science*, *76*(9), 2812–2823. https://doi.org/10.3168/jds.S0022-0302(93)77620-1

Mojgani, N., Bagheri, M., Moharrami, M., Toutiaee, S., & Rousta, H. M. (2023). Bioactive metabolites from autochthonous lactic acid bacteria inhibit the growth of *Melissococcus plutonius*, causal agent of European foulbrood disease in honey bees. *Entomologia Experimentalis et Applicata*, *171*(5), 386–394. https://doi.org/10.1111/eea.13295

Mojgani, N., Bagheri, M., & Vaseji, N. (2024). Invitro and invivo analysis of human milk lactic acid bacteria isolates for their anti-hypercholesterolemia actions. *Indian Journal of Microbiology*, *64*(1), 175–185. https://doi.org/10.1007/s12088-023-01150-0

Moradi, M., Kousheh, S. A., Almasi, H., Alizadeh, A., Guimarães, J. T., Yılmaz, N., & Lotfi, A. (2020). Postbiotics produced by lactic acid bacteria: The next frontier in food safety. *Comprehensive Reviews in Food Science and Food Safety*, *19*(6), 3390–3415. https://doi.org/10.1111/1541-4337.12613

Mosca, A., Abreu, Y., Abreu, A. T., Gwee, K. A., Ianiro, G., Tack, J., Nguyen, T. V. H., & Hill, C. (2022). The clinical evidence for postbiotics as microbial therapeutics. *Gut Microbes*, *14*(1). https://doi.org/10.1080/19490976.2022.2117508

Mowat, A. M., & Viney, J. L. (1997). The anatomical basis of intestinal immunity. *Immunological Reviews*, *156*(1), 145–166. https://doi.org/10.1111/j.1600-065X.1997.tb00966.x

Nakamura, F., Ishida, Y., Sawada, D., Ashida, N., Sugawara, T., Sakai, M., … Fujiwara, S. (2016). Fragmented lactic acid bacterial cells activate peroxisome proliferator-activated receptors and ameliorate dyslipidemia in obese mice. *Journal of Agricultural and Food Chemistry*, *64*(12), 2549–2559. https://doi.org/10.1021/acs.jafc.5b05827

Nataraj, B. H., Ali, S. A., Behare, P. V., & Yadav, H. (2020). Postbiotics-parabiotics: The new horizons in microbial biotherapy and functional foods. *Microbial Cell Factories*, *19*, 1–22. https://doi.org/10.1186/s12934-020-01426-w

Peluzio, M. D. C. G., Martinez, J. A., & Milagro, F. I. (2021). Postbiotics: Metabolites and mechanisms involved in microbiota–host interactions. *Trends in Food Science & Technology*, *108*, 11–26. https://doi.org/10.1016/j.tifs.2020.12.004

Rad, A. H., Hosseini, S., & Pourjafar, H. (2022). Postbiotics as dynamic biological molecules for antimicrobial activity: A mini-review. *Biointerface Research in Applied Chemistry*, *12*(5), 6543–6556. https://doi.org/10.33263/BRIAC125.65436556

Rad, A. H., Aghebati-Maleki, L., Kafil, H. S., Gilani, N., Abbasi, A., & Khani, N. (2021). Postbiotics, as dynamic biomolecules, and their promising role in promoting food safety. *Biointerface Research in Applied Chemistry*, *11*(6), 14529–14544. https://doi.org/10.33263/BRIAC116.1452914544

Rocchetti, M. T., Russo, P., De Simone, N., Capozzi, V., Spano, G., & Fiocco, D. (2024). Immunomodulatory activity on human macrophages by cell-free supernatants to explore the probiotic and postbiotic potential of *L. plantarum* strains of plant origin. *Probiotics and Antimicrobial Proteins*, *16*(3), 911–926. https://doi.org/10.1007/s12602-023-10084-4

Sabahi, S., Homayouni Rad, A., Aghebati-Maleki, L., Sangtarash, N., Ozma, M. A., Karimi, A., & Abbasi, A. (2023). Postbiotics as the new frontier in food and pharmaceutical research. *Critical Reviews in Food Science and Nutrition*, *63*(26), 8375–8402. https://doi.org/10.1080/10408398.2022.2056727

Salminen, S., Collado, M. C., Endo, A., Hill, C., Lebeer, S., Quigley, E. M., … Vinderola, G. (2021). The International Scientific Association of Probiotics and Prebiotics (ISAPP) consensus statement on the definition and scope of postbiotics. *Nature Reviews Gastroenterology & Hepatology*, *18*(9), 649–667. https://doi.org/10.1038/s41575-021-00440-6

Schogor, A. L. B., Palin, M. F., dos Santos, G. T., Benchaar, C., Lacasse, P., & Petit, H. V. (2013). Mammary gene expression and activity of antioxidant enzymes and oxidative indicators in the blood, milk, mammary tissue and ruminal fluid of dairy cows fed flax meal. *British Journal of Nutrition*, *110*(10), 1743–1750. https://doi.org/10.1017/S0007114513001220

Sharifi-Rad, J., Rodrigues, C. F., Stojanović-Radić, Z., Dimitrijević, M., Aleksić, A., Neffe-Skocińska, K., … Calina, D. (2020). Probiotics: Versatile bioactive components in promoting human health. *Medicina*, *56*(9), 433. https://doi.org/10.3390/medicina56090433

Siedler, S., Balti, R., & Neves, A. R. (2019). Bioprotective mechanisms of lactic acid bacteria against fungal spoilage of food. *Current Opinion in Biotechnology*, *56*, 138–146. https://doi.org/10.1016/j.copbio.2018.11.015

Stoyanova, L. G., Ustyugova, E. A., & Netrusov, A. I. (2012). Antibacterial metabolites of lactic acid bacteria: Their diversity and properties. *Applied Biochemistry and Microbiology*, *48*, 229–243. https://doi.org/10.1134/S0003683812030143

Sugahara, M., Tanaka, T., & Nangaku, M. (2017). Prolyl hydroxylase domain inhibitors as a novel therapeutic approach against anemia in chronic kidney disease. *Kidney International*, *92*(2), 306–312. https://doi.org/10.1016/j.kint.2017.02.035

Tarique, M., Abdalla, A., Masad, R., Al-Sbiei, A., Kizhakkayil, J., Osaili, T., … Ayyash, M. (2022). Potential probiotics and postbiotic characteristics including immunomodulatory effects of lactic acid bacteria isolated from traditional yogurt-like products. *LWT*, *159*, 113207. https://doi.org/10.1016/j.lwt.2022.113207

Tiptiri-Kourpeti, A., Spyridopoulou, K., Santarmaki, V., Aindelis, G., Tompoulidou, E., Lamprianidou, E. E., … Chlichlia, K. (2016). *Lactobacillus casei* exerts anti-proliferative effects accompanied by apoptotic cell death and up-regulation of TRAIL in colon carcinoma cells. *PloS One*, *11*(2), e0147960. https://doi.org/10.1371/journal.pone.0147960

Veiga, P., Suez, J., Derrien, M., & Elinav, E. (2020). Moving from probiotics to precision probiotics. *Nature Microbiology*, *5*(7), 878–880. https://doi.org/10.1038/s41564-020-0721-1

Vinderola, G., Sanders, M. E., & Salminen, S. (2022). The concept of postbiotics. *Foods*, *11*(8), 1077. https://doi.org/10.3390/foods11081077

Wang, Y., Wu, Y., Wang, Y., Xu, H., Mei, X., Yu, D., … Li, W. (2017). Antioxidant properties of probiotic bacteria. *Nutrients*, *9*(5), 521. https://doi.org/10.3390/nu9050521

Whitman, W. B., Coleman, D. C., & Wiebe, W. J. (1998). Prokaryotes: The unseen majority. *Proceedings of the National Academy of Sciences*, *95*(12), 6578–6583. https://doi.org/10.1073/pnas.95.12.6578

Wu, J., He, T., Wang, Z., Mao, J., & Sha, R. (2024). The dynamic analysis of non-targeted metabolomics and antioxidant activity of *Dendrobium officinale* Kimura et Migo by the synergistic fermentation of bacteria and enzymes. *LWT*, *203*, 116354. https://doi.org/10.1016/j.lwt.2024.116354

Yeşilyurt, N., Yılmaz, B., Ağagündüz, D., & Capasso, R. (2021). Involvement of probiotics and postbiotics in the immune system modulation. *Biologics*, *1*(2), 89–110. https://doi.org/10.3390/biologics1020006

Yilmaz, N., Özogul, F., Moradi, M., Fadiloglu, E. E., Šimat, V., & Rocha, J. M. (2022). Reduction of biogenic amines formation by foodborne pathogens using postbiotics in lysine-decarboxylase broth. *Journal of Biotechnology*, *358*, 118–127. https://doi.org/10.1016/j.jbiotec.2022.09.003

Zulkhairi Amin, F. A., Sabri, S., Ismail, M., Chan, K. W., Ismail, N., Mohd Esa, N., … Zawawi, N. (2020). Probiotic properties of Bacillus strains isolated from stingless bee (*Heterotrigona itama*) honey collected across Malaysia. *International Journal of Environmental Research and Public Health*, *17*(1), 278. https://doi.org/10.3390/ijerph17010278

12 Postbiotics in Managing Insulin Resistance and Glucose Metabolism

Aline Dias Paiva and Ana Beatriz Marcari
Federal University of Triângulo Mineiro, Uberaba - MG, Brazil

João Batista Camargo Neto
Ribeirão Preto School of Medicine, University of São Paulo, Ribeirão Preto-SP, Brazil

Maria Emilia Soares Martins dos Santos
Federal University of Triângulo Mineiro, Uberaba - MG, Brazil

12.1 INTRODUCTION/OVERVIEW

Carbohydrates comprise a large part of the human diet, and the primary hormone regulating carbohydrate metabolism is insulin. Insulin resistance (IR) can be defined as a decrease in insulin sensitivity in target tissues, which diminishes glucose utilization. An important cause associated with IR is obesity. In addition to obesity, dysbiosis—a disruption in the balance of gut microbiota—has also been linked to the development of IR. This chapter will address the importance of glucose, insulin, and the pathological changes resulting from dysfunction of this hormone, as well as how postbiotics could play a role in promoting insulin action and modulating carbohydrate metabolism.

12.2 IMPORTANCE OF GLUCOSE AND GLUCOSE METABOLISM

Carbohydrates comprise a large part of the human diet, representing about 45% to 65% of daily caloric intake [1]. Primarily consumed in the form of polysaccharides, they are hydrolyzed in the digestive tract and released as monosaccharides for intestinal absorption. Glucose is the primary monosaccharide used for energy generation and biomolecule synthesis in the human body [2]. Although glucose circulates freely in the bloodstream and represents a rapid energy source for cells, glucose metabolism depends primarily on its entry into cells through specialized glucose transporter proteins (GLUTs). In energy metabolism, GLUT-2 and GLUT-4 are of utmost importance. GLUT-2 is mainly found in hepatocytes, while GLUT-4 is present in white adipose tissue (WAT) and skeletal muscles, constituting about 60% of body mass [3]. The main difference between GLUT-2 and GLUT-4 is their dependence on insulin action for their expression on the plasma membrane, a key point in glucose metabolism [4]. Regardless of the glucose transporter used, upon entering the cell, glucose undergoes glycolysis. Through its complete metabolism in the glycolytic pathway, followed by the citric acid cycle and oxidative phosphorylation, glucose provides cells with appreciable amounts of energy. When an excess supply of glucose and the cellular energy demands have been met, this energy substrate is directed toward storage. The two primary forms of energy storage from excess glucose intake are glycogen synthesis in the liver and skeletal muscle and triglyceride

DOI: 10.1201/9781003607779-12

synthesis, primarily in WAT. These processes are highly regulated, ensuring energetic and glycemic homeostasis so that any alteration in the induction of glucose uptake and utilization may lead to the pathological increase of glucose levels in the bloodstream, a condition known as hyperglycemia [5].

In chronic hyperglycemia, protein glycation may occur. Glycation is a nonenzymatic process that promotes the appearance of highly reactive molecules called advanced glycation end products (AGEs). The accumulation of AGEs leads to binding to their specific receptors (RAGEs), which are widely expressed in immune and endothelial cells. This event triggers a signaling cascade that promotes the production of pro-inflammatory cytokines, recruiting immune cells and releasing more inflammatory cytokines. The accumulation of these cytokines, combined with processes that induce cell death, leads to endothelial injury and cardiovascular alterations [6]. Thus, it becomes clear that the regulation of glucose metabolism must operate in a highly controlled manner, maintaining its uptake and utilization to ensure glycemic and cellular homeostasis.

12.2.1 Insulin: Structure, Functions, and Mechanism of Action

The primary hormone regulating carbohydrate metabolism is insulin. Expressed by pancreatic beta cells, insulin is initially synthesized as preproinsulin. This precursor is cleaved to form proinsulin. Later, proinsulin is cleaved in its α and β chains, releasing them from the C-peptide. The α and β chains, bound by disulfide bridges, form the active insulin and are stored in granules along with the C-peptide in the pancreatic beta cells [7]. Insulin secretion occurs via a mechanism dependent on increased blood glucose levels. When glucose enters pancreatic beta cells, mediated by GLUT-2 in response to an increase in blood glucose (after a meal, for example), glucose is metabolized, generating ATP. The increase in ATP concentrations promotes the depolarization of the pancreatic beta cell through the closure of ATP-sensitive potassium channels. The depolarization of the pancreatic beta cell induces the opening of voltage-gated calcium channels, increasing intracellular calcium levels and promoting insulin secretion from storage granules [8]. Insulin circulates freely in the blood and has a relatively short half-life, the time needed for its action on metabolism [9]. When insulin binds to its tyrosine kinase receptors on the plasma membrane and is widely expressed in cells, the phosphorylation of tyrosine residues occurs on the intracellular subunits of this receptor. This event brings the insulin receptor substrates (IRS) closer to the receptor, where they are also phosphorylated. The phosphorylated IRS serves as anchors for a signaling cascade that induces kinases and phosphatases that mediate insulin's effects on glucose uptake, metabolism, and lipid and protein synthesis [10]. Glucose utilization in the glycolytic pathway is induced by insulin by stimulating enzymes in this metabolic pathway. In cases of excess glucose, the synthesis of hepatic and muscular glycogen is favored by this hormone. Also in the liver, but primarily in WAT, insulin stimulates the synthesis of lipids from excess metabolized glucose, inhibiting the mobilization of fats for energy generation. Specifically in skeletal muscle, insulin stimulates protein synthesis and inhibits protein degradation, increasing muscle mass. Furthermore, inhibiting endogenous hepatic glucose production is also one of insulin's primary functions, ensuring the metabolism of excess circulating glucose [11]. In summary, insulin ensures glucose uptake and metabolism, maintaining glycemic homeostasis. Through its anabolic functions in lipid and protein synthesis, this hormone contributes to the structure of WAT and skeletal muscles, which are essential for life maintenance [12].

12.2.2 Definition and Possible Cellular Mechanisms of Insulin Resistance Installation

The importance of an adequate insulin response in achieving glycemic homeostasis is well known. The association between IR and the development of type 2 diabetes mellitus (T2DM) grows exponentially each year. In 2021, the International Diabetes Federation (IDF) reported that approximately 537 million adults (20–79 years) were living with diabetes, of which more than 90% had T2DM. Alarmingly, this number could reach 643 million in 2030 and 783 million in 2045 [13].

IR can be defined as a reduced response to insulin action in target tissues, such as the muscle, liver, and adipose tissue. In skeletal muscle, the reduced insulin sensitivity diminishes glucose uptake and storage as glycogen. In the liver, IR impairs the suppression of gluconeogenesis, contributing to elevated blood glucose levels. Upon altered response to insulin, the pancreas compensates by producing more insulin, resulting in hyperinsulinemia. Over time, the sustained demand for insulin production can lead to pancreatic β-cell dysfunction and ultimately reduce insulin secretion, worsening glucose metabolism. This cycle of IR and hyperglycemia is central to the development of metabolic syndrome, obesity, and cardiovascular diseases associated with T2DM [14].

As mentioned before, an important cause associated with IR is obesity, strongly driven by environmental factors, especially lifestyle. Imbalanced diets and sedentary behavior favor the establishment of obesity and IR [14]. Although the molecular mechanisms involved in this association are not fully understood, research conducted in the past decades has contributed to its partial understanding. In general, increased caloric intake, associated with lower energy expenditure, favors the storage of excess energy in the form of triglycerides in WAT. Consequently, adipocytes undergo hypertrophy to increase their storage capacity, which promotes weight gain. In this scenario, the physiological oxygen supply is insufficient to meet the high energy demand of these hypertrophied adipocytes, leading to local hypoxia and a pathophysiological remodeling of adipocytes. This microenvironment favors the release of pro-inflammatory cytokines and the recruitment of immune cells with an inflammatory phenotype, leading to the accumulation of circulating inflammatory cytokines, especially interleukin-6 (IL-6), tumor necrosis factor alpha (TNF-α), and interleukin-1β (IL-1β), resulting in a state of chronic low-grade inflammation [15–17].

In addition to their pleiotropic effects on the immune system, these inflammatory cytokines favor the mobilization of triglycerides from hypertrophied adipocytes, increasing the levels of fatty acids in the bloodstream [18]. With sedentary behavior combined with obesity, these fatty acids are not metabolized for energy generation but are captured and stored ectopically, especially in the liver (hepatic steatosis) and skeletal muscle [19]. These fatty acids, captured by highly expressed transporters in obesity (CD36), esterify and accumulate diacylglycerols (DAG) in the cytoplasm [20]. Insulin signaling is impaired in this inflammatory microenvironment, and DAG accumulation is in the cytosol. Inflammatory cytokines and DAGs favor serine and/or threonine phosphorylation on IRS, which are then directed for degradation [21]. Reducing IRS prevents insulin from recruiting its signaling pathway, decreasing glucose uptake and utilization, which promotes hyperglycemia [22]. In response, pancreatic beta cells secrete more insulin, but as the inflammatory and obesity conditions progress, these cells become exhausted, entering cell death and releasing more pro-inflammatory cytokines [23]. The failure of pancreatic beta cells to secrete insulin, combined with resistance to this hormone, leads to the establishment of hyperglycemia that becomes difficult to reverse, constituting the condition of T2DM [24]. In this context, therapies aimed at mitigating the development of insulin resistance associated with obesity are urgent and highly relevant.

12.3 GUT MICROBIOTA: EUBIOSIS AND DYSBIOSIS

The gut microbiota is a dynamic and diverse microbial community. In healthy individuals, a predominance of bacteria from the Bacteroidetes and Firmicutes phyla, along with smaller proportions of Proteobacteria and Actinobacteria, are evidenced [25]. In a state of eubiosis, defined as a balanced microbial ecosystem, beneficial microorganisms prevail. These microorganisms establish a harmonious relationship with the host, preventing the proliferation of potentially pathogenic microbes and performing essential functions. These include synthesizing compounds beneficial to the host (such as vitamins and short-chain fatty acids [SCFAs]), digesting dietary components (especially fibers), modulating the immune system, and promoting trophic effects on the intestinal epithelium [26–29]. In contrast, dysbiosis, a disruption in the gut microbiota balance, results in reduced biodiversity, decline in important bacterial species, altered metabolic capacity, and proliferation of pathogenic species. Consequently, dysbiosis severely impacts the host's physiology and energy metabolism [27, 30]. Among the factors influencing the composition of the gut microbiota (e.g., exposure to toxic substances, genetics, age, and medication used (including antibiotics), dietary patterns play a crucial role. Diet significantly contributes to establishing and maintaining eubiosis and dysbiosis, thus affecting health and disease [31]. The Western diet, characterized by high consumption of refined carbohydrates and saturated fats, besides low intake of fruits and vegetables, is closely associated with gut microbiota alterations, leading to dysbiosis and increased intestinal permeability [29].

12.3.1 Dysbiosis and Insulin Resistance

Dysbiosis has been associated with the development of various metabolic conditions, including IR. Several studies have shown that IR could appear due to increased intestinal permeability, which allows bacterial lipopolysaccharides (LPS) to enter the bloodstream, triggering chronic low-grade inflammation. This inflammation produces cytokines, such as IL-1β, IL-6, and TNF-α, which interfere with insulin signaling in the liver, muscles, and adipose tissue, resulting in IR. In this way, IR and chronic inflammation exacerbate each other, worsening dysbiosis and accelerating the clinical progression of diabetes, affecting both nerves and blood vessels and impacting the intestinal epithelium [32]. Moreover, dysbiosis can influence the production of metabolites like SCFAs and secondary bile acids, essential for insulin sensitivity [33]. For instance, butyrate has been shown to enhance insulin sensitivity in peripheral tissues and may help control hepatic glucose production [34–37]. Propionate can stimulate the release of glucagon-like peptide-1 (GLP-1) and peptide YY, which improve insulin sensitivity and help with weight regulation [33, 38]. Secondary bile acids, formed from primary bile acids through gut microbiota, can activate G protein-coupled receptors, increasing cAMP production, improving insulin sensitivity, and interacting with the Farnesoid X receptor (FXR), Pregnane X receptor (PXR), and vitamin D receptors to regulate lipid and glucose metabolism [37, 39].

12.4 POSTBIOTICS: DEFINITION AND METABOLIC EFFECTS

Recent research has shown that modulating the gut microbiota by increasing beneficial microorganisms could be an interesting therapeutic alternative for combating the aforementioned diseases. In this context, prebiotics, probiotics, symbiotics, and postbiotics have been proposed as supplements to restore the gut microbiota [40, 41]. More specifically, postbiotics have been considered a relevant therapeutic strategy, particularly for preventing and treating IR and T2DM. Postbiotics are "preparations of inanimate microorganisms and/or components of these microorganisms that confer beneficial effects to the host." Generally, postbiotics can be classified into inactivated microorganisms, microbial lysates, microbial cell components, and metabolites produced by microorganisms [42]. This heterogeneous group of molecules shows different biological activities than their viable counterparts and exhibits some advantages over traditional probiotics [43]. Among the studied benefits of postbiotics, their immunomodulatory, anti-inflammatory, and anticancer effects stand out [44, 45]. Although the use of probiotics seems promising for the treatment of metabolic diseases, postbiotics' safety and potential risks have not yet been fully understood [46].

12.4.1 Main Postbiotics Already Tested for Treating Insulin Resistance

The relationship between glucose metabolism, insulin resistance, and the use of postbiotics is an area of growing scientific interest, particularly in managing metabolic disorders (Figure 12.1).

Modulating gut microbiota composition by postbiotics is mainly achieved by increasing the production of SCFAs, which enhance glucose homeostasis by improving insulin signaling and reducing inflammation. Additionally, postbiotics exhibit anti-inflammatory properties that can mitigate the chronic low-grade inflammation often present in insulin-resistant states [47]. Through their

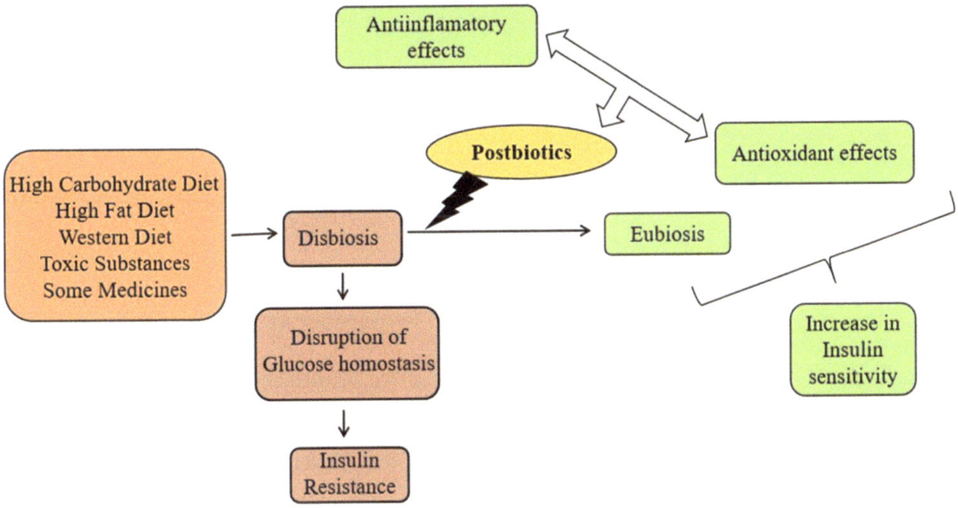

FIGURE 12.1 Interplay between glucose metabolism, insulin resistance, and the use of postbiotics.

immunomodulatory effects, postbiotics can down-regulate pro-inflammatory cytokine production, thus potentially improving insulin responsiveness and glucose metabolism [48]. Furthermore, postbiotics have shown promise in enhancing gut barrier integrity, reducing endotoxin translocation and subsequent systemic inflammation [49]. Postbiotics help maintain normal insulin signaling and reduce insulin resistance by preventing endotoxemia. Specific postbiotic compounds, such as extracellular vesicles or bacterial lysates, directly interact with insulin signaling pathways, enhancing glucose uptake in muscle and adipose tissues [50] (Table 12.1).

12.4.1.1 Inactivated Microorganisms

One of the first studies showing the effect of inactivated microorganisms on glucose metabolism was published in 2013. In this study, Shi and colleagues (2013) showed that treatment with *L. gasseri* TMC0356 in an experimental model of metabolic syndrome decreased the fasting glucose levels in these animals (51). Subsequently, Hsieh and colleagues (2016) showed that treatment with inactivated *L. reuteri* GMNL-263 (Lr263) was able to attenuate metabolic abnormalities induced in rats by a high-fat diet (HFD) (52). Supporting these findings, Caimari and colleagues (2017) showed that animals on an obesogenic diet that led to IR showed improvement in insulin action under treatment with inactivated *B. animalis* subsp. *lactis* CECT 8145 [53]. The use of *L. plantarum* L-137, *B. longum*, *A. muciniphila*, *S. thermophilus* MN-ZLW-002, *L. brevis* KB290, *L. casei* 01, *B. animalis*, and *L. plantarum* LRCC5314 showed similar results regarding the regulation of carbohydrate metabolism. In these studies, it was generally observed that the administration of these inactivated microorganisms improved fasting glucose, postprandial glucose, enhanced insulin sensitivity, and regulated insulin signaling in tissues [54–62]. Based on these initial findings, Companys and colleagues (2022) aimed to investigate the beneficial effects of consuming seafood sticks made with white fish enriched with *B. animalis* subsp. *lactis* CECT 8145, inulin, and omega-3 in individuals with abdominal obesity. They observed that individuals with abdominal obesity who consumed the enriched seafood sticks showed reduced insulin levels and Homeostatic Model Assessment for Insulin Resistance (HOMA-IR) compared to individuals who did not consume them. The authors believe that the results may be partially due to changes in the gut microbiota composition [63]. More recently, Kim and colleagues (2023) evaluated the efficacy of the postbiotic *L. plantarum* LRCC5314 supplemented in powdered milk (MP-LRCC5314) in a model of mice with stress-induced type II diabetes (stress-T2D). They observed that postbiotic administration reduced corticosterone levels,

TABLE 12.1
Postbiotics in Managing Insulin Resistance and Glucose Metabolism

Microorganism Used	Model	Duration	Observed Data	Reference
		Inactivated Microorganisms		
L. gasseri TMC0356	Metabolic syndrome SD rats induced by high-fat and high-salt diet	15 weeks	Decreased fasting blood glucose	[51]
L. reuteri GMNL-263	HFD-fed SD rats	12 weeks	Decreased blood glucose and insulin	[52]
B. animalis subsp. *Lactis* CECT 8145	Cafeteria diet-induced obese Wistar rats	84 days	Improved insulin sensitivity	[53]
L. plantarum L-137	Metabolic syndrome rats (DS)/obese rats	5 weeks	Improved insulin sensitivity; regulated insulin signaling in visceral and subcutaneous adipose tissue	[54]
B. longum	TSOD mice	5 weeks	Decreased blood glucose	[55]
A. muciniphila	HFD-induced obese C57BL/6 mice	5 weeks	Decreased blood glucose	[56]
L. plantarum L-137	HFD-induced obese C57BL/6J mice	16 weeks	Decreased blood glucose	[57]
S. thermophilus MN-ZLW-002	HFD-fed C57BL/6 mice	4–12 weeks	Reduced fasting blood glucose	[58]
L. brevis KB290	HFD-induced obese C57BL/6J mice	8 weeks	Decreased serum glucose, insulin, and HOMA-IR	[59]
L. casei 01	Healthy individuals (BMI, 18–25 kg/m^2)	7 days	Reduced postprandial glycemia	[60]
B. animalis	Healthy Wistar rats	15 days	Reduced serum glucose	[61]
L. plantarum LRCC5314	Cold-induced stress–diabetes mellitus type II C57BL/6 mice	12 weeks	Reduced fasting glucose	[62]
B. animalis subsp lactis CECT 8145 + Inulin + omega-3	Obese subjects	12 weeks	Reduced insulin level and HOMA-IR	[63]
L. plantarum LRCC5314 supplemented in powdered milk	Cold-induced stress–diabetes mellitus type II C57BL/6J mice	8–12 weeks	Improved glucose metabolism and insulin sensitivity	[64]
		Microbial Cell Components		
Exopolysaccharides *B. subtilis* sp. *suppress*	Streptozotocin induced diabetic SD rats	4 weeks	Reduced glucose and increased insulin levels	[65]
MDP	HFD-induced obese C57BL/6J mice	5 weeks	Reduced glucose intolerance and increased hepatic insulin sensitization	[66]
MDP	Murine and human L-cells Chow-fed C57BL/6 mice	2 hours 30 min	Increased GLP-1 secretion	[67]
MDP	HFD-induced obesity in male and female adipocyte-specific IRF4 knockout mice (AdipoIRF4fl/fl)	3 days	Reduced glucose intolerance and increased insulin sensitization in wild-type mice	[68]
SLPs from PLAB	HFD-induced obese C57BL/6J mice	6 weeks	Improved insulin resistance	[69]

(*Continued*)

TABLE 12.1
(Continued)

Microorganism Used	Model	Duration	Observed Data	Reference
		Microbial Cell Components		
Intestinal extracts derived from male and female mice fed a CD or HFD	Lean male and female C57BL/6J mice	5 weeks	Improved glucose and insulin tolerance	[70]
Fecal transplant from mice fed a resveratrol diet	High-fat, high-sucrose-induced obese C57BL/6N mice	1, 3, and 5 days	Improved glucose homeostasis	[71]
		Microbial Metabolites		
Butyrate	HFD-induced obese C57BL/6J mice	4 weeks	Decreased blood glucose, fasting insulin, and insulin resistance	[72]
Isobutyric and isovaleric acids	Primary rat and human adipocytes	-	Potentiated insulin-stimulated glucose uptake	[73]
Acetate	HFD-induced obesity in Wistar rats	12 weeks	Reversed insulin resistance with reduction in HOMA-IR; Nevertheless, blood glucose levels did not change.	[74]
Butyrate	HFD-induced obese C57BL/6J mice	16 weeks	Improved glucose tolerance and insulin sensitivity	[75]
Acetate and butyrate	NOD	5 weeks	Preserved pancreatic endocrine cell	[76]

Legend: *L. gasseri*: *Lactobacillus gasseri*; *L. reuteri*: *Limosilactobacillus reuteri*; *B. animalis*: *Bifidobacterium animalis*; *L. plantarum*: *Lactiplantibacillus plantarum*; *B. longum*: *Bifidobacterium longum*; *A. muciniphila*: *Akkermansia muciniphila*; *S. thermophilus*: *Streptococcus thermophilus*; *L. brevis*: *Levilactobacillus brevis*; *L. casei*: *Lacticaseibacillus casei*; *B. subtilis*: *Bacillus subtilis*; SD rats: Spargue Dawley rats; (DS)/obese rats: Salt sensitive and Zucker rats; TSOD mice: Tsumura, Suzuki, obese diabetes; MDP: Muramyl dipeptide; SLP: Surface layer proteins; PLAB: Paraprobiotic kefir lactic acid bacteria; NOD mice: Non-obese-diabetic mice.

regulated the immune response (decreasing pro-inflammatory cytokines and increasing anti-inflammatory cytokines), and improved glucose and insulin sensitivity [64].

12.4.1.2 Microbial Cell Components

The work of Ghonein and collaborators (2016) was one of the early studies that evaluated the ability of bacterial cellular components to cause beneficial effects on metabolism. They showed that diabetic rats induced with streptozotocin and treated with exopolysaccharides from *Bacillus subtilis* sp. showed a reduction in glucose levels and an increase in insulin levels compared to untreated rats [65]. Subsequently, several studies were conducted evaluating the effects of different bacterial components. Peptidoglycan (PG) (consisting of muramyl dipeptides [MDP]) is a component of the bacterial cell wall responsible for maintaining its structural integrity. MDPs are constantly released from PG degradation and are recognized by various receptors, including nucleotide-binding oligomerization domain (NOD) proteins [77]. Cavallari and colleagues (2017) showed that the injection of MDP in ob/ob mice or mice fed with HFD improved insulin sensitivity through the activation of NOD2. The researchers believe this effect is mediated by interferon regulatory factor 4 (IRF4), which reduces the inflammatory effects observed in obesity and endotoxemia [78]. Later, it was also found that RIPK2 (receptor-interacting serine/threonine-protein kinase 2) in nonhematopoietic cells mediates the reduction of glucose and the anti-inflammatory effects of NOD2-activating postbiotics in animals consuming HFD [66]. Willians and colleagues (2020) also evaluated the effect of MDP on glucagon-like peptide-1 (GLP-1) secretion. GLP-1 is a hormone produced by intestinal L-cells known for its action on pancreatic β-cells, increasing insulin secretion. The researchers observed that mice fed a control diet treated with MDP significantly increased GLP-1. However, this did not lead to improved oral glucose tolerance. On the other hand, when the mice were fed a high-fat diet, which can cause hyperglycemia, they lost the MDP-induced GLP-1 release. These findings suggest that MDP may be important in increasing GLP-1 during normal glycemic conditions, but this ability is lost under hyperglycemia [67].

More recently, Duggan and colleagues (2022) also observed that the effect of MDP and its action on insulin sensitivity appears to be IRF4- and sex-dependent. This study evaluated male and female mice with adipocyte-specific knockout of IRF4 (AdipoIRF4fl/fl) on HFD and during endotoxemia (LPS injection). They found that IRF4 in adipocytes is necessary for the effects of MDP on glucose reduction in male mice receiving HFD and during endotoxemia. However, MDP did not alter glucose levels in female wild-type (WTfl/fl) mice or AdipoIRF4fl/fl/HFD mice during endotoxemia [68]. Surface-layer proteins (SLPs) are components of the cell envelope composed of glycoprotein subunits, usually found in microorganisms that mediate binding to host epithelial cells. SLPs have been associated with anti-inflammatory properties [79]. In this context, Kim and colleagues (2021) observed a significant improvement in insulin resistance in mice fed with HFD upon administration of SLPs from kefir probiotic lactic acid bacteria. The authors believe this effect may be related to the immunomodulatory action of SLPs [69].

Duggan and colleagues (2022) explored the action of another type of bacterial component. The researchers injected bacterial extract derived from the small intestine (BIE) of male and female mice fed with chow diet (CD) or HFD into lean mice of both sexes. They found that this administration reduced glucose levels and improved insulin resistance in all lean mice, with no changes in body mass. Vaccination of NOD1-/-, NOD2-/-, and RIPK2-/- mice with BIE from male WT mice fed with CD showed that these innate immune system proteins were necessary for the bacterial extract injection to improve blood glucose control. Furthermore, the researchers also found that lean mice only required a single injection of small intestine–derived bacterial extract. In contrast, obese mice fed an HFD required prime-boost injections of the bacterial extract to improve blood glucose control [70]. Corroborating the data mentioned previously, a group of researchers also found that sterile fecal filtrates containing nonliving components from mice that received resveratrol (a polyphenolic compound found in various plant foods that have shown promising results in preventing insulin resistance caused by obesity) were also sufficient to improve glucose homeostasis in obese mice [71, 80].

12.4.1.3 Microbial Metabolites

As previously stated, the SCFAs are the primary microbial metabolites with recognized beneficial effects on the host metabolism [33]. Gao and collaborators (2009) showed that the administration of sodium butyrate prevented the development of insulin resistance and obesity in mice fed with HFD. Fasting glucose, fasting insulin, and insulin tolerance were preserved in the HFD-treated animals compared to the untreated ones [72]. Since then, several other studies have been published emphasizing the importance of SCFAs in regulating glycemia and insulin action. For example, the administration of isobutyric and isovaleric acids (branched SCFAs) has a beneficial effect on glucose metabolism in adipocytes in both rats and humans. Notably, isobutyric acid at different concentrations enhanced insulin-stimulated glucose uptake in rat adipocytes [73]. Olaniyi and colleagues (2021) also observed a significant decrease in insulin and HOMA-IR in HFD-supplemented animals with sodium acetate compared to the untreated HFD group [74]. Kushwaha and colleagues (2022) also showed that obese mice receiving sodium butyrate improved insulin sensitivity, glucose tolerance, and adipose tissue inflammation. Additionally, butyrate mitigated metabolic disorders and intestinal epithelial dysfunction in type 2 diabetic mice, promoting insulin secretion without compensatory hyperplasia in pancreatic beta cells [75]. Complementing these findings, Vandenbempt and colleagues (2024) also showed that dietary supplements containing acetate and butyrate improved glycemic control in type 1 diabetic mice by increasing the expression of functional β-cell genes and insulin production [76].

12.5 CONCLUSIONS

Postbiotics present a promising therapeutic approach for addressing insulin resistance by modulating glucose metabolism through different mechanisms, including altering gut microbiota, reducing systemic inflammation, and enhancing insulin signaling pathways. As research progresses, understanding the specific molecular mechanisms of postbiotic actions may lead to more targeted interventions for managing insulin resistance and associated metabolic disorders.

REFERENCES

1. World Health Organization. (2023). *Carbohydrate intake for adults and children: WHO guideline*. World Health Organization.
2. Dashty, M. (2013). A quick look at biochemistry: Carbohydrate metabolism. *Clinical Biochemistry*, 46(15), 1339–1352. https://doi.org/10.1016/j.clinbiochem.2013.04.027
3. Zierler, K. (1999). Whole body glucose metabolism. *American Journal of Physiology-Endocrinology and Metabolism*, 276(3), E409–E426. https://doi.org/10.1152/ajpendo.1999.276.3.E409
4. Moore, M. C., et al. (2012). Regulation of hepatic glucose uptake and storage in vivo. *Advances in Nutrition*, 3(3), 286–294. https://doi.org/10.3945/an.112.002089
5. Roden, M., & Bernroider, E. (2003). Hepatic glucose metabolism in humans—its role in health and disease. *Best Practice & Research Clinical Endocrinology & Metabolism*, 17(3), 365–383. https://doi.org/10.1016/s1521-690x(03)00031-9
6. Nowotny, K., et al. (2015). Advanced glycation end products and oxidative stress in type 2 diabetes mellitus. *Biomolecules*, 5(1), 194–222. https://doi.org/10.3390/biom5010194
7. Mayer, J. P., et al. (2007). Insulin structure and function. *Peptide Science: Original Research on Biomolecules*, 88(5), 687–713. https://doi.org/10.1002/bip.20734
8. Komatsu, M. T., et al., (2013). Glucose-stimulated insulin secretion: A newer perspective. *Journal of Diabetes Investigation*, 4(6), 511–516. https://doi.org/10.1111/jdi.12094
9. Konturek, S. J., et al. (1990). Comparison of organ uptake and disappearance half-time of human epidermal growth factor and insulin. *Regulatory Peptides*, 30(2), 137–148. https://doi.org/10.1016/0167-0115(90)90054-z
10. Saltiel, A. R. (2021). Insulin signaling in health and disease. *The Journal of Clinical Investigation*, 131(1). https://doi.org/10.1172/JCI142241
11. Rahman, M. S., et al. (2021). Role of insulin in health and disease: An update. *International Journal of Molecular Sciences*, 22(12), 6403. https://doi.org/10.3390/ijms22126403
12. Norton, L., et al. (2022). Insulin: The master regulator of glucose metabolism. *Metabolism*, 129, 155142. https://doi.org/10.1016/j.metabol.2022.155142
13. International Diabetes Federation. (2021). *IDF diabetes atlas* (10th ed.). Brussels: IDF. https://diabetesatlas.org
14. Martyn, J. A. J., et al. (2008). Obesity-induced insulin resistance and hyperglycemia: Etiologic factors and molecular mechanisms. *The Journal of the American Society of Anesthesiologists*, 109(1), 137–148. https://doi.org/10.1097/ALN.0b013e3181799d45
15. Kahn, S. E., et al. (2006). Mechanisms linking obesity to insulin resistance and type 2 diabetes. *Nature*, 444(7121), 840–846. https://doi.org/10.1038/nature05482
16. Ye, J. (2009). Emerging role of adipose tissue hypoxia in obesity and insulin resistance. *International Journal of Obesity*, 33(1), 54–66. https://doi.org/10.1038/ijo.2008.229
17. Blüher, M. (2016). Adipose tissue inflammation: A cause or consequence of obesity-related insulin resistance? *Clinical Science*, 130(18), 1603–1614. https://doi.org/10.1042/CS20160005
18. Kang, Y. E., et al. (2016). The roles of adipokines, proinflammatory cytokines, and adipose tissue macrophages in obesity-associated insulin resistance in modest obesity and early metabolic dysfunction. *PLoS One*, 11(4), e0154003. https://doi.org/10.1371/journal.pone.0154003

19. Alkhouri, N., et al. (2010). Adipocyte apoptosis, a link between obesity, insulin resistance, and hepatic steatosis. *Journal of Biological Chemistry*, 285(5), 3428–3438. https://doi.org/10.1074/jbc.M109.074252
20. Miquilena-Colina, M. E., et al. (2011). Hepatic fatty acid translocase CD36 upregulation is associated with insulin resistance, hyperinsulinaemia and increased steatosis in non-alcoholic steatohepatitis and chronic hepatitis C. *Gut*, 60(10), 1394–1402. https://doi.org/10.1136/gut.2010.222844
21. Jornayvaz, F. R., & Shulman, G. I. (2012). Diacylglycerol activation of protein kinase Cε and hepatic insulin resistance. *Cell Metabolism*, 15(5), 574–584. https://doi.org/10.1016/j.cmet.2012.03.005
22. An Gerwen, J. V., et al. (2023). Insulin signalling and GLUT4 trafficking in insulin resistance. *Biochemical Society Transactions*, 51(3), 1057–1069. https://doi.org/10.1042/BST20221066
23. Cantley, J., & Ashcroft, F. M. (2015). Q&A: Insulin secretion and type 2 diabetes: Why do β-cells fail?. *BMC Biology*, 13, 1–7. https://doi.org/10.1186/s12915-015-0140-6
24. Kahn, B. B. (1998). Type 2 diabetes: When insulin secretion fails to compensate for insulin resistance. *Cell*, 92(5), 593–596. https://doi.org/10.1016/s0092-8674(00)81125-3
25. Belizário, J. E., et al. (2018). Gut microbiome dysbiosis and immunometabolism: New frontiers for treatment of metabolic diseases. *Mediators of Inflammation*, 2037838.https://doi.org/10.1155/2018/2037838
26. Mukherjee S., et al.. (2018). Gut microbes as future therapeutics in treating inflammatory and infectious diseases: Lessons from recent findings. *The Journal of Nutritional Biochemistry*, 61, 111–128. https://doi.org/10.1016/j.jnutbio.2018.07.010
27. Riccio, P., & Rossano, R. (2019). Undigested food and gut microbiota may cooperate in the pathogenesis of neuroinflammatory diseases: A matter of barriers and a proposal on the origin of organ specificity. *Nutrients*, 11(11), 2714. https://doi.org/10.3390/nu11112714
28. Portincasa P., et al. (2021). Intestinal barrier and permeability in health, obesity and NAFLD. *Biomedicines*,10(1), 83. https://doi.org/10.3390/biomedicines10010083
29. Malesza, I. J., et al. (2021, November 14). Western-style diet, systemic inflammation, and gut microbiota: A narrative review. *Cells*, *10*(11), 3164. https://doi.org/10.3390/cells10113164
30. Bishehsari, F., et al. (2021). Abnormal food timing and predisposition to weight gain: Role of barrier dysfunction and microbiota. *Translational Research*,231, 113–123. https://doi.org/10.1016/j.trsl.2020.11.007
31. Croci, S., et al. (2021). Dietary strategies for management of metabolic syndrome: Role of gut microbiota metabolites. *Nutrients*, 13(5), 1389. https://doi.org/10.3390/nu13051389. PMID: 33919016; PMCID: PMC8142993
32. Crudele, L., et al. (2023). Gut microbiota in the pathogenesis and therapeutic approaches of diabetes. *EBioMedicine*, 97, 104821. https://doi.org/10.1016/j.ebiom.2023.104821
33. Kant, R., et al. (2022). Gut microbiota interactions with anti-diabetic medications and pathogenesis of type 2 diabetes mellitus. *World Journal of Methodology*, 12, 246–257. https://doi.org/10.5662/wjm.v12.i4.246
34. Hamer, H. M., et al. (2009). Butyrate modulates oxidative stress in the colonic mucosa of healthy humans. *Clinical Nutrition*, 28, 88–93. https://doi.org/10.1016/j.clnu.2008.11.002
35. Tang, R., & Li, L. (2021). Modulation of short-chain fatty acids as potential therapy method for type 2 diabetes mellitus. *Canadian Journal of Infectious Diseases and Medical Microbiology*, 6632266. https://doi.org/10.1155/2021/6632266. Erratum in: https://doi.org/10.1155/2021/9756586
36. Fusco, W., et al. (2023, May 6). Short-chain fatty-acid-producing bacteria: Key components of the human gut microbiota. *Nutrients*, 15(9), 2211. https://doi.org/10.3390/nu15092211
37. Zhou, Z., et al. (2022). Gut microbiota: An important player in type 2 diabetes mellitus. *Frontiers in Cellular and Infection Microbiology*, 12, 834485 https://doi.org/10.3389/fcimb.2022.834485
38. Canfora, E. E., et al. (2015). Short-chain fatty acids in control of body weight and insulin sensitivity. *Nature Reviews Endocrinology*, 11, 577–591. https://doi.org/10.1038/nrendo.2015.128
39. Massey, W., & Brown, J. M. (2021). The gut microbial endocrine organ in type 2 diabetes. *Endocrinology*, 162. https://doi.org/10.1210/endocr/bqaa235
40. Li, H. Y., et al. (2021). Effects and mechanisms of probiotics, prebiotics, synbiotics, and postbiotics on metabolic diseases targeting gut microbiota: A narrative review. *Nutrients*, 13. https://doi.org/10.3390/nu13093211.
41. Kataria, J., et al. (2009). Probiotic microbes: Do they need to be alive to be beneficial? *Nutrition Reviews*, *67*: 546–550. https://doi.org/10.1111/j.1753-4887.2009.00226.x

42. Salminen, S. M. C., et al. (2021). The International Scientific Association of Probiotics and Prebiotics (ISAPP) consensus statement on the definition and scope of postbiotics. *Nature Reviews: Gastroenterology & Hepatology*, 18(9), 649–667. https://doi.org/10.1038/s41575-021-00440-6
43. Cabello-Olmo, M., et al. (2021). Role of postbiotics in diabetes mellitus: Current knowledge and future perspectives. *Foods*, 10(7), 1590. https://doi.org/10.3390/foods10071590
44. Zolkiewicz J., et al. (2020). Postbiotics: A step beyond pre- and probiotics. *Nutrients*, 12(8), 2189. https://doi.org/10.3390/nu12082189
45. Daliri, E. B., et al. (2019). Safety of probiotics in health and disease. In *The role of functional food security in global health* (pp. 603–622). Academic Press.
46. Zhong, Y., et al. (2024). Recent advances and potentiality of postbiotics in the food industry: Composition, inactivation methods, current applications in metabolic syndrome, and future trends. *Critical Reviews in Food Science and Nutrition*, 64(17), 5768–5792. https://doi.org/10.1080/10408398.2022.2158174
47. Püschel, G. P., et al. (2022). Macrophages, low-grade inflammation, insulin resistance and hyperinsulinemia: A mutual ambiguous relationship in the development of metabolic diseases. *Journal of Clinical Medicine*, 11(15), 4358. https://doi.org/10.3390/jcm11154358
48. Mehta, J. P., et al. (2023). The potential of paraprobiotics and postbiotics to modulate the immune system: A review. *Microbiological Research*, 275, 127449. https://doi.org/10.1016/j.micres.2023.127449
49. Scott, E., et al. (2022). Postbiotics and their health modulatory biomolecules. *Biomolecules*, 12(11), 1640. https://doi.org/10.3390/biom12111640
50. Lakhter, A. J., & Sims, E. K. (2015). Minireview: Emerging roles for extracellular vesicles in diabetes and related metabolic disorders. *Molecular Endocrinology*, 29, 1535–1548. https://doi.org/10.1210/me.2015-1206
51. Shi, L., et al. (2013). Effects of heat-inactivated *Lactobacillus gasseri* TMC0356 on metabolic characteristics and immunity of rats with the metabolic syndrome. *The British Journal of Nutrition*, 109(2), 263–272. https://doi.org/10.1017/S000711451200116X
52. Hsieh, F. C., et al. (2016). Heat-killed and live *Lactobacillus reuteri* GMNL-263 exhibit similar effects on improving metabolic functions in high-fat diet-induced obese rats. *Food & Function*, 7(5), 2374–2388. https://doi.org/10.1039/c5fo01396h
53. Caimari, A., et al. (2017). Heat-killed *Bifidobacterium animalis* subsp. *Lactis* CECT 8145 increases lean mass and ameliorates metabolic syndrome in cafeteria-fed obese rats. *Journal of Functional Foods*, 38, 251–263. https://doi.org/10.1016/j.jff.2017.09.029
54. Uchinaka, A., et al. (2018). Anti-inflammatory effects of heat-killed *Lactobacillus plantarum* L-137 on cardiac and adipose tissue in rats with metabolic syndrome. *Scientific Reports*, 8(1), 8156. https://doi.org/10.1038/s41598-018-26588-x
55. Othman, M. B., & Sakamoto, K. (2020). Effect of inactivated *Bifidobacterium longum* intake on obese diabetes model mice (TSOD). *Food Research International (Ottawa, Ont.)*, 129, 108792. https://doi.org/10.1016/j.foodres.2019.108792
56. Ashrafian, F., et al. (2021). Extracellular vesicles and pasteurized cells derived from *Akkermansia muciniphila* protect against high-fat induced obesity in mice. *Microbial Cell Factories*, 20(1), 219. https://doi.org/10.1186/s12934-021-01709-w
57. Yoshitake, R., et al. (2021). Heat-killed *Lactobacillus plantarum* L-137 attenuates obesity and associated metabolic abnormalities in C57BL/6 J mice on a high-fat diet. *Bioscience of Microbiota, Food and Health*, 40(2), 84–91. https://doi.org/10.12938/bmfh.2020-040
58. Kang, X., et al. (2021). Anti-adipogenesis and metabolism-regulating effects of heat-inactivated *Streptococcus thermophilus* MN-ZLW-002. *Letters in Applied Microbiology*, 72(6), 677–687. https://doi.org/10.1111/lam.13398
59. Watanabe, J., et al. (2021). Heat-killed *Lactobacillus brevis* KB290 attenuates visceral fat accumulation induced by high-fat diet in mice. *Journal of Applied Microbiology*, 131(4), 1998–2009. https://doi.org/10.1111/jam.15079
60. Barros, C. P., et al. (2021). Paraprobiotic obtained by ohmic heating added in 5786 whey-grape juice drink is effective to control postprandial glycemia in healthy adults. *Food Research International*, 140, 109905. https://doi.org/10.1016/j.foodres.2020.109905
61. Almada, C.N., et al. (2021). Obtaining paraprobiotics from *Lactobacilus acidophilus, Lacticaseibacillus casei* and *Bifidobacterium animalis* using six inactivation methods: Impacts on the cultivability, integrity, physiology, and morphology. *Journal of Functional Foods*, 87, 104826. https://doi.org/10.1016/j.jff.2021.104826

62. Nam, Y., et al. (2022). Heat-killed *Lactiplantibacillus plantarum* LRCC5314 mitigates the effects of stress-related type 2 diabetes in mice via gut microbiome modulation. *Journal of Microbiolog and Biotechnology*, 32(3), 324–332. https://doi.org/10.4014/jmb.2111.11008
63. Companys, J., et al. (2022). Effects of enriched seafood sticks (heat-inactivated *B. animalis* subsp. lactis CECT 8145, inulin, omega-3) on cardiometabolic risk factors and gut microbiota in abdominally obese subjects: randomized controlled trial. *European Journal of Nutrition*, 61(7), 3597–3611. https://doi.org/10.1007/s00394-022-02904-0
64. Kim, J.H., et al. (2023). Effect of postbiotic *Lactiplantibacillus plantarum* LRCC5314 supplemented in powdered milk on type 2 diabetes in mice. *Journal of the Dairy Science*,107(8), 5301–5315. https://doi.org/10.3168/jds.2023-24103
65. Ghoneim, M.A., et al.,(2016). Effect of polysaccharide from *Bacillus subtilis* sp. on cardio vascular diseases and atherogenic indices in diabetic rats. *BMC Complementary and Alternative Medicine*, 16, 112. https://doi.org/10.1186/s12906-016-1093-1
66. Cavallari, J.F., et al. (2020). Postbiotics for NOD2 require nonhematopoietic RIPK2 to improve blood glucose and metabolic inflammation in mice. *American Journal of Physiology-Endocrinology and Metabolism*, 318: E579–E585. https://doi.org/10.1152/ajpendo.00033.2020
67. Williams, L., et al. (2020). The role of the bacterial muramyl dipeptide in the regulation of GLP-1 and glycemia. *International Journal of Molecular Sciences*, 21, 5252. https://doi.org/10.3390/ijms21155252
68. Duggan, B.M., et al. (2022a). Postbiotics engage IRF4 in adipocytes to promote sex-dependent changes in blood glucose during obesity. *Physiological Reports*, 10(16), e15439. https://doi.org/10.14814/phy2.15439
69. Kim, E., et al. (2021). Effect of surface layer proteins derived from paraprobiotic kefir lactic acid bacteria on inflammation and high-fat diet-induced obesity. *Journal of Agricultural and Food Chemistry*, 69, 15157–15164. https://doi.org/10.1021/acs.jafc.1c05037
70. Duggan, B. M., et al. (2022b, January). Gut microbiota-based vaccination engages innate immunity to improve blood glucose control in obese mice. *Molecular Metabolism*, *55*, 101404. https://doi.org/10.1016/j.molmet.2021.101404
71. Kim, T. T., et al. (2018). Fecal transplant from resveratrol-fed donors improves glycaemia and cardiovascular features of the metabolic syndrome in mice. *American Journal of Physiology-Endocrinology and Metabolism*, 315, E511–E519. https://doi.org/10.1152/ajpendo.00471.2017
72. Gao, Z. et al., (2009). Butyrate improves insulin sensitivity and increases energy expenditure in mice. *Diabetes*, 58(7), 1509–1517. https://doi.org/10.2337/db08-1637
73. Heimann, E., et al. (2016). Branched short-chain fatty acids modulate glucose and lipid metabolism in primary adipocytes. *Adipocyte*, 5(4), 359–368. https://doi.org/10.1080/21623945.2016.1252011
74. Olaniyi, K. S., et al. (2021). Acetate supplementation restores testicular function by modulating Nrf2/PPAR-gamma in high fat diet-induced obesity in Wistar rats. *Journal of Diabetes and Metabolic Disorders*, 20 (2), 1685–1696. https://doi.org/10.1007/s40200-021-00924-x
75. Kushwaha, V., et al. (2022).Sodium butyrate reduces endoplasmic reticulum stress by modulating CHOP and empowers favorable anti-inflammatory adipose tissue immune-metabolism in HFD fed mice model of obesity. *Food Chemistry*, 4, Article 100079. https://doi.org/10.1016/j.fochms.2022.100079
76. Vanderbilt, V., et al. (2024). HAMSAB diet ameliorates dysfunctional signaling in pancreatic islets in autoimmune diabetes. *iScience*, 27, 108694. https://doi.org/10.1016/j.isci.2023.108694
77. Molinaro, R., et al. (2019). Trace levels of peptidoglycan in serum underlie the NOD dependent cytokine response to endoplasmic reticulum stress. *Journal of Biological Chemistry*, 294(22), 9007–9015. https://doi.org/10.1074/jbc.RA119.007997
78. Cavallari, J.F., et al. (2017). Muramyl dipeptide-based postbiotics mitigate obesity-induced insulin resistance via IRF4. *Cell Metabolism*, 25(5), 1063–1074.e3. https://doi.org/10.1016/j.cmet.2017.03.021
79. Malamud, M., et al. (2019). S-Layer glycoprotein from *Lactobacillus kefiri* exerts its immunostimulatory activity through glycan recognition by mincle. *Frontiers in Immunology*, 1422. https://doi.org/10.3389/fimmu.2019.01422
80. Baur, J. A., et al. (2006). Resveratrol improves health and survival of mice on a high-calorie diet. *Nature*, 444, 337–342. https://doi.org/10.1038/nature05354

13 Postbiotics

In Metabolic Endotoxemia and Inflammation in Type 2 Diabetes

Muhammad Usman
Faculty of Veterinary Science, University of Agriculture, Faisalabad, Pakistan.

Muhammad Kasib Khan and Zaheer Abbas
Department of Veterinary Parasitology, Faculty of Veterinary Science, University of Agriculture, Faisalabad, Pakistan.

Sultan Ali and Muhammad Shahid Mehmood
Institute of Microbiology, Faculty of Veterinary Science, University of Agriculture, Faisalabad, Pakistan.

13.1 INTRODUCTION

Probiotics are everlasting architects of gut harmony, showing their potential significance in gut health. But their negative effects on health are alarming, as they can enhance the risk of infection and disturb the microbial balance in the gut. So postbiotics are a healthy and safer alternative to probiotics, as they have long shelf life and are more stable than probiotics. Postbiotics are nonviable metabolites or soluble factors produced from probiotics. They might be fragments of microorganisms like secreted proteins and extracellular polysaccharides. Some probiotics having potential importance in fermentation like *Lactobacillus*, *Bifidobacterium*, and *Bacillus coagulans* are common producers of postbiotics [1]. Fermented foods like kombucha, yogurt, and sauerkraut are rich in postbiotics due to these probiotics [2–4]. They are of great commercial importance as they are used as supplements in food because they play important roles in regulating gut health, barrier functions, and metabolic syndromes (diabetes) and act as anti-inflammatory mediators. Peptidoglycan, teichoic acids, polysaccharides (EPS), cell-free supernatants, extracellular vesicles, short-chain fatty acids (SCFAs), enzymes, and bacteriocins are some common examples of postbiotics.

13.2 HISTORY OF POSTBIOTICS

Over 2,000 years ago, Hippocrates said that all the diseases begin in the gut. Modern science also agrees with this statement that gut health is very important for causing or preventing many diseases. A Russian microbiologist, Elie Metchnikoff, started work on gut health in the 1900s. He analyzed that people living in Bulgaria unusually live for more than 100 years. He demonstrated that toxins produced due to harmful biota are a major cause of aging. He also described that eating healthy and fermented food improves gut health, stops growth of harmful microbes, and may prevent aging. He was titled as the "founder of probiotics." This research opened new doors for researchers and lead to discovery of postbiotics. This leads to uncovering the health benefits of all the probiotics, postbiotics, and metabolites.

DOI: 10.1201/9781003607779-13

13.3 CLASSIFICATION OF POSTBIOTICS

Postbiotics can be classified on the basis of factors such as microorganism type, structure, and functions. Extracellular and intracellular probiotic bacteria are also used to produce many of the postbiotic compounds. For instance, muropeptides can be obtained from peptidoglycans, teichoic acids, fimbriae, pili or flagella, secreted proteins/peptides, bacteriocins, neurotransmitters, and biosurfactants [5]. They are classified into different categories like peptidoglycans, bacteriocins, enzymes, cell-free supernatants, vitamins, inactivated and dead probiotics, teichoic acids, short-chain fatty acids, and exopolysaccharides on the basis of their different chemical, functional, and physical characteristics.

13.3.1 Dead and Inactivated Probiotics

Dead or inactivated probiotics are synthesized by various methods like tyndallization, gamma or UV radiation, chemical treatment, and sonication, but they are most commonly prepared by heat. This inactivation results in cellular changes and functions. Experiments have proved that traits such as reducing inflammatory indicators and the ability to scavenge oxygen radicals are retained by nonviable products from their viable counterparts [6]. Caco 2 cell cultures adhered with live and heat-killed strains of *Lactobacillus reuteri* prevented attachment of *Enterococcus faecalis*, *E. coli*, *Listeria monocytogenes*, and *Salmonella typhi* to cell culture [7].

13.3.2 Fragments of Cell Wall

The components of bacterial cell wall like lipoteichoic acid and teichoic acid, comprising 60% of cell mass, can be used as immunogenic components [8]. Lipoteichoic acid and teichoic acid having distinct structures are characterized for specific functions in the body. It has been studied that lipoteichoic acid from the K88, K8, K55-5, and K5-5 strains of *Lactobacillus plantarum* having different structure result in different immunological responses in immune cells [9]. Teichoic acid and lipoteichoic acid are beneficial because of their anticancer, immunomodulatory, and antioxidant activities. Teichoic acid is responsible for development of antibiotic resistance.

13.3.3 Exopolysaccharides

A large number of homopolysaccharides and heteropolysaccharides such as glucans and uronic acid are derived from lactobacilli [10]. These are exopolysaccharides that adhere to bacterial surface as a slime layer or capsule. These structures perform beneficial roles for bacteria like prevention of phagocytosis and lipid metabolism. In *Lactobacillus paraplantarum* these structures enhance anti-inflammatory activity. In *Paenibacillus polymyxa* and *Lactobacillus fermentum* cultures, these macromolecules show antioxidant property and treat various disorders like rheumatoid arthritis and atherosclerosis. *S. aureus* and *E. coli* isolated EPSs play a key role in suppressing tumor growth and inflammation. To isolate EPSs, first centrifuge them; this also includes filtration, acid protein removal, dialysis, cold ethanol precipitation, and lyophilization. They exhibit antitumor, immunomodulatory, antioxidant, antimutagenicity, antibacterial, antiviral, cholesterol-lowering, antihypertensive, antigastrointestinal, and anti-inflammatory properties [11]. EPS isolated from *Lactococcus lactis* have shown antioxidant efficacy by enhancing the activities of antioxidant enzymes such as glutathione peroxidase, catalase, and superoxide dismutase and decreasing the peroxidation of lipid in serum and mice livers [12]. They are used as water-binding agents, stabilizers, and emulsifiers.

13.3.4 Enzymes

Enzymes catalyze biological reactions and are divided into six primary groups on the basis of their functions. Transferases, oxidoreductases, lyases, hydrolases, and isomerases are some common enzymes [13, 14]. Some strains of bacteria and fungi such as *Bacillus licheniformis* and *B. subtilis*, *Aspergillus niger*, *Bacillus licheniformis*, and *A. oryzae* are also source of enzymes. *Lactobacillus fermentum* produced glutathione peroxidase possesses antioxidant capabilities. Some stable proteolytic enzymes are also produced from *Bacillus* species under the influence of unfavourable pH, temperature, oxidizing agents, detergents, and organic solvents. Catalase isolated from genetically mutated strain of *Lactobacillus lactis* can shield mice against chemically induced colon cancer [15].

13.3.5 Short-Chain Fatty Acids (SCFAs)

Polysaccharide metabolizing gut bacteria like Bacteroides and Firmicutes usually produce a significant class of chemicals called SCFAs [16]. SCFAs like propionate, butyrate, and acetate are produced by fermentation of prebiotics like inulin and fructooligosaccharides. They reside in the feces and colon, facilitating intestinal epithelial renewal [17, 18]. They promote the blood flow in the colon and proliferation of epithelial cells [19]. SCFAs lowered the risk of colorectal diseases. Acetate helps in cholesterol regulation. SCFAs like propionate and butyrate play a key role in gluconeogenesis. This energy is used by colonocytes and epithelial cells to increase apoptosis of colon cancer cells.

13.3.6 Bacteriocins

These are basically ribosomal synthesized peptides or proteins obtained from lactic acid bacteria, eubacteria, and archaebacteria. Bacteriocins have ability to kill or inhibit bacterial growth. They do so by creating small pores in their cell membranes, preventing the proper formation of cell walls, and inhibiting enzymes. They are widely used in tumor therapy, medicine, cosmetics, food, and veterinary medicine.

13.3.7 Vitamins

Vitamins are involved in numerous body functions such as DNA replication, repair, and methylation, brain function, blood clotting, and bone health. They can be given exogenously. Vitamins have wide range of sources. Vitamins like MK-8, MK-9, and MK-6 are synthesized by *Bacteroides fragilis*, while *Eubacterium lentum*, *Lactococcus lactis*, and *Lactococcus lactiscremoris* are sources of MK-10, MK-11, and MK-12 [20]. Gut microbiota are also responsible for production of polyphenols.

Lactic acid and *Bifidobacterium* also give rise to a large number of vitamins like nicotinic acid, niacin, thymine, folate, cobalamin, riboflavin, and pyridoxine. Vitamins are also rich in fermented foods like yogurt, cheese, and milk [21]. The gut microbiome is also able to synthesize B_2, B_{12}, B_6, vitamin K, and B_9. Water-soluble vitamin B_{12} is important for hematopoiesis and neuronal health. Riboflavin works as a hydrogen carrier for oxidation–reduction reactions. Vitamin K plays a crucial role in blood clotting by working as a cofactor of gamma carboxylase.

13.3.8 Extracellular Vesicles

Extracellular vesicles are spherical, lipid bilayer, membrane-bound particles playing crucial roles in the body. They contain DNA, proteins, RNA, polysaccharides, glycolipids, toxins, and enzymes. They have a wide range of functions in the body, which are described as follows.

- They are responsible for horizontal genetic transfer.
- They maintain intestinal homeostasis by modulating permeability of the gut barrier and signaling pathways, facilitate communication between the gut and brain and improve lipid profile in blood [22, 23].
- They play a crucial role in regulating pathogenesis and immunomodulation in body.
- They cross mucosal barrier and lower the risk of sepsis. *Akkermansia muciniphila* synthesized extracellular vesicles play a crucial role in reducing fat accumulation and affect adipocytes, thus resulting in reduced chances of obesity, leading to less risk of endotoxemia.
- EVs obtained from *Propionibacterium freudenreichii* can alleviate inflammation [24–26]. These vesicles possess antibacterial and antibiofilm effects.
- *Bacillus licheniformis* produced extracellular vesicles exhibit anticancer effects against breast and lung cancer.

13.4 METABOLIC ENDOTOXEMIA

Metabolic disorders cover a wide range of conditions like osteoarthritis, diabetes mellitus, and obesity. These disorders are usually manifested by insulin resistance, inflammation, hyperglycemia, and oxidative stress [27, 28]. In addition, disruption of microbial balance in the gut also leads to metabolic disorders [29, 30]. So, the regulation of gut microbiota is necessary for preventing metabolic diseases [31, 32]. Some bioactive molecules present in gut microbiota are continuously secreted in the lumen of the bowel, while some are entered into systemic circulation by crossing the intestinal barrier. The intestinal barrier is an epithelial monolayer barrier found between circulation and the environment of the gut. It normally allows the entry of water, nutrients, and healthy bioactive particles and resists entry of harmful substances from gut into circulation. There are several factors that cause loss of intestinal integrity, leading to entry of harmful substances (endotoxins) in circulation. Lipopolysaccharide (LPS) is one of the most common endotoxins leading to metabolic endotoxemia and is produced in gut microbiota as it is a component in the outer membrane of Gram-negative bacteria [33]. It is usually produced by gram negative bacteria upon lysis.

Usually, LPS is not able to cross the intestinal barrier in normal conditions. But the intestinal barrier loses its integrity and allows LPS to enter blood circulation in the presence of soybean lecithin, sodium taurocholate, cholesterol, 2-oleoylglycerol, and oleic acid, which are constituents of a high fat diet [34]. Intake of a high fat diet may cause high LPS, resulting in metabolic endotoxemia in the following ways.

- It may lead to increase in LPS-producing bacteria like Firmicutes and Proteobacteria.
- It may weaken the junctions in gut wall, which lead to high LPS levels in the bloodstream.
- It may weaken the intestinal barrier by inhibiting the production of proteins like occludin and ZO-1, which are necessary for sealing gaps between gut cells [35].

This results in high levels of LPS in the bloodstream. LPS usually consists of lipid A, repeating oligosaccharide units, and a core oligosaccharide [36]. Lipid A is the toxic portion of LPS, which triggers inflammatory and immune responses in the body.

LPS has a short half-life in systemic circulation, so it combines with LPS binding protein (LBP) [37], resulting in LPS-LBP complex to enhance its half-life [38]. This complex binds with the CD14 expressed by immune cells, mainly neutrophils and macrophages. TLR-4 receptor facilitates CD14 in immune cells to initiate downstream signaling, triggering immune and inflammatory responses like NF-κ B in the body [39]. These immune and inflammatory responses play crucial roles in the pathogenesis of various chronic metabolic disorders like diabetes mellitus and obesity [40]. Type 2 diabetes is characterized by high LPS levels, less SCFAs, insulin resistance, and sensitivity [41, 42].

13.5 TYPE 2 DIABETES

Diabetes mellitus is a long-term group of diseases that are very common all over the world. It is characterized by high blood glucose level [43]. Following are two kinds of diabetes mellitus:

- Type 1 diabetes (T1D): This is an autoimmune disorder in which immune cells attack and damage insulin producing cells in pancreas.
- Type 2 diabetes (T2D): Abnormal responses of normal body cells to insulin or decrease production of insulin are two possible abnormalities in this disease [44].

An analysis in last decade has demonstrated that gut microbiota, intestinal barrier, and T2D are closely related. According to this analysis, T2D patients experience imbalanced gut microbiota. These imbalanced gut microbiota lead to insulin resistance and disturbance in glucose metabolism. This imbalance results in a decrease in butyrate producing bacteria, which is a potential cause of insulin resistance.

Various factors like dysbiosis or lysis of Gram-negative bacteria are leading cause of changing permeability of intestinal barrier which results in high levels of toxins (LPS) in blood. These toxins cause systemic inflammation by producing inflammatory mediators like IL-6, TNF-γ, IL-1β, and TNF-α. This inflammation causes phosphorylation of insulin receptor substrate (IRS) [45]. This results in disturbance in the normal working of insulin in the body. In this condition, the body becomes less responsive to insulin (insulin resistance). Insulin resistance affects the production of insulin by damaging beta-cells in the pancreas and decreasing the activity of the PDX1 gene, which is essential for normal working of beta cells.

13.6 ROLE OF POSTBIOTICS IN METABOLIC ENDOTOXEMIA AND TYPE 2 DIABETES

Metabolic endotoxemia and T2D are caused due to high blood LPS levels, which might be due to weak intestinal barrier or changes in permeability of the intestinal barrier. So, a healthy gut is necessary for prevention of metabolic endotoxemia and T2D. Researchers are exploring the beneficial role of synbiotics and postbiotics in preventing metabolic disorders by regulating the balance of microbiota in the gut [46–48]. Postbiotics are used in different ways to regulate the integrity of intestinal barrier.

13.6.1 Short-Chain Fatty Acids

Short-chain fatty acids are the molecules produced from bacteria when they undergo breakdown of fiber [49]. They are the energy-regulating molecules in the human body. They play an essential role in preventing T2D in different ways, which are described as follows.

- They enhance the production of insulin and reduce the release of glucagon in blood by attaching with L cells in the intestine with the help of free fatty acid receptor (FFAR2 or FFAR3). This attachment triggers the release of glucagon-like peptide-1 (GLP-1) and peptide YY (PYY), which facilitates insulin production [50–53].
- Short-chain fatty acids play a key role in strengthening the intestinal barrier by triggering claudin-1, which leads to production of occludin and ZO-1. SCFA triggers the claudin-1 producing gene by interacting with SP1 transcription factor [54].
- SCFA also plays role in the production of MUC-2 by activating the MUC-2 promoter region. This leads to the production of thick mucus, which enhances the integrity of the intestinal barrier [55].
- SCFA plays a key role in reducing LPS-produced inflammation. SCFAs like butyrate have an anti-inflammatory role in the body. They enhance the production of regulatory T-cells, which leads to reduced inflammation [56–59]. In the same way, SCFAs like propionate play a key role in anti-inflammation by triggering the production of dendritic cells and macrophages [60].

13.6.2 Glucose Tolerance and Insulin Resistance

High LPS levels in blood promote the insulin resistance and inflammation leading to T2D. Muramyl dipeptide (MDP) is a bacteria-produced postbiotic thaT plays an important role in reducing inflammation and improving glucose tolerance in T2D. It plays a crucial role in metabolic endotoxemia and T2D in the following ways:

- It reduces harmful effects of LPS and improves glucose tolerance.
- It reduces insulin resistance during endotoxemia through activation of NOD2 receptors.
- It has insulin-sensitizing and anti-inflammatory effects that are facilitated by transcription factor IRF4.

13.6.3 Regulating the Composition of Gut Microbiota

The composition of gut microbiota is very crucial for preventing metabolic diseases. It does so by increasing SCFA producing bacteria and strengthening intestinal barrier [35, 61]. Butyrate is produced by bacteria in the gut when they break down complex dietary fiber. It plays a preventive role in metabolic endotoxemia and T2D in the following ways:

- It upregulates the growth of beneficial bacteria like *Lachnospiraceae* and *Proteobacteria* while it downregulates the growth of many harmful bacteria like *Clostridiaceae* [62].
- It plays a crucial role in decreasing inflammation due to high blood LPS levels which is caused by Gram-negative bacteria like *Citrobacter rodentium*.
- Butyrate administration leads to an increase in butyrate producing bacteria like *Blautia* [63].

In addition, there are many potential postbiotic substances and polysaccharides that play an important role in maintaining health by downregulating the blood LPS level by reducing the growth of many Gram-negative bacteria like *Yersinia*.

13.6.4 Regulation of Microbial Metabolites in Gut

Different microbial metabolites present in the gut are playing necessary role in maintaining gut health. Various polysaccharides have significance in maintaining microbial metabolites in the gut. Some examples of postbiotics having importance in maintaining gut health are described as follows.

- *Lactobacillus plantarum* produced exopolysaccharide has an ability to control the lipid profile in blood by inhibiting the differentiation of immature cells into mature adipocytes by triggering the TLR2-AMPK pathway [64].
- *Lactobacillus brevis* produced long-chain polyphosphate plays a crucial role in upregulating extracellular regulated protein kinase pathways, which helps in improving intestinal barrier function [65].
- Potential postbiotic substances and polysaccharides lead to the synthesis of beneficial metabolites like butyrate, which results in low LPS levels. In addition, it plays a necessary role in suppressing various inflammatory responses like the NF-κB and TNF-α pathway [66].

13.6.5 Regulating Integrity of Intestinal Barrier

Imbalance in the gut microbiota (dysbiosis) leads to damage of the intestinal barrier. This damage results in the entry of harmful bacteria and their metabolites into circulation. These harmful substances lead to systemic inflammation resulting in insulin resistance. So there is a normal pathway in the body to regulate the integrity of the intestinal barrier to prevent insulin resistance. Postbiotics have the potential to prevent metabolic syndrome by strengthening the gut barrier [67].

Harmful microbes in the gut are usually recognized by certain protein receptors named nucleotide-binding oligomerization domain-like receptors (NLRs) [68]. Overactivation of these receptors may lead to damage of the intestinal barrier. But in contrast, laboratory studies have demonstrated that an increase in number of specific NLR protein (NLRC3) plays a critical role in improving the permeability of gut barrier. This increase of NLRC3 is crucial for boosting the activity of TRAF6. TRAF6 leads to the increased production of proteins like occludin and ZO-1 which are necessary for sealing gaps between gut cells, resulting a strong intestinal barrier [69].

- Butyrate is an SCFA that helps in maintaining the strength of the intestinal barrier. It enhances the activity of NLRC3 by activating GPR43 receptors in gut cells. This results in more production of occludin and ZO-1 proteins and strengthens the intestinal barrier [63] (Figure 13.1).

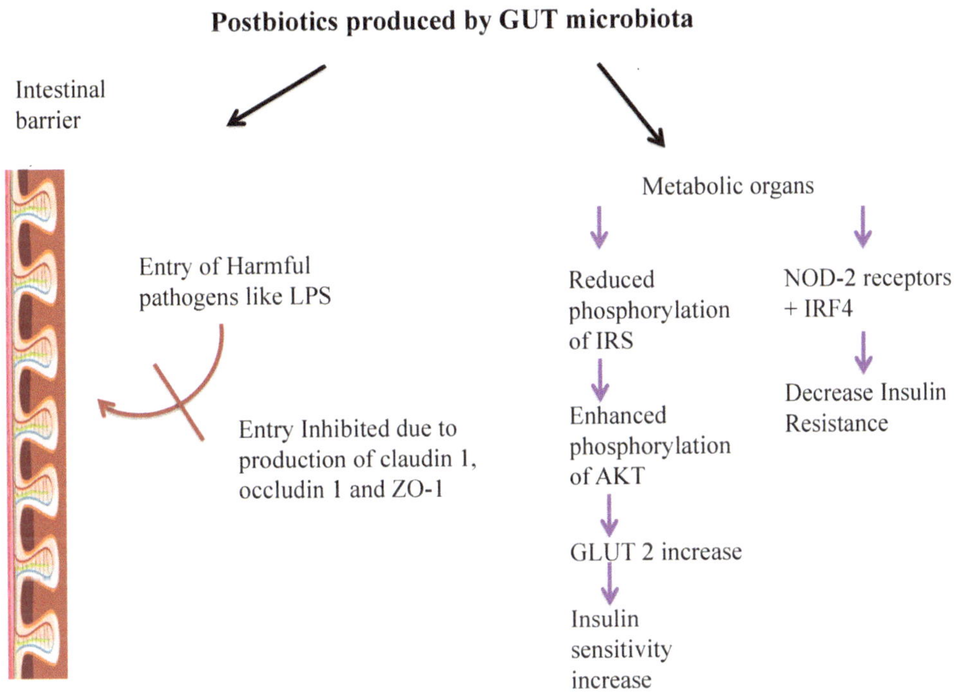

FIGURE 13.1 Role of postbiotics in preventing entry of pathogens and T2D. GLUT2, glucose transporter 2; IRS1, insulin receptor substrate 1; Akt, protein kinase B; NOD2, oligomerization domain-containing protein 2; IRF4, interferon regulatory factor 4; ZO 1, zonula occludens-1 [70]

13.7 CONCLUSION

Postbiotics are nonviable substances produced in the gut as metabolites of probiotics and prebiotics. They play a critical role in maintaining strength of the intestinal barrier by increasing production of claudin-1, occluding-1, and ZO-1. This prevents from entry of harmful pathogens in circulation. They also reduce insulin resistance and prevent from T2D.

REFERENCES

1. Arshad, F., et al., Lactobacilli as probiotics and their isolation from different sources. *British Journal of Research*, 2018. **5**(3): p. 43. https://doi.org/10.21767/2394-3718.100043
2. Aguilar-Toalá, J., et al., Postbiotics: An evolving term within the functional foods field. *Trends in Food Science & Technology*, 2018. **75**: p. 105–114. https://doi.org/10.1016/j.tifs.2018.03.009
3. Amores-Arrocha, A., et al., Evaluation of the use of multiflora bee pollen on the volatile compounds and sensorial profile of Palomino fino and Riesling white young wines. *Food Research International*, 2018. **105**: p. 197–209. https://doi.org/10.1016/j.foodres.2017.11.013
4. Barros, C.P., et al., Ohmic heating as a method of obtaining paraprobiotics: Impacts on cell structure and viability by flow cytometry. *Food Research International*, 2021. **140**: p. 110061. https://doi.org/10.1016/j.foodres.2020.110061
5. Nataraj, B.H., et al., Postbiotics-parabiotics: The new horizons in microbial biotherapy and functional foods. *Microbial Cell Factories*, 2020. **19**: p. 1–22. https://doi.org/10.1186/s12934-020-01426-w
6. Sugahara, M., T. Tanaka, and M. Nangaku, Prolyl hydroxylase domain inhibitors as a novel therapeutic approach against anemia in chronic kidney disease. *Kidney International*, 2017. **92**(2): p. 306–312. https://doi.org/10.1016/j.kint.2017.02.035
7. Singh, T.P., et al., Antagonistic activity of *Lactobacillus reuteri* strains on the adhesion characteristics of selected pathogens. *Frontiers in Microbiology*, 2017. **8**: p. 486. https://doi.org/10.3389/fmicb.2017.00486

8. van der Es, D., et al., Teichoic acids: Synthesis and applications. *Chemical Society Reviews*, 2017. **46**(5): p. 1464–1482. https://doi.org/10.1039/C6CS00270F
9. Jung, B.-J., H. Kim, and D.-K. Chung, Differential immunostimulatory effects of lipoteichoic acids isolated from four strains of *Lactiplantibacillus plantarum*. *Applied Sciences*, 2022. **12**(3): p. 954. https://doi.org/10.3390/app12030954
10. Caggianiello, G., M. Kleerebezem, and G. Spano, Exopolysaccharides produced by lactic acid bacteria: From health-promoting benefits to stress tolerance mechanisms. *Applied Microbiology and Biotechnology*, 2016. **100**: p. 3877–3886. https://doi.org/10.1007/s00253-016-7471-2
11. Gezginç, Y., et al., Health promoting benefits of postbiotics produced by lactic acid bacteria: Exopolysaccharide. *Biotech Studies*, 2022. **31**(2): p. 61–70. https://doi.org/10.38042/biotechstudies.1159166
12. Gao, Q., et al., SLC27A5 deficiency activates NRF2/TXNRD1 pathway by increased lipid peroxidation in HCC. *Cell Death & Differentiation*, 2020. **27**(3): p. 1086–1104. https://doi.org/10.1038/s41418-019-0399-1
13. Singhania, R.R., et al., Genetic modification: A tool for enhancing beta-glucosidase production for biofuel application. *Bioresource Technology*, 2017. **245**: p. 1352–1361. https://doi.org/10.1016/j.biortech.2017.05.126
14. Kumar, B., V. Prakash Gupta, and V. Kumar, A perspective on monoamine oxidase enzyme as drug target: Challenges and opportunities. *Current Drug Targets*, 2017. **18**(1): p. 87–97.
15. Contesini, F.J., R.R. de Melo, and H.H. Sato, An overview of *Bacillus* proteases: From production to application. *Critical Reviews in Biotechnology*, 2018. **38**(3): p. 321–334. https://doi.org/10.1080/07388551.2017.1354354
16. Kim, M., et al., Microbial metabolites, short-chain fatty acids, restrain tissue bacterial load, chronic inflammation, and associated cancer in the colon of mice. *European Journal of Immunology*, 2018. **48**(7): p. 1235–1247. https://doi.org/10.1002/eji.201747122
17. Żółkiewicz, J., et al., Postbiotics: A step beyond pre- and probiotics. *Nutrients*, 2020. **12**(8): p. 2189.
18. Takagi, R., et al., A single-batch fermentation system to simulate human colonic microbiota for high-throughput evaluation of prebiotics. *PLoS One*, 2016. **11**(8): p. e0160533. https://doi.org/10.1371/journal.pone.0160533
19. Markowiak-Kopeć, P., and K. Śliżewska, The effect of probiotics on the production of short-chain fatty acids by human intestinal microbiome. *Nutrients*, 2020. **12**(4): p. 1107. https://doi.org/10.3390/nu12041107
20. Camelo-Castillo, A., et al., Gut microbiota and the quality of oral anticoagulation in vitamin K antagonists users: A review of potential implications. *Journal of Clinical Medicine*, 2021. **10**(4): p. 715. https://doi.org/10.3390/jcm10040715
21. LeBlanc, J.G., et al., Bacteria as vitamin suppliers to their host: A gut microbiota perspective. *Current Opinion in Biotechnology*, 2013. **24**(2): p. 160–168. https://doi.org/10.1016/j.copbio.2012.08.005
22. Ahmadi Badi, S., et al., Microbiota-derived extracellular vesicles as new systemic regulators. *Frontiers in Microbiology*, 2017. **8**: p. 1610. https://doi.org/10.3389/fmicb.2017.01610
23. Chelakkot, C., et al., *Akkermansia muciniphila*-derived extracellular vesicles influence gut permeability through the regulation of tight junctions. *Experimental & Molecular Medicine*, 2018. **50**(2): p. e450–e450. https://doi.org/10.1038/emm.2017.282
24. Ashrafian, F., et al., Extracellular vesicles and pasteurized cells derived from Akkermansia muciniphila protect against high-fat induced obesity in mice. *Microbial Cell Factories*, 2021. **20**: p. 1–17. https://doi.org/10.1186/s12934-021-01709-w
25. Gurunathan, S., et al., Antibacterial and antibiofilm effects of Pseudomonas aeruginosa derived outer membrane vesicles against Streptococcus mutans. *Heliyon*, 2023. **9**(12).
26. Keshavarz Azizi Raftar, S., et al., The protective effects of live and pasteurized Akkermansia muciniphila and its extracellular vesicles against HFD/CCl4-induced liver injury. *Microbiology Spectrum*, 2021. **9**(2): p. e00484–21. https://doi.org/10.1128/Spectrum.00484-21
27. Gowd, V., et al., Dietary polyphenols to combat the metabolic diseases via altering gut microbiota. *Trends in Food Science & Technology*, 2019. **93**: p. 81–93. https://doi.org/10.1016/j.tifs.2019.09.005
28. Li, H.-Y., et al., Plant-based foods and their bioactive compounds on fatty liver disease: Effects, mechanisms, and clinical application. *Oxidative Medicine and Cellular Longevity*, 2021. **2021**(1): p. 6621644. https://doi.org/10.1155/2021/6621644

29. Zhang, Y.-J., et al., Impacts of gut bacteria on human health and diseases. *International Journal of Molecular Sciences*, 2015. **16**(4): p. 7493–7519. https://doi.org/10.3390/ijms16047493
30. Stewart, C.J., et al., Temporal bacterial and metabolic development of the preterm gut reveals specific signatures in health and disease. *Microbiome*, 2016. **4**: p. 1–10. https://doi.org/10.1186/s40168-016-0216-8
31. Zou, J., et al., Fiber-mediated nourishment of gut microbiota protects against diet-induced obesity by restoring IL-22-mediated colonic health. *Cell Host & Microbe*, 2018. **23**(1): p. 41–53.
32. Régnier, M., et al., Gut microbiome, endocrine control of gut barrier function and metabolic diseases. *Journal of Endocrinology*, 2021. **248**(2): p. R67–R82.
33. Fuke, N., et al., Regulation of gut microbiota and metabolic endotoxemia with dietary factors. *Nutrients*, 2019. **11**(10): p. 2277. https://doi.org/10.3390/nu11102277
34. Laugerette, F., et al., Emulsified lipids increase endotoxemia: possible role in early postprandial low-grade inflammation. *The Journal of Nutritional Biochemistry*, 2011. **22**(1): p. 53–59. https://doi.org/10.1016/j.jnutbio.2009.11.011
35. Cani, P.D., et al., Changes in gut microbiota control metabolic endotoxemia-induced inflammation in high-fat diet–induced obesity and diabetes in mice. *Diabetes*, 2008. **57**(6): p. 1470–1481. https://doi.org/10.2337/db07-1403
36. Raetz, C.R. and C. Whitfield, Lipopolysaccharide endotoxins. *Annual Review of Biochemistry*, 2002. **71**(1): p. 635–700. https://doi.org/10.1146/annurev.biochem.71.110601.135414
37. Vreugdenhil, A.C., et al., Lipopolysaccharide (LPS)-binding protein mediates LPS detoxification by chylomicrons. *The Journal of Immunology*, 2003. **170**(3): p. 1399–1405. https://doi.org/10.4049/jimmunol.170.3.1399
38. Gomes, J.M.G., J. de Assis Costa, and R.D.C.G. Alfenas, Metabolic endotoxemia and diabetes mellitus: A systematic review. *Metabolism*, 2017. **68**: p. 133–144. https://doi.org/10.1016/j.metabol.2016.12.009
39. Liu, X., et al., Lipopolysaccharide binding protein, obesity status and incidence of metabolic syndrome: a prospective study among middle-aged and older Chinese. *Diabetologia*, 2014. **57**: p. 1834–1841. https://doi.org/10.1007/s00125-014-3288-7
40. Liang, H., et al., Effect of lipopolysaccharide on inflammation and insulin action in human muscle. *PloS One*, 2013. **8**(5): p. e63983. https://doi.org/10.1371/journal.pone.0063983
41. Zimmet, P., et al., Epidemic T2DM, early development and epigenetics: Implications of the Chinese famine. *Nature Reviews Endocrinology*, 2018. **14**(12): p. 738–746. https://doi.org/10.1038/s41574-018-0106-1
42. Qin, J., et al., A metagenome-wide association study of gut microbiota in type 2 diabetes. *Nature*, 2012. **490**(7418): p. 55–60. https://doi.org/10.1038/nature11450
43. Zhang, L., et al., Gut microbiota and type 2 diabetes mellitus: Association, mechanism, and translational applications. *Mediators of Inflammation*, 2021. **2021**(1): p. 5110276. https://doi.org/10.1155/2021/5110276
44. Zhao, J., et al., A potential synbiotic strategy for the prevention of type 2 diabetes: *Lactobacillus paracasei* JY062 and exopolysaccharide isolated from *Lactobacillus plantarum* JY039. *Nutrients*, 2022. **14**(2): p. 377. https://doi.org/10.3390/nu14020377
45. Creely, S.J., et al., Lipopolysaccharide activates an innate immune system response in human adipose tissue in obesity and type 2 diabetes. *American Journal of Physiology-Endocrinology and Metabolism*, 2007. **292**(3): p. E740–E747. https://doi.org/10.1152/ajpendo.00302.2006
46. Vallianou, N., et al., Probiotics, prebiotics, synbiotics, postbiotics, and obesity: Current evidence, controversies, and perspectives. *Current Obesity Reports*, 2020. **9**: p. 179–192. https://doi.org/10.1007/s13679-020-00379-w
47. Wu, J., et al., The role of the gut microbiome and its metabolites in metabolic diseases. *Protein & Cell*, 2021. **12**(5): p. 360–373. https://doi.org/10.1007/s13238-020-00814-7
48. Salminen, S., et al., The International Scientific Association of Probiotics and Prebiotics (ISAPP) consensus statement on the definition and scope of postbiotics. *Nature Reviews Gastroenterology & Hepatology*, 2021. **18**(9): p. 649–667. https://doi.org/10.1038/s41575-021-00440-6
49. Cummings, J.H., et al., Short chain fatty acids in human large intestine, portal, hepatic and venous blood. *Gut*, 1987. **28**(10): p. 1221–1227. https://doi.org/10.1136/gut.28.10.1221
50. Wang, Y., et al., Fourteen composite probiotics alleviate type 2 diabetes through modulating gut microbiota and modifying M1/M2 phenotype macrophage in db/db mice. *Pharmacological Research*, 2020. **161**: p. 105150. https://doi.org/10.1016/j.phrs.2020.105150

51. Tolhurst, G., et al., Short-chain fatty acids stimulate glucagon-like peptide-1 secretion via the G-protein–coupled receptor FFAR2. *Diabetes*, 2012. **61**(2): p. 364–371. https://doi.org/10.2337/db11-1019
52. Lin, H.V., et al., Butyrate and propionate protect against diet-induced obesity and regulate gut hormones via free fatty acid receptor 3-independent mechanisms. *PloS One*, 2012. **7**(4): p. e35240. https://doi.org/10.1371/journal.pone.0035240
53. Psichas, A., et al., The short chain fatty acid propionate stimulates GLP-1 and PYY secretion via free fatty acid receptor 2 in rodents. *International Journal of Obesity*, 2015. **39**(3): p. 424–429. https://doi.org/10.1038/ijo.2014.153
54. Wang, H.-B., et al., Butyrate enhances intestinal epithelial barrier function via up-regulation of tight junction protein Claudin-1 transcription. *Digestive Diseases and Sciences*, 2012. **57**: p. 3126–3135. https://doi.org/10.1007/s10620-012-2259-4
55. Burger-van Paassen, N., et al., The regulation of intestinal mucin MUC2 expression by short-chain fatty acids: Implications for epithelial protection. *Biochemical Journal*, 2009. **420**(2): p. 211–219. https://doi.org/10.1042/BJ20082222
56. Takahashi, D., et al., Microbiota-derived butyrate limits the autoimmune response by promoting the differentiation of follicular regulatory T cells. *EBioMedicine*, 2020. 58.
57. Arpaia, N., et al., Metabolites produced by commensal bacteria promote peripheral regulatory T-cell generation. *Nature*, 2013. **504**(7480): p. 451–455. https://doi.org/10.1038/nature12726
58. Smith, P.M., et al., The microbial metabolites, short-chain fatty acids, regulate colonic Treg cell homeostasis. *Science*, 2013. **341**(6145): p. 569–573. https://doi.org/10.1126/science.1241165
59. Furusawa, Y., et al., Commensal microbe-derived butyrate induces the differentiation of colonic regulatory T cells. *Nature*, 2013. **504**(7480): p. 446–450. https://doi.org/10.1038/nature12721
60. Trompette, A., et al., Gut microbiota metabolism of dietary fiber influences allergic airway disease and hematopoiesis. *Nature Medicine*, 2014. **20**(2): p. 159–166. https://doi.org/10.1038/nm.3444
61. Sharma, S., and P. Tripathi, Gut microbiome and type 2 diabetes: Where we are and where to go? *The Journal of Nutritional Biochemistry*, 2019. **63**: p. 101–108. https://doi.org/10.1016/j.jnutbio.2018.10.003
62. Jiminez, J.A., et al., Butyrate supplementation at high concentrations alters enteric bacterial communities and reduces intestinal inflammation in mice infected with Citrobacter rodentium. *MSphere*, 2017. **2**(4). https://doi.org/10.1128/msphere.00243-17
63. Zhou, D., et al., Sodium butyrate attenuates high-fat diet-induced steatohepatitis in mice by improving gut microbiota and gastrointestinal barrier. *World Journal of Gastroenterology*, 2017. **23**(1): p. 60. https://doi.org/10.3748/wjg.v23.i1.60
64. Thiennimitr, P., et al., *Lactobacillus paracasei* HII01, xylooligosaccharides, and synbiotics reduce gut disturbance in obese rats. *Nutrition*, 2018. **54**: p. 40–47. https://doi.org/10.1016/j.nut.2018.03.005
65. Isozaki, S., et al., Probiotic-derived polyphosphate accelerates intestinal epithelia wound healing through inducing platelet-derived mediators. *Mediators of Inflammation*, 2021. **2021**(1): p. 5582943. https://doi.org/10.1155/2021/5582943
66. Li, H.-Y., et al., Effects and mechanisms of probiotics, prebiotics, synbiotics, and postbiotics on metabolic diseases targeting gut microbiota: A narrative review. *Nutrients*, 2021. **13**(9): p. 3211. https://doi.org/10.3390/nu13093211
67. Liu, W., et al., A bridge for short-chain fatty acids to affect inflammatory bowel disease, type 1 diabetes, and non-alcoholic fatty liver disease positively: By changing gut barrier. *European Journal of Nutrition*, 2021. **60**: p. 2317–2330. https://doi.org/10.1007/s00394-020-02431-w
68. Cheng, D., et al., Butyrate ameliorated-NLRC3 protects the intestinal barrier in a GPR43-dependent manner. *Experimental Cell Research*, 2018. **368**(1): p. 101–110. https://doi.org/10.1016/j.yexcr.2018.04.018
69. Mei, X., et al., Insulin sensitivity-enhancing activity of phlorizin is associated with lipopolysaccharide decrease and gut microbiota changes in obese and type 2 diabetes (db/db) mice. *Journal of Agricultural and Food Chemistry*, 2016. **64**(40): p. 7502–7511. https://doi.org/10.1021/acs.jafc.6b03474
70. Gurunathan, S., P. Thangaraj, and J.-H. Kim, Postbiotics: Functional food materials and therapeutic agents for cancer, diabetes, and inflammatory diseases. *Foods*, 2023. **13**(1): p. 89. https://doi.org/10.3390/foods13010089

14 Postbiotics

In Nonalcoholic Fatty Liver Disease

Susmita Chakrabarty and Bangmayee Dash
School of Paramedical and Allied Sciences, DRIEMS University, Odisha, India

Krishnendu Adhikary
Department of Interdisciplinary Science, Centurion University of Technology and Management, Odisha, India

Bhabani Shankar Nayak
KIIT School of Pharmacy, KIIT Deemed to be University, Bhubaneswar, Odisha, India

14.1 INTRODUCTION

During previous decades, expeditious growth in productivity and industrialization occurred, consequently leading to some of the worst challenges in the quality of ecosystems, like air and water pollution and heavy metal pollution as well as the mental stress associated with this hectic life schedule (Salminen, 2021). In the 21st century, with the advancement of time and more people advocating compressive health and functional foods for eliminating sub-healthy conditions, probiotics and postbiotics have gradually received the focus of scientific and nutrition communities. Microorganisms are known to modify an inoculated food's microbiological, physiochemical, and sensory properties, which are well documented. Among the most interesting organisms commonly used in industrial foods due to their health advantages are those secreting several substances called postbiotics, which ensure food safety (Bourebaba, 2022). Postbiotics are a type of bioactive compounds or nonviable bacterial products derived from probiotics. These are finding their way as promising agents for handling nonalcoholic fatty liver disease (NAFLD). This is one of the most common liver diseases in developed countries that is primarily based on key factors such as obesity, diabetes, insulin resistance, sedentary lifestyle, and Western diet and is becoming a serious global health burden as its prevalence has risen rapidly (Korinkova, 2020; Hass et al., 2016).

14.2 TYPES OF POSTBIOTICS AND THEIR MECHANISMS

- *Lactobacillus*-derived postbiotics: Postbiotics prepared from *Lactobacillus* species have proven to be helpful in combating NAFLD by maintaining the gut microbiota and enhancing the liver's metabolism. According to experiments, such postbiotics can diminish hepatic steatosis and inflammation alongside improving the sensitivity of insulin (Pan et al., 2022).
- Short-chain fatty acids (SCFAs): SCFAs, including butyrate, propionate, and acetate, are products of fermentation of dietary fiber by gut bacteria. SCFAs play a very crucial role in maintaining the gut's health and are believed to possess anti-inflammatory actions that could help protect from damage occurring at the liver due to NAFLD (Grander et al., 2023)

DOI: 10.1201/9781003607779-14

- Peptides and exopolysaccharides: Exopolysaccharides enhance intestinal barrier integrity, which reduces microbial translocation that leads to inflammation of the liver, while some of the peptides produced during fermentation possess antioxidant properties (Kaufmann et al., 2023).

14.3 PATHOPHYSIOLOGY OF NAFLD

NAFLD etiology is complex. There are several factors—environmental, genetic, and metabolic—that encourage hepatocyte fat formation. The theory that was best supported in the last ten years of the 20th century was the "two hit pathogenesis" (Marchisello,2019). NFLD is marked by an excessive accumulation of fat in the liver, which may lead to a more dangerous disease than simple steatosis: nonalcoholic steatohepatitis (NASH), fibrosis, and cirrhosis. NAFLD's pathophysiology is quite complex, involving multiple factors including dysregulation of metabolism, alterations in gut microbiota, and inflammatory processes.

14.4 NAFLD

NAFLD is worthy of attention and is a relevant human disorder related to altered ways of survival. This term is associated with a wide range of pathophysiology-related liver disorders linked to alcohol but becomes visible in those who intake little alcohol (Brunt, 2007). In 1980, Ludwig et al. used the term "nonalcoholic steatohepatitis" (NASH) to outline a condition of severely overweight females having adult-onset diabetes along with alcoholic hepatitis and without any consumption of alcohol. Nowadays it has become evident that nonalcoholic fatty liver is a diapason of disorders. This disease is indirectly connected with obesity and visceral fat deposition (Clark, 2002). Nevertheless, not all overweight people come into having NAFLD. Based on research data, moreover, 30% of the overweight males and 40% of overweight females had NAFLD and had lipodystrophy, a biochemical process of lack of adipose tissues in opposition to insulin resistance. It is also known as metabolic dysfunction-associated steatotic liver disease (MASLD), which is a major global chronic liver problem; 30% of the population suffers from this chronic disease. This chronic disease can lead to remarkable health problems such as cirrhosis, hepatocellular carcinoma (HCC), excess fluid in the abdomen, improper brain functions due to abnormalities in the bile acid metabolism, and other various biochemical processes that are associated with the liver, as it is the metabolism center for all biomolecules such as lipids, amino acids, protein, carbohydrates, and vitamins. Primarily nonalcoholic fatty liver disease is a multifaceted disease associated with cardiometabolic and environmental risk factors (Chen, 2021). NAFLD is an extensive term used to cover a large number of states that are distinguished by affirmation of steatotic liver disease on immunohistochemistry and absence of other causes of fatty liver such as intake of excess alcohol or persistent intake of various medicines that are responsible for diseases. The notable intake of alcohol has not been observed. For NAFLD clinical trials, it can be explained as the continuing or recent use of more than 14 standard drinks in a week in females, and in the case of males, it is about 21 in 7 days. It is most frequently recognized in passing on imaging or in other complicated conditions. The widespread presence of NAFLD in Western countries is approximately 20% to 30% (Gong, 2024). It is contemplated to be the liver incorporation of various metabolic syndromes. Apart from this, NAFLD is common in the case of sugar patients. Various development stages of this disease are simple steatosis, steatohepatitis, fibrosis, and cirrhosis, and ultimately could even progress to hepatocellular carcinoma. NASH constitutes liver injuries with inflammation with or without fibrosis. The major causes of NAFLD development are obesity, diabetes, dyslipidemias, insulin resistance, and metabolic syndrome. Another reason for the formation of NAFLD is exposure to arsenic and enhancement in the alanine aminotransferase enzyme, which is responsible for various heart diseases, cirrhosis, and hepatocellular carcinoma. Based on the epidemiology, 80% to 90% of adults with obesity, 30% to 50% with diabetes mellitus (DM), and 95% of people with high amounts of lipid (hyperlipidemia) suffer from this metabolic syndrome. In the case of NAFLD, the amount of fat tissue increases in the liver without any symptoms of inflammation. NAFLD is of two types: Type I deals with metabolic syndrome, and Type II is associated with human immunodeficiency virus (HIV) and hepatitis C. The biochemical mechanism of the infectious NAFLD is based on the "multiple-hit" hypothesis, which includes the deposition of triacylglycerol in the liver tissue in opposition to insulin, enhanced antioxidants, and abnormal formation of lipids in the body. NAFLD formation is associated with local and systemic inflammation and abnormalities in the adipose tissues. Factors affecting NAFLD are numerous extrahepatic occurrences, such as malnutrition, intake of unhealthy food, and abnormalities in the liver–gut axis. Intrahepatic factors such as fatty acid oxidation, endoplasmic reticulum stress, and abnormalities in mitochondria, Kupffer cells, and lymphocytes are associated with NAFLD and NASH (Rinella & Sookoian, 2024).

14.5 SYMPTOMS AND CAUSES

NAFLD is a hepatic issue that exert influence on people who consume zero alcohol. The primary reason for this disease is an excess accumulation of fat in the liver of those who are overweight or obese, swelling and harming it. Major symptoms are fatigue, excess pain in the upper belly area, malaise, skin irritation, difficulty breathing, increased size of the spleen, and blood vessels that look like spiders under the skin. Other factors that are indirectly associated with NAFLD are opposition resistance, excess cholesterol, excess triacylglycerol (TAG), high blood pressure, overnutrition, and a sedentary lifestyle with hepatic steatosis, which are linked with NAFLD but do not result in liver inflammation and fibrosis (Malnick, 2003). Major lifestyle-associated parameters are excess amounts of glucose, fructose, and single-bond-containing fatty acids, and infection in the adipose tissues. Other factors connected with it include ageing, and genetics might have a major role. Based on epigenetics and genome analysis reports, PNPLA3, TM6SF2, MBOAT7, GCKR, and HSD17B13 are responsible for the progression of this epidemic. Major functions of these genes are the formation of TAG from lipids, synthesis of very-low-density lipoprotein, reconstruction of hepatic phosphatidyl inositol acyl chains, synthesis of lipids, and so forth. The risk factors of NAFLD are shown in Figure 14.1 (Cuevas, 2020).

14.6 BIOCHEMICAL MECHANISM

Various biochemical processes involved in NAFLD are explained one by one. One process, autophagy, regulates intracellular recycling and energy homeostasis by destroying the different cellular compartments. The major signaling pathway in this biochemical stage is macroautophagy, based on

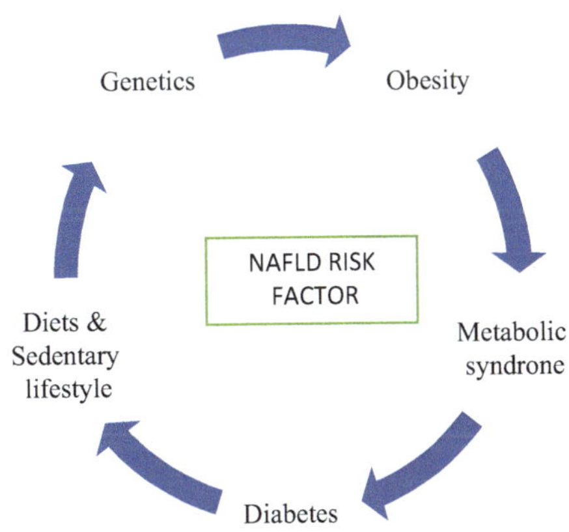

FIGURE 14.1 Various risk factors associated with NAFLD.

chaperone-mediated and microautophagy. Autophagosome vesicles are two-layered with large cytosolic areas that contain ATG7 protein. This protein is essential for the formation of autophagosomes, which are the major biochemical process of macroautophagy. The process is based on chaperone-mediated biochemical cell signaling pathways, which depend on the identification of various amino acids and polypeptide groups, by the Hsp 70 chaperone, which is essential for movements of protein molecules to the lysosome. Finally, in the case of microautophagy, membrane invagination occurs for intake of lysosomal engulfment. These are major biochemical processes for the regulation of lipid metabolism, excess deposition of fat in the liver, and primary factors for initiating NAFLD. The next mechanism is oxidative stress, which is associated with the transformation of steatosis to steatohepatitis due to the accumulation of high amounts of sustenance in the liver tissues coordinated with the electron transport chain, affecting the formation of free radicals and denaturation of mitochondrial protein, and enhancing the oxidative stress in NAFLD. The lipoxygenase cell-signaling pathway plays a vital role by connecting major mechanisms associated with NAFLD. By analyzing the lipidomics studies on mouse models having a high-fat diet, Hall et al. observed that excess amounts of triacyl glycerol and fatty acids 18:1 are major factors in NAFLD (Liang, 2024). Based on this report, their metabolism and abnormalities in their structure are the primary factors. Lipoxygenase biocatalysts convert polyunsaturated fatty acid (PUFA) in the plasma membrane such as linoleic acid and arachidonic acid to synthesize various intermediates for proinflammation. The endoplasmic reticulum secretes various peptides and proteins that are essential for the transportation of fatty acid in various parts of the body, with the enhancement of synthesis of protein across the endoplasmic reticulum's (ER's) capacity. The misfolded protein response is triggered by the deposition of mismatched proteins inside the lumen. The activation of ER stress sensors occurs due to the presence of protein folding complex molecules, which increases the unfolded protein mount. Unfolded protein response started with various cell signaling sensor pathways such as protein kinase R (PERK), which is known as double-stranded (RNA-associated protein kinase), initiating transcription factor 6, and inositol required element-1 alpha. Fats containing single bonds like palmitic acid have been observed to change the integrity properties of the membrane of ER with the help of phosphatidylcholine and TAG. With the help of iron response element (IRE) and PERK, NF-Kb promotes ER stress, which is one of the major biochemical processes. The combinatorial role of ER stress and oxidative stress is a promoting factor in the formation of NASH (Idilman, 2016).

14.7 LIVER INFLAMMATION

Two primary hypotheses that are involved in liver inflammation are intake of inadequate calories and lack of nutrition associated with the initiation of NAFLD.it was observed that unhealthy foods along with their composition play a role. One observational study on mice evaluated the concept of tissue-proven steatosis and inflammation, which are related to the intake of bad foods and their duration. After 40 weeks of observation, enhancement of inflammation of the liver and in the hepatocytes was assessed. Methionine and choline in adequate foods are used to reduce liver inflammation. Inflammation was reversible with 16 weeks of methionine–choline deficient (MCD) diets but irreversible after 20 weeks (Food containing trans-unsaturated fatty acid enhanced the synthesis of cholesterol by using the sterol regulatory element binding protein pathway by the utilization of sterol regulatory element binding protein (SREBP) cleavage-activating protein. antisense oligonucleotides are used to reduce insulin signaling with the presence of SREBP-1c. Carbohydrate reactive compounds such as binding proteins are involved in the intake of sugars. Fructose enhances the NAFLD dependently on carbohydrate response element binding protein (ChREBP), and liver inflammation is suppressed by the presence of hepatic carbon binding protein along with SREBP2. The formation of hepatic inflammation is led by the systemic knockout of ChREBP. It was researched that fructose combined with different amino acids and polypeptides helps form glycation-end compounds along with the nuclear factor kappa B cell signaling pathway. By targeting the nuclear factor kappa

B cell signaling pathway, researchers have tried to reduce the progression of liver inflammation in the critical condition of NAFLD. The oxidase catalyst of NADPH alleviates the impact of fructose. Another factor affecting liver inflammation is fibroblast growth factor 21 (FGF21) in combination with carbohydrate-binding protein. FGF21 and carbohydrate-binding proteins are responsible for the pathogenesis of NAFLD. Astaxanthin is one of the promising factors for the treatment of this hepatic disorder by developing the mitochondrial function by catalyzing the peroxisome proliferator-activated receptor gamma (PPAR-) coactivator-1 (PGC-1), which is synthesized by FGF21. Circulatory elements of free fatty acid are altered by saturated fatty acid and are responsible for the formation of inflammation in NAFLD (Powell, 2021).

14.8 IMPACT OF BILE ACID

Bile acids are formed primarily in the liver with the help of cholesterol by involving two cell signaling pathways. The first is classical cell signaling, and the second is an alternative pathway, which includes a total of 17 biocatalysts. The enzyme associated with the initiation of the classical pathway is cholesterol 7 a-hydroxylase and with the combination of sterol-12α hydroxylase, essential for the metabolism of bile acid in NAFLD. In the case of alternative pathways, required enzymes are chenodeoxycholic acid, cholic acid, and beta-muricholic acid. These play an important role in the conversion of cholic acid to chenodeoxycholic acid. Bile acids are combined with glycine, a simple amino acid, which helps to form a conjugated form of bile, which can flow in the bloodstream, which is pumped out from the hepatocytes with the help of bile salt export pump (BSEP) and protein, which are connected to antibiotic resistance. In the second phase, bile acids are converted into deoxycholic acid and lithocholic acid in the stomach by the process of deconjugation. The bile acid transporter is used to reabsorb approximately 90% of bile acids and released into the portal vein and reabsorbed by the liver. Sodium-taurocholate cotransport polypeptide (NTCP) is used for the movement of bile acids from the liver. Lipid digestion and nutrient absorption are important for bile acids homeostasis, which includes the formation and metabolism of bile acids. Bile acid receptor molecules are used to regulate the levels of bile acid by feedback metabolism. The formation and concentration of bile acids in the liver are indirectly proportional to the concentration of intrahepatic bile acids. An abnormality in the concentration of bile acid or homeostasis is linked with various hepatic diseases. Examples of these disorders are NAFLD, cholestasis, other disorders such as obesity and diabetes. The last part of the gastrointestinal (GI) tract contains apical sodium–dependent bile acid transporter (ASBT)-blocking molecules, which are used to reabsorb bile acids and used to enhance the removal of bile acid. These are the major factors used to improve this hepatic disease by reducing by intake of a high-fat diet. Various receptors involved in the absorption of bile acids are NR1H4, GPCR molecules, and farnesoid X receptors. The major function of bile acids signaling molecule breakdown of fats and carbohydrates. The gut microbiota and liver are closely connected to NAFLD. In the crosstalk of gut microbiota and bile acid (BA) are intestinal FXR-FGF15/1, which showed protection in opposition to hepatic steatosis. Recent clinical research has been investigated to modify the treatment potentials of molecules such as FGF19 for the treatment of nonalcoholic fatty liver. To regulate the intestinal cholesterol and absorption of lipids, the required properties are hydrophobic nature and conjugation of bile acids. These are the biochemical investigations of bile acids in NAFLD (Wong, 2023).

14.8.1 Signature of Postbiotics in NAFLD

Postbiotics can be explained by novel research that deals with various bioactive compounds. These bioactive compounds are synthesized from probiotics microorganisms, which are inactivated microbial cells and associated with various biological activities directly or indirectly used for health development. *Lactobacillus*, *Bifidobacterium*, *Saccharomyces*, *Bacillus*, *Streptococcus*, or *Faecali* bacterium genera offer highly efficacious postbiotics derived from cytoplasmic extracts and cell wall components (Koning, 2023). Various microorganisms synthesize metabolites that are responsible for an important role in shaping the host and stomach microbiome associated with different bacteria. These formulated metabolites are essential for the possible diagnosis and treatment of NAFLD. The International Scientific Association of Probiotics and Prebiotics has published the modern definition of postbiotics in which not all microorganisms are added. The meaning of postbiotics is biologically active molecules and the synthesis of nonliving compounds from various pathogens and associated components, which bestow health benefits on the host. The formation, however, is not living, turning aside the possibilities of colonization. Hence, the health advantages bestowed by postbiotics are based on the well-ordered intake of biochemically active biomolecules. Examples of some vital metabolites are phenylacetic acid, bile acids, trimethylamine oxide (TMAO), tryptophan derivatives, imidazole propionate, ethanol, and SCFAs and also pasteurized A. Phenylacetic acid (PAA) is a bacterial metabolite, which is synthesized by using a meta-organismal pathway; in this biochemical process phenylalanine acts as a nutrient precursor. The case of NAFLD involves various complex proteins and metabolism of enhanced levels of PAA. In research, it was observed that the therapeutic application of butyrate prodrug tributyrin reduced insulin resistance and hepatic steatosis. Bacteria-producing SCFAs are essential for the treatment of NAFLD. The breakdown of tryptophan can occur by both biochemical pathways of microorganisms; in these pathways, tryptophan converts into indole, which enters into the biochemical pathways and helps to form compounds having kynurenine or serotonergic molecules and oxindoles. In the case of NAFLD, researchers observed a decrease in the parameters of tryptophan synthesis indoleamine 2,3-dioxygenase 1 and 2 (IDO1 and IDO2) and tryptophan-2,3-dioxygenase (TDO), which are related with an enhanced risk of inflammation and NAFLD. Ethanol is converted into acetaldehyde in the presence of the enzyme alcohol dehydrogenase (ADH), CYP2E1, and catalase. Acetaldehyde is responsible for DNA damage. Various phases are used in the treatment of NAFLD. Various advantages of postbiotics in NAFLD are described in Table 14.1 (Kerksic, 2024). Peptidoglycan is one of the major components of the bacterial cell wall, which has anti-inflammatory and antiproliferative properties. Insulin-sensitive postbiotics that contain MDP are activated in the presence of NOD2 and can reduce the inflammation of adipose tissues. Lactic acid bacteria and yeasts form some active metabolites that contain organic acids, di-diacetylene, CO_2, and bacteriocins, which are known as cell-free supernatants. These obtained materials have significant antioxidant capacity. These are the major factors for lipid metabolism. In the gastrointestinal tract, anaerobic bacteria produce primarily butyric, propionic, and acetic acids, which are volatile and are used the reduce the chances of NAFLD.

TABLE 14.1
Impact of Postbiotics in NAFLD (Kenkik, 2024)

Microorganisms	Constituents	Advantages
L. plantarum *L. paracasei* *L. rhamnosus*	Teichoic acids (Polyglycerol phosphate unit)	Immunomodulatory property Antiobesogenic action Anti-inflammatory property in opposition to NAFLD
Lactobacillus acidophilus *L. casei* *L. reuteri* *L. lactis*	Cell-free supernatant	Antioxidant property Anti-inflammatory effect
L. delbrueckii *L. plantarum*	Exopolysaccharides	Antioxidant property In opposition to insulin To regulate adult-onset diabetes

14.9 CLINICAL SIGNIFICANCE, LIMITATIONS, AND CHALLENGES

Over the last few decades, the frequency of metabolic disorders and NAFLD has been perilously enhanced due to modern lifestyles. The inhibitory and healing impacts have been set up. Postbiotics are a novel dietary intermediation to reduce the chances of obesity. Some major clinical significance of postbiotics is due to varieties of metabolites from lysates and cell membrane components, and some major components are muramyl dipeptide (MDP), substrate-level phosphorylation (SLP), lipoteichoic acid (LTA), extracellular polymeric substances (EPS), and short-chain fatty acids. Based on the comparative studies between the probiotics and postbiotics, postbiotics product generally have less impact on the modulation of intestinal metabolism. The safety and potential of postbiotic treatments have not been analyzed properly yet. Postbiotics need to prevail over various hurdles (Park, 2023).

14.10 FUTURE PERSPECTIVES

NAFLD is a prevalent condition worldwide, driven by lifestyle factors such as poor diet and sedentary behavior. Its progression can lead to severe complications like NASH, cirrhosis, and liver failure. Currently, there are limited pharmacological treatments for NAFLD, which has prompted interest in alternative approaches, such as postbiotics. Postbiotics are bioactive compounds produced by the gut microbiota through fermentation processes, and they have shown potential in modulating metabolic pathways, inflammation, and oxidative stress. These compounds, which include SCFAs, peptides, and other metabolites, could offer novel therapeutic strategies for managing NAFLD. However, the mechanisms through which postbiotics affect liver function remain largely unexplored and require further research.

A key area for future investigation is in understanding the specific mechanisms by which postbiotics affect liver health, particularly through the gut–liver axis. The gut microbiota play a pivotal role in the production of postbiotics, and alterations in microbiome composition have been associated with NAFLD. Postbiotics can influence lipid metabolism, reduce inflammation, and modulate oxidative stress in liver cells, which are key factors in the development and progression of NAFLD. Exploring these pathways could help identify targeted therapeutic interventions for the disease. Furthermore, optimizing the delivery of postbiotics to the liver is a critical challenge. The bioavailability of these compounds can be affected by factors such as digestion, absorption, and metabolism. Developing novel drug delivery systems, such as encapsulation technologies or nanoparticle-based carriers, could enhance the therapeutic effectiveness of postbiotics. These systems could help ensure that postbiotics reach the liver in sufficient concentrations to exert their beneficial effects.

Research also needs to focus on identifying and characterizing a broader range of postbiotics that could influence liver function. While SCFAs like butyrate have shown promise in early studies, other postbiotics, such as antimicrobial peptides and polyphenols, might also contribute to improved liver health. A more comprehensive understanding of how different classes of postbiotics interact with the liver could lead to the development of combination therapies that target multiple pathways involved in NAFLD. Additionally, the role of postbiotics in advanced stages of NAFLD, such as NASH and cirrhosis, remains underexplored. While postbiotics have shown potential in preventing liver fat accumulation, their ability to reverse inflammation and fibrosis in advanced stages requires further investigation. Clinical trials are essential to evaluate the safety, efficacy, and long-term effects of postbiotics in NAFLD patients, particularly in those with more severe forms of the disease. Given the association between NAFLD and metabolic syndrome, future research should also explore the broader effects of postbiotics on obesity, insulin resistance, and other components of metabolic dysfunction. Understanding how postbiotics influence these comorbidities could help develop more comprehensive treatment strategies for NAFLD. Additionally, personalized approaches based on individual microbiome profiles and dietary habits could enhance the effectiveness of postbiotic-based therapies. Finally, postbiotics present a promising therapeutic strategy for NAFLD, with the potential to address various aspects of the disease, from lipid metabolism to inflammation. However, extensive research is required to optimize their use, understand their mechanisms, and assess their safety and efficacy in clinical settings. By expanding the understanding of postbiotics and their effects on liver health, researchers could contribute to the development of novel, targeted therapies for NAFLD.

14.11 CONCLUSION

In conclusion, nonalcoholic fatty liver disease (NAFLD) is a growing health concern characterized by the accumulation of fat in liver cells, not related to excessive alcohol consumption. The pathophysiology of NAFLD involves a complex interplay of genetic, environmental, and metabolic factors that lead to liver steatosis, inflammation, and, in severe cases, fibrosis or cirrhosis. The disease can be classified into simple fatty liver (NAFL) and its more advanced form, nonalcoholic steatohepatitis (NASH), where inflammation and liver damage occur. NAFLD has various causes, including poor dietary habits, obesity, insulin resistance, and genetic predisposition, while its symptoms may remain asymptomatic or manifest as fatigue, abdominal discomfort, and liver enzyme abnormalities. Biochemical mechanisms underlying the disease involve alterations in lipid metabolism, oxidative stress, and dysregulated inflammatory pathways contributing to hepatocyte damage. Liver inflammation plays a crucial role in the progression from simple steatosis to more severe forms of NAFLD, with immune responses exacerbating tissue injury. Additionally, bile acids, which help in lipid digestion and absorption, can have a dual role in liver health, with their dysregulation potentially contributing to NAFLD progression.

In this context, postbiotics are bioactive compounds produced by gut microbiota during fermentation, which have emerged as a potential therapeutic approach for NAFLD. Postbiotics, such as short-chain fatty acids (SCFAs) and other microbial metabolites, may help regulate lipid metabolism, reduce liver inflammation, and modulate oxidative stress, thereby offering a novel treatment avenue. However, while postbiotics hold promise, their clinical significance remains limited by challenges related to bioavailability, optimal formulation, and the need for more robust clinical trials. The effectiveness of postbiotics in treating NAFLD will depend on overcoming these limitations and addressing the underlying complexities of the disease, including the role of the gut–liver axis, microbiome diversity, and the individual patient's metabolic state. In summary, postbiotics represent a promising frontier in NAFLD management, but further research is required to fully understand their potential, refine therapeutic strategies, and ensure their clinical applicability.

REFERENCES

Bourebaba, Y., Marycz, K., Mularczyk, M., & Bourebaba, L. (2022). Postbiotics as potential new therapeutic agents for metabolic disorders management. *Biomedicine & Pharmacotherapy*, *153*, 113138.

Brunt, E. M. (2007). Pathology of fatty liver disease. *Modern Pathology*, *153*, 40–48.

Chen, Y.-Y. (2021). Non-alcoholic fatty liver disease: A review with clinical and pathological correlation. *Journal of the Formosan Medical Association*, *120*, 68–77.

Clark, J. M. (2002). Nonalcoholic fatty liver disease. *Gastroenterology*, *122*, 1649–1657.

Cuevas, P. F. (2020). Postbiotics and parabiotics: From concepts to applications, *Food Research International*, *136*, 109502.

Powell, E. E. (2021). Non-alcoholic fatty liver disease. *Lancet*, *397*, 2212–2224.

Feng, G. (2024). Recompensation in cirrhosis: Unraveling the evolving natural history of nonalcoholic fatty liver disease. *Nature Reviews Gastroenterology & Hepatology 21*, 46–56.

Grander, C., Grabherr, F., & Tilg, H. (2023). Non-alcoholic fatty liver disease: Pathophysiological concepts and treatment options. *Cardiovascular Research*, *119*(9), 1787–1798.

Gong, H. (2024). Associations between systemic inflammation indicators and nonalcoholic fatty liver disease: Evidence from a prospective study. *Frontier in Immunology*, *15*, 1389967.

Haas, J. T., Francque, S., & Staels, B. (2016). Pathophysiology and mechanisms of nonalcoholic fatty liver disease. *Annual Review of Physiology*, *78*(1), 181–205.

Idilman, Ilkay S (2016). *Hepatic steatosis: Etiology, patterns, and quantification.* Elsevier.

Kaufmann, B., Seyfried, N., Hartmann, D., & Hartmann, P. (2023). Probiotics, prebiotics, and synbiotics in nonalcoholic fatty liver disease and alcohol-associated liver disease. *American Journal of Physiology-Gastrointestinal and Liver Physiology*, *325*(1), G42–G61.

Kořínková, L., Pražienková, V., Černá, L., Karnošová, A., Železná, B., Kuneš, J., & Maletínská, L. (2020). Pathophysiology of NAFLD and NASH in experimental models: The role of food intake regulating peptides. *Frontiers in Endocrinology*, *11*, 597583.

Kerksick. (2024). It's dead! Can postbiotics really help performance and recovery? A systematic review. *Nutrients 16*, 720. https://doi.org/10.3390/nu16050720

Koning. (2023). Targeting nonalcoholic fatty liver disease via gut microbiome-centered therapies. *Gut Microbes*, 151, 2226922. https://doi.org/10.1080/19490976.2023.2226922

Malnick, S. (2003). Non-alcoholic fatty liver: A common manifestation of a metabolic disorder. *Quarterly Journal of Medicin*, 75, 699–709.

Rinella, M. E., & Sookoian, S. (2024). From NAFLD to MASLD: Updated naming and diagnosis criteria for fatty liver disease. *Journal of Lipid Research*, *65*(1).

Marchisello, S., Di Pino, A., Scicali, R., Urbano, F., Piro, S., Purrello, F., & Rabuazzo, A. M. (2019). Pathophysiological, molecular and therapeutic issues of nonalcoholic fatty liver disease: An overview. *International Journal of Molecular Sciences*, *20*(8), 1948.

Pan, Z., Mao, B., Zhang, Q., Tang, X., Yang, B., Zhao, J., … Zhang, H. (2022). Postbiotics prepared using *Lactobacillus paracasei* CCFM1224 prevent nonalcoholic fatty liver disease by modulating the gut microbiota and liver metabolism. *International Journal of Molecular Sciences*, *23*(21), 13522.

Park, S.-J., Sharma, A., & Lee, H.-J. (2023). Postbiotics against obesity: Perception and overview based on pre-clinical and clinical studies. *International Journal of Molecular Sciences 24*, 6414. https://doi.org/10.3390/ijms24076414

Pan, Y. (2024). Efficacy of probiotics, prebiotics, and synbiotics on liver enzymes, lipid profiles, and inflammation in patients with non-alcoholic fatty liver disease: A systematic review and meta-analysis of randomized controlled trials. *BMC Gastroenterology*, *24*, 283.

Paulino, P. (2024). Non alcoholic fatty liver disease; disease burden, management, and future perspectives. *International Research Journal of Gastroenterology and Hepatology*, 7, 1–13

Parola, M. (2024). Liver fibrosis in NAFLD/NASH: From pathophysiology towards diagnostic and therapeutic strategies, *Molecular Aspects of Medicine*, *95*, 101231.

Petrescu, M. (2022). Chronic inflammation—A link between nonalcoholic fatty liver disease (NAFLD) and dysfunctional adipose tissue. *Medicina*, *58*, 641. https://doi.org/10.3390/medicina58050641

Salminen, S., Collado, M. C., Endo, A., Hill, C., Lebeer, S., Quigley, E. M., … Vinderola, G. (2021). The International Scientific Association of Probiotics and Prebiotics (ISAPP) consensus statement on the definition and scope of postbiotics. *Nature Reviews Gastroenterology & Hepatology*, *18*(9), 649–667.

Samala, N. (2017). Molecular mechanisms of nonalcoholic fatty liver disease: Potential role for 12-lipoxygenase. *Journal of Diabetes and Its Complications*, *31*(11), 1630–1637.

Wei, M. (2024). Regulating bile acids signaling for NAFLD: Molecular insights and novel therapeutic interventions. *Frontiers in Microbiology*, *15*, 1341938.

Liang, Z. (2024). A critical evaluation of the role of iron overload in fatty liver disease. *Journal of Gastroenterology and Hepatology*, *37*(10), 1873–1883.

Wong, V. (2023). Changing epidemiology, global trends and implications for outcomes of NAFLD. *Journal of Hepatology*, *95*, 842–852.

15 Postbiotics

Regulating Lipid Profiles and Hypertension

Pratyush Porel, Navpreet Kaur, Rohit and Khadga Raj Aran
ISF College of Pharmacy, Moga, Punjab, India

15.1 INTRODUCTION

Gut microbiota (GM) are a complex biological community of abundant microorganisms, consisting of bacteria, fungi, archaea, protozoa, bacteriophages, and eukaryotic viruses, which imposes an enormous threat to gut health. The GM profile begins to develop in all individuals with its uniqueness and plays a pivotal role in host physiology. Prebiotics are nondigestible supplements that are used for improving host health by promoting the growth and division of beneficial gut bacteria. Several microbial species, like *Bifidobacteria* and *Lactobacilli*, produce numerous metabolites that are intended to improve gut integrity, suppress inflammation, and maintain overall host health [1]. Postbiotics, synthesized as metabolic byproducts of probiotics and other microbes, have been reported to be an essential modulator of host physiology in recent studies. These studies have shown an increasing recognition of the dynamic relationship between gut health and cardiovascular health, emphasizing the substantial impact of GM on metabolic processes in the last decades. The ability of these bioactive substances, including peptides, short-chain fatty acids (SCFAs), bacterial cell wall fragments, and other metabolites, to regulate blood pressure and lipid metabolism is a crucial factor in the development of several cardiovascular diseases (CVDs) [2]. Hypertension, commonly known as the silent killer disease, is a major risk factor for developing CVDs, and postbiotics exert their effects by diverse mechanisms in these diseases [3]. GM produces SCFAs, such as acetate, propionate, and butyrate, by the fermentation process of dietary fibers, from which the postbiotics are formed. It also regulates lipid homeostasis by alterations in hepatic lipid production, improved lipoprotein clearance, and decreased triglyceride (TG) buildup in peripheral organs. Specifically, butyrate enhances fatty acid oxidation, regulates adipocyte differentiation, and suppresses inflammation, all of which lead to a healthy lipid profile [4]. Additionally, bacteriocins, peptides, and cell wall fragments have shown a significant impact on lipid metabolism by altering intestinal cholesterol absorption, reducing the enzymatic activities for lipid synthesis, and boosting lipoprotein clearance. In addition to lipid metabolism, postbiotics have a significant impact on blood pressure regulation by an interplay with arterial tone and sodium balance, namely the renin–angiotensin system (RAS) [5]. Several studies on SCFAs indicate that butyrate may encourage vasodilation and reduce blood pressure by modulating gene expression implicated in RAS signaling. Furthermore, postbiotics can change endothelial function and lower systemic inflammation, which plays a pivotal role in the pathogenesis of hypertension. Postbiotics inhibit angiotensin-converting enzyme (ACE), a key component of the RAS, which also led to their identification as possible natural antihypertensive medicines. Further evidence is required to find the exact mechanism of postbiotics in regulating lipid metabolism that has been associated with the management of hypertension [6].

Current clinical research is supported by these preclinical results, with significant evidence of the interrelation between GM and hypertension suggesting that postbiotics would be a novel and promising approach to the regulation of lipid metabolism and blood pressure, offering potential benefits

DOI: 10.1201/9781003607779-15

for the prevention and management of CVDs. Although the precise mechanisms underlying these benefits remain to be explored, an intriguing area of research suggests that postbiotics may be employed as a preventative or additional therapy for CVDs. The chapter aims to present a comprehensive analysis of the available data on postbiotic function in controlling lipid profiles and hypertension. It will also cover the challenges and future aspects of postbiotic research, such as the requirement for more extensive clinical trials, the discovery of certain postbiotics with therapeutic efficacy, and the creation of focused therapies for the prevention and treatment of hypertension.

15.2 MECHANISM OF ACTION

15.2.1 Influence of Postbiotics on Gut Microbiota

A diverse variety of microbial communities are present in the human gut, which are collectively called GM. This ecosystem includes bacteria, viruses, parasites, eukaryotes, and archaea, with 90% of bacteria belonging to the Gram-positive Firmicutes and Gram-negative Bacteroidetes phyla [7]. In 1681, Antonie van Leeuwenhoek first reported the presence of little living organisms in his stool sample, and the later isolation of *E. coli* from the gastrointestinal (GI) tract was the first evidence of GM in 1855 [8]. Then dramatic developments in highly sophisticated techniques including high throughput screening, metabolomics, and artificial intelligence resulted in a rapid advancement in the research associated with the interplay between GM and vital host organs, such as the brain, heart, liver, and kidney. Recent studies show that GM plays a crucial role in metabolic and immune functions, aiding in the digestion of macronutrients, preventing pathogen colonization, and promoting gut immunity [9].

Between healthy and diseased individuals, there are significant differences in GM composition and gut health that lead to several diseases, including inflammatory and metabolic disorders. Various postbiotic compounds, including SCFAs like butyrate, propionate, and acetate, help to promote the growth of beneficial bacteria while hindering harmful bacteria. For instance, by lowering gut pH, SCFAs provide a favorable environment for *Lactobacillus* and *Bifidobacterium*, which are considered beneficial microbes, while suppressing the colonization of harmful bacteria [10]. By strengthening the intestinal barrier and enhancing tight junctions in adhering gut cells, postbiotics prevent "leaky gut," which reduces gut inflammation, a crucial factor for developing inflammatory bowel disease (IBD) and irritable bowel syndrome (IBS). Additionally, postbiotics show protective actions by interacting with gut-associated lymphoid tissue, present in the GI tract, endorsing a homeostasis between immune responses and factors of autoimmune diseases [11].

15.2.2 Modulation of Cholesterol Homeostasis

Cholesterol serves as a chief structural component of cell membranes, modulating intracellular signaling and neuronal conduction and acting as a precursor to steroid hormones, vitamin D, oxysterols, and bile acids (BAs). About one-third of cholesterol is dietary, while the rest is synthesized in body cells. After esterification, the liver eliminates cholesterol by secreting it into bile in the packed form of lipoproteins or converting it into BAs, thus maintaining cholesterol homeostasis through coordination with the intestines, as shown in Figure 15.1 [12]. These BAs, synthesized from cholesterol, are stored in the gallbladder, released during digestion, and reabsorbed in the intestines, maintaining balance via enterohepatic circulation. GM plays a vital role in this process, impacting cholesterol levels and hypertension risk. Approximately 1 g of cholesterol from bile, diet, and desquamated cells enters the colon daily, where bacteria convert cholesterol to coprostanol, influencing blood cholesterol levels and cardiovascular disease risk [13]. Additionally, intestinal bacteria transform cholesterol into BAs through deconjugation, oxidation, epimerization, and 7α-dihydroxylation, carried out by several bacteria, such as *Clostridium*, *Bifidobacterium*, *Lactobacillus*, and *Eubacterium*. This biotransformation modulates BA toxicity and reabsorption, influencing cholesterol excretion [14].

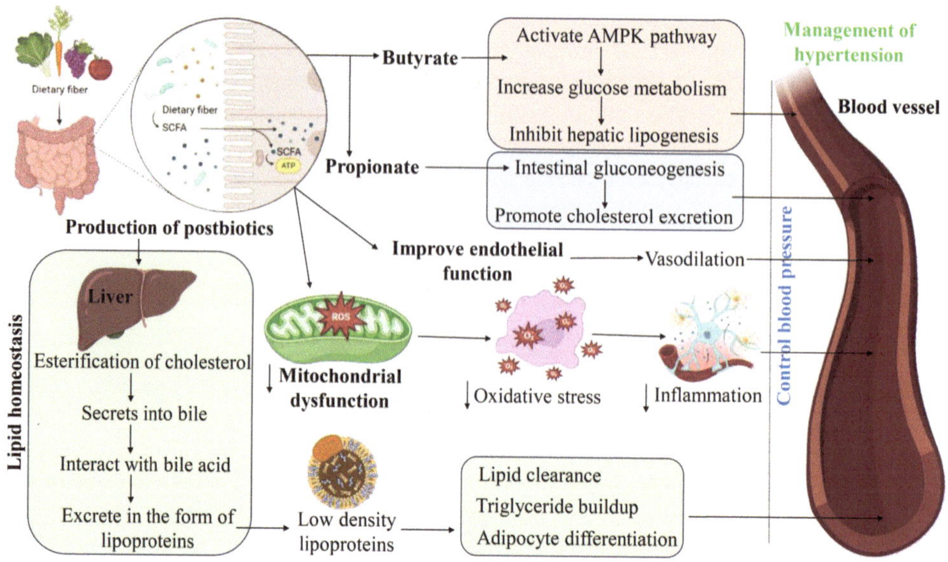

FIGURE 15.1 Mechanisms of postbiotics in hypertension management.

On the other side, the accumulation of cholesterol molecules inside of the arterial walls forms plaques or atheromas that make the blood vessels narrower and stiffen, which contributes to a reduction in blood flow. Reduced blood flow will increase vascular resistance, followed by an elevation of blood pressure. If it persists for a longer time, it leads to hypertension called cholesterol-mediated hypertension [15]. Postbiotics also stimulate the synthesis of bile acids in the liver, which can upregulate the body's cholesterol level, leading to hypercholesteremia. 3-hydroxy-3-methyl-glutaryl coenzyme-A (HMG CoA) reductase is the rate-limiting enzyme in the mevalonate pathway, responsible for the conversion of HMG CoA to mevalonic acid, a precursor of cholesterol synthesis [16]. Some studies suggest that postbiotics might act as HMG CoA reductase inhibitors by modulating intracellular signaling pathways like adenosine monophosphate-activated protein kinase (AMPK). This effect of postbiotics establishes its role in cholesterol homeostasis endogenously by reducing the risk of developing atherosclerosis and cholesterol-mediated hypertension [17].

15.2.3 Production of Short-Chain Fatty Acids (SCFAs)

SCFAs such as acetate, propionate, and butyrate are produced by GM through the fermentation of dietary fiber in the colon. SCFAs are crucial for cholesterol metabolism, with butyrate playing a significant role in maintaining cholesterol homeostasis. SCFAs lower serum levels of LDL-C by inhibiting HMG-CoA reductase, suppressing cholesterol synthesis, enhancing LDL receptor expression, and promoting LDL absorption from the circulation. This regulation reduces cholesterol levels, lowering the risk of hypertension and CVDs, diagrammatically illustrated in Figure 15.1 [18].

15.2.4 Anti-Inflammatory Properties

Postbiotics are also permissible in the field of developing novel therapeutics as an anti-inflammatory agent. Propionate, another SCFA, promotes the growth of peripheral regulatory T cells (Tregs) in the intestine, while butyrate contributes to the induction of variety among regulatory T cells. By increasing colonic function, reducing pH, and encouraging the growth of epithelial cells and blood flow in the colon, SCFA has a positive impact on health [19]. Furthermore, in vitro research shows that the supernatant of a *Bifidobacterium breve* culture inhibits tumor necrosis factor-α (TNF-α) secretion while increasing interleukin-10 (IL-10) production, which supports dendritic cell maturation and survival. The decrease in Th1-mediated responses and the rise in Th2-mediated responses could be explained by these traits [20]. According to research, probiotics have a crucial function in controlling immunological activity in both healthy people and infected patients by modifying immune cell behavior and activating vital immune signaling pathways. Notably, more research is required to determine how probiotics might be used as a substitute supplement to treat diseases and basic functions, as shown in Figure 15.1 [21].

15.3 POSTBIOTICS AND LIPID METABOLISM

15.3.1 Effects on Cholesterol Levels

Postbiotics are bioactive substances produced by microbial metabolism that show promising effects on cholesterol regulation and lipid metabolism. One primary mechanism is the synthesis of SCFAs, which regulate lipid oxidation and energy expenditure. SCFAs lower visceral fat and triglyceride levels by increasing mitochondrial fatty acid oxidation and energy consumption [22]. Butyrate activates AMPK, which controls lipid metabolism and inhibits hepatic lipogenesis. Propionate influences intestinal gluconeogenesis and promotes cholesterol excretion by regulating bile acid synthesis and metabolism, as shown in Figure 15.1. By altering BA compositions, postbiotics improve the deconjugation and conversion of bile acids into secondary forms, facilitating cholesterol excretion and enhancing lipid homeostasis [23].

15.4 REDUCTION OF TOTAL CHOLESTEROL

Postbiotics reduce overall cholesterol levels through several interconnected mechanisms. Bacterial bile salt hydrolase (BSH) catalyzes the deconjugation of bile acids, producing free bile acids that are excreted in feces. This process requires the liver to use cholesterol to synthesize new bile acids, reducing blood cholesterol [24]. Additionally, SCFAs produced by postbiotic bacteria inhibit HMG-CoA reductase, decreasing the liver's cholesterol synthesis. SCFAs also promote cholesterol utilization for energy, improving lipid oxidation and export. By enhancing gut health, postbiotics strengthen the intestinal barrier and reduce lipopolysaccharide translocation, preventing systemic inflammation and dysregulation of lipid metabolism [25].

15.5 IMPACT ON LDL AND HDL LEVELS

Postbiotics improve lipid profiles and lower the risk of CVD by beneficially affecting both LDL and high-density lipoprotein (HDL) cholesterol levels. Postbiotics increase the clearance and decrease the oxidative modification of LDL cholesterol. Postbiotic-produced SCFAs promote the expression of hepatic LDL receptors, enhancing the absorption and breakdown of LDL from the circulation, which lowers plasma LDL levels and reduces atherogenicity. They also reduce the production of oxidized LDL, a primary cause of atherosclerosis, through antioxidant effects mediated by exopolysaccharides and peptides. These compounds scavenge reactive oxygen species (ROS) and prevent oxidative damage to LDL particles, thereby reducing vascular inflammation and preventing plaque and foam cell formation in artery walls [26].

Additionally, postbiotics enhance HDL cholesterol levels and function through several mechanisms. By facilitating reverse cholesterol transport (RCT), which moves cholesterol from peripheral tissues to the liver for elimination, HDL protects cardiovascular health. Postbiotics increase the activity of cholesterol transporters like ATP-binding cassette transporters A1 (ABCA1) and G1

(ABCG1), improving this process [27]. These transporters help macrophages and other peripheral cells release cholesterol into developing HDL particles, which then transport cholesterol to the liver. By enhancing RCT efficiency and reducing systemic inflammation, postbiotics indirectly improve HDL function. Chronic inflammation hampers HDL's capacity to function in lipid metabolism, but by lowering inflammatory cytokines such as TNF-α, interleukin-1β (IL-1β), and interleukin-6 (IL-6), postbiotics enhance HDL particles' antioxidant capacity and structural integrity, protecting against lipid peroxidation [28].

By altering the composition of the GM, postbiotics also indirectly impact LDL and HDL levels, influencing lipid transport and metabolism. A healthier GM composition encouraged by postbiotic activity promotes the generation of bioactive compounds that improve fat utilization and transport, ultimately reducing the risk of atherosclerosis and CVD by ensuring a positive lipid profile with decreased LDL levels and enhanced HDL activity [29].

15.6 INSULIN SENSITIVITY ENHANCEMENT

Postbiotics play a crucial role in improving insulin sensitivity through various intricate mechanisms, and modification of gut flora is one of the most important of all. Postbiotics nurture a healthy and diverse GM, increasing the growth of beneficial bacteria while decreasing the prevalence of harmful bacteria, which is crucial for metabolic health and enhanced insulin sensitivity [30]. Postbiotics also have anti-inflammatory properties, important due to the interlink between chronic low-grade inflammation and insulin resistance. By reducing inflammation, postbiotics improve insulin sensitivity and overall metabolic performance. Postbiotics directly affect several metabolic pathways linked to insulin signaling and glucose metabolism. They increase the sensitivity of insulin receptors, stimulate cells to glucose uptake, and alter the activity of enzymes involved in glucose metabolism, which helps improve glucose management and insulin sensitivity. Additionally, postbiotics might regulate the release of adipokines and lipid metabolism in adipose tissue, promoting healthy adipose tissue activity to further enhance insulin sensitivity and metabolic health [31].

The synthesis of SCFAs via postbiotic fermentation of dietary fibers is another key factor in this context. Acetate, propionate, butyrate, and other SCFAs have been linked to better glucose homeostasis and increased insulin sensitivity. SCFAs are a significant connection between postbiotics, GM, and metabolic health. Postbiotics also help maintain the integrity of the intestinal barrier, preventing the translocation of pathogens and harmful chemicals, which can affect insulin sensitivity and systemic inflammation [32].

15.7 IMPLICATIONS FOR METABOLIC DISORDERS

Postbiotics have broad implications for metabolic disorders, offering potential treatments for conditions such as type 2 diabetes, insulin resistance, and other metabolic abnormalities. Postbiotics provide a novel approach to developing targeted therapies to enhance insulin sensitivity and glycemic control. Researchers and healthcare professionals could explore new ways to treat insulin resistance and metabolic disorders by leveraging postbiotics' multiple mechanisms of action [33].

Postbiotics' capacity to alter the GM and reduce inflammation can influence the underlying pathophysiology of metabolic disorders. Additionally, postbiotics' ability to affect metabolic pathways linked to insulin signaling and glucose metabolism can impact the development of targeted therapies. Precision medicine strategies that exploit postbiotics' varied metabolic effects to target specific features of metabolic illnesses may be developed. Furthermore, postbiotics' potential to regulate adipose tissue function and promote SCFA synthesis highlights their importance in managing metabolic health. By targeting adipose tissue function and SCFA synthesis, postbiotics can offer a comprehensive approach to addressing metabolic abnormalities associated with insulin resistance and type 2 diabetes [34].

15.8 POSTBIOTICS AND HYPERTENSION MANAGEMENT

15.8.1 Endothelial Function Improvement

Postbiotics have drawn interest due to their ability to control hypertension and modify endothelium functions. Several mechanisms are used by these bioactive substances, such as polyphenolic derivatives, peptidoglycans, and SCFAs, to produce their advantageous effects. First, postbiotics impact the expression of endothelial nitric oxide synthase (eNOS), which increases the generation of nitric oxide (NO). By encouraging endothelial relaxation and lowering vascular resistance, NO is an essential vasodilator that lowers blood pressure [35]. Postbiotics may also alter inflammatory pathways by inhibiting the synthesis of pro-inflammatory cytokines such as TNF-α and IL-6, which are known to cause endothelial dysfunction and elevated vascular tone in hypertension conditions. Additionally, postbiotics support tight junction protein production, which contributes to vascular homeostasis and improves the integrity of the endothelial barrier. They also affect the gut–vascular axis, where metabolites produced by the GM can control the release of vasoactive chemicals, affecting the tone of the systemic arteries. Postbiotics are a viable supplementary therapy via these multifactorial pathways for controlling hypertension, enhancing endothelial function, and maybe lowering the risk of CVDs [36].

15.8.2 Mechanisms of Blood Pressure Regulation

Postbiotics, particularly SCFAs, are crucial for regulating blood pressure and maintaining cardiovascular health. Research shows that *Lactobacillus*, a beneficial bacterium, can increase HDL cholesterol while decreasing LDL and fatty acid cholesterol in obese mice. This improvement in lipid profiles, brought on by fragmented *Lactobacillus* bacteria, stimulates the peroxisome proliferator-activated kinase receptor (PPAR), a target for cholesterol-lowering drugs [37]. SCFAs like propionate reduce cholesterol formation by blocking the condensation of cholesterol precursors, similar to statins. The composition of the GM during early life stages significantly influences the development of hypertension in children, linked to maternal high sugar intake. Research has investigated the protective effects of butyrate or propionate supplementation for mothers on their offspring's hypertension in rat models. Results show that supplementing with these postbiotics during pregnancy mitigates the negative impact of a high-fat diet on offspring blood pressure [38]. Both butyrate and propionate treatments increase plasma levels of beneficial SCFAs, enhance lipid metabolism, and affect gut-flora composition. While propionate treatment alters the abundance of specific gut bacterial taxa, butyrate has a major effect on trimethylamine N-oxide metabolism and nitric oxide–related metrics. After reviewing all the studies, researchers are also suggesting that postbiotics can be implemented as a therapeutic approach to treat hypertension [39].

15.8.3 Clinical Evidence and Studies

Postbiotics have shown encouraging promise in recent preclinical research for the treatment of several illnesses, including hypertension. These probiotic-derived bioactive chemicals have shown positive results in lab settings, indicating that they may be important for blood pressure regulation and cardiovascular health enhancement. Nevertheless, more research is required to validate these findings in human populations, even though the preclinical data are intriguing. The precise mechanisms and therapeutic potential of postbiotics in clinical practice require further preclinical trials as well

TABLE 15.1
List of Some Case Studies and Clinical Trials That Aim to Explore the Various Potential Roles of Postbiotics

S. no.	Title of Study	Location	Type of Study	Diseases / Conditions	Study Outcomes	References
1.	Clinical trials of the effects of postbiotic supplementation on inflammation, oxidative stress, and clinical outcomes in patients with CVA	Imam-Reza Hospital in Kermanshah, Iran	Single-center, randomized, controlled trial	Cerebrovascular accident (CVA)	A promising treatment strategy for lowering oxidative stress and inflammation and enhancing outcomes in CVA patients is postbiotic supplementation.	[41]
2.	Effects of postbiotics on chronic diarrhea in young adults: a randomized, double-blind, placebo-controlled crossover trial assessing clinical symptoms, gut microbiota, and metabolite profiles		Randomized, double-blind, placebo-controlled crossover trial	Chronic diarrhea	Chronic diarrhea can be efficiently treated with postbiotics by increasing gut flora, decreasing the frequency of defecation, and improving stool consistency.	[42]
3.	A randomized controlled trial of postbiotic administration during antibiotic treatment increases microbiome diversity and enriches health-associated taxa		Single-center, randomized placebo-controlled trial	Antibiotic treatment	In addition to increasing helpful bacteria and decreasing detrimental pathobionts, postbiotics can potentially reverse the damage that antibiotics do due to the GM.	[43]
4.	Efficacy and safety of postbiotic contained inactivated *Lactobacillus reuteri* (*Limosilactobacillus reuteri*) *DSM*17648 as adjuvant therapy in the eradication of *Helicobacter pylori* in adults with functional dyspepsia: A randomized double-blind placebo-controlled trial	Sechenov University, Russia	Randomized, double-blind, placebo-controlled, multicenter, parallel study	Dyspepsia	Postbiotics improve quality of life, decrease GI side effects, and increase the efficacy of *H. pylori* eradication therapy in patients with functional dyspepsia	[44]
5.	Home oral care of periodontal patients using antimicrobial gel with postbiotics, lactoferrin, and *Aloe barbadensis* leaf juice powder versus conventional chlorhexidine gel: A split-mouth randomized clinical trial	Department of Clinical, Surgical, Diagnostic and Paediatric Sciences of the University of Pavia, Italy	Single-center, randomized controlled trial	Periodontitis	According to the study's findings, there was no discernible difference in the efficacy of postbiotic gel and conventional chlorhexidine gel in treating periodontitis. This suggests that postbiotics could be a viable natural substitute for traditional chlorhexidine treatment for people with a history of periodontal disease.	[45]

6.	Impact of the food-grade heat-killed probiotic and postbiotic oral lozenges in oral hygiene	Taiwan	Randomized controlled trial (RCT)	Effect on IgG and pathogenic microbes	The study evaluated how the oral and intestinal health of participants with an age range of 20–40, was affected by lozenges containing postbiotics from *Lactobacillus salivarius*, *Lactobacillus paracasei*, and *Lactobacillus plantarum* that were heat-killed and supernatant extract. Both postbiotics increased salivary IgG A, inhibited the growth of harmful oral bacteria, and had antibacterial properties.	[46]
7.	Urolithin A improves muscle strength, exercise performance, and biomarkers of mitochondrial health in a randomized trial in middle-aged adults	-	Randomized, double-blind, placebo-controlled study	Skeletal system	The study evaluated the effects of *A. muciniphila* postbiotics on immunological responses, lipid and glucose metabolism, and obesity. The postbiotics increased metabolic health, decreased inflammation and obesity, and improved glucose tolerance.	[47]
8.	The moisturizing efficacy of a proprietary dermo-cosmetic product (CLS02021) versus placebo in a 4-week application period	Sarajevo, Bosnia, and Herzegovina	Randomized, intra-individual, double-blind controlled study	De-moisturization	According to the study, postbiotic treatment had a corticosteroid-sparing effect, suggesting that it could be used as an adjuvant treatment, even if it did not significantly lessen the severity of atopic dermatitis (AD) when compared to steroid therapy.	[48]
9.	A randomized trial of *Lactobacillus rhamnosus* IDCC 3201 tyndallizate (RHT3201) for treating atopic dermatitis	Department of Pediatrics, Ajou University	Randomized, double-blind, placebo-controlled parallel study	Atopic dermatitis	According to the study, the shampoo containing heat-killed *L. paracasei* GMNL-653 postbiotics enhanced hair growth, reduced dandruff, controlled sebum production, and modulated the scalp microbiota, demonstrating better performance in certain areas than a commercial product.	[49]

as carefully planned clinical studies. Postbiotics may have comparable therapeutic potential in real-world situations as they do in preclinical research, according to recently released data showing that clinical trial results have matched laboratory results [40]. Key findings from preclinical and clinical studies are highlighted in Table 15.1, which provides a summary of these findings.

15.9 FUTURE DIRECTIONS

There are several areas where future research is necessary, especially on how postbiotics affect cardiovascular health and lipid metabolism. First of all, it is necessary to explore the molecular mechanism of postbiotics in lipid metabolism, cholesterol homeostasis, and blood pressure regulation; afterward it can be determined as a therapeutic approach. Further, validation of the preclinical data, safety, and efficacy of postbiotic supplementation, along with pharmacokinetic and pharmacodynamic profiling and their impact on chronic use in humans still require additional clinical studies. Also, there is a potential for creating concrete postbiotic therapies pertinent to CVD, including hypertension and atherosclerosis. Understanding how postbiotics affect GM and metabolic diseases is another major area of research interest as it may provide direction on developing customized approaches. However, more research is required to understand the potential of postbiotics as a novel therapy for CVD, including hypertension, and its associated complications. Last but not least, understanding the interaction between postbiotics and other cardiovascular medications may reveal a new method for combination therapy to improve therapeutic outcomes, and establish that postbiotics can be utilized alone or in combination with other medicaments for managing cardio complications.

15.10 CONCLUSION

Postbiotics have emerged as a new horizon in understanding and preventing the pathogenesis of CVD and metabolic diseases considering the complex interaction between GM and the host organism. These bioactive compounds produced from postbiotics include SCFA, peptides, cell wall fragments, and bacterial metabolites, which play pivotal roles in the regulation of lipid metabolism, cholesterol homeostasis, and blood pressure. They lower LDL cholesterol, increase the efficiency of HDL, and promote reverse cholesterol transport, thus lowering the risk of atherosclerosis and cholesterol-induced hypertension. Furthermore, postbiotics modulate bile acid metabolism, enhance the mitochondrial β-oxidation of fatty acids, and inhibit ACE, thereby exhibiting antihypertensive activity. Postbiotics also promote cardiovascular health due to their anti-inflammatory and antioxidant properties. Although they showcase promising results in preclinical research and certain clinical studies, information regarding their molecular mechanisms and long-term safety in humans is still poorly explored. Further studies should be directed toward the evaluation of their safety and efficacy in larger patient groups, particularly concerning the use in combination with recent standard therapies, as well as toward the identification of formulations suitable for disease states. From a broader perspective, postbiotics can be viewed as prospective novel microbiota-targeted approaches to treat CVD within the framework of the concept of individualized medicine.

ABBREVIATIONS

ACE Angiotensin-converting enzyme
CVD Cardiovascular diseases
eNOS Nitric oxide synthase
GM Gut microbiota
HDL High-density lipoprotein
IL Interleukin

LDL Low-density lipoprotein
NO` Nitric oxide
RAS Renin–angiotensin system
SCFAs Short-chain fatty acids
TNF-α Tumor necrosis factor-α

REFERENCES

1. Thursby E, Juge N. Introduction to the human gut microbiota. *Biochemical Journal*. 2017 Jun 1;474(11):1823–36. https://doi.org/10.1042/BCJ20160510
2. Pelton R. Postbiotic metabolites: How probiotics regulate health. *Integrative Medicine: A Clinician's Journal*. 2020 Feb;19(1):25.
3. Poznyak AV, Sadykhov NK, Kartuesov AG, Borisov EE, Melnichenko AA, Grechko AV, Orekhov AN. Hypertension as a risk factor for atherosclerosis: Cardiovascular risk assessment. *Frontiers in Cardiovascular Medicine*. 2022 Aug 22;9:959285. https://doi.org/10.3389/fcvm.2022.959285
4. Fusco W, Lorenzo MB, Cintoni M, Porcari S, Rinninella E, Kaitsas F, Lener E, Mele MC, Gasbarrini A, Collado MC, Cammarota G. Short-chain fatty-acid-producing bacteria: key components of the human gut microbiota. *Nutrients*. 2023 May 6;15(9):2211. https://doi.org/10.3390/nu15092211
5. Jaworska K, Koper M, Ufnal M. Gut microbiota and renin–angiotensin system: a complex interplay at local and systemic levels. *American Journal of Physiology-Gastrointestinal and Liver Physiology*. 2021 Oct 1;321(4):G355–66. https://doi.org/10.1152/ajpgi.00099.2021
6. Den Besten G, Van Eunen K, Groen AK, Venema K, Reijngoud DJ, Bakker BM. The role of short-chain fatty acids in the interplay between diet, gut microbiota, and host energy metabolism. *Journal of Lipid Research*. 2013 Sep 1;54(9):2325–40. https://doi.org/10.1194/jlr.R036012
7. Chen Y, Xiao L, Zhou M, Zhang H. The microbiota: a crucial mediator in gut homeostasis and colonization resistance. *Frontiers in Microbiology*. 2024 Aug 6;15:1417864. https://doi.org/10.3389/fmicb.2024.1417864
8. Kutschera U. Antonie van Leeuwenhoek (1632–1723): master of fleas and father of microbiology. *Microorganisms*. 2023 Aug 2;11(8):1994. https://doi.org/10.3390/microorganisms11081994
9. Chi J, Shu J, Li M, Mudappathi R, Jin Y, Lewis F, Boon A, Qin X, Liu L, Gu H. Artificial intelligence in metabolomics: a current review. *TrAC Trends in Analytical Chemistry*. 2024 Jul 3;178:117852. https://doi.org/10.1016/j.trac.2024.117852
10. Zhang YJ, Li S, Gan RY, Zhou T, Xu DP, Li HB. Impacts of gut bacteria on human health and diseases. *International Journal of Molecular Sciences*. 2015 Apr;16(4):7493–519. https://doi.org/10.3390/ijms16047493
11. Chen Y, Zhou J, Wang L. Role and mechanism of gut microbiota in human disease. *Frontiers in Cellular and Infection Microbiology*. 2021 Mar 17;11:625913. https://doi.org/10.1080/17474124.2017.1343143
12. Paukner K, Králová Lesná I, Poledne R. Cholesterol in the cell membrane—an emerging player in atherogenesis. *International Journal of Molecular Sciences*. 2022 Jan 4;23(1):533. https://doi.org/10.3390/ijms23010533
13. Kriaa A, Bourgin M, Potiron A, Mkaouar H, Jablaoui A, Gérard P, Maguin E, Rhimi M. Microbial impact on cholesterol and bile acid metabolism: current status and prospects. *Journal of Lipid Research*. 2019 Feb 1;60(2):323–32. https://doi.org/10.1194/jlr.R088989
14. Grüner N, Mattner J. Bile acids and microbiota: multifaceted and versatile regulators of the liver gut axis. *International Journal of Molecular Sciences*. 2021 Jan 30;22(3):1397. https://doi.org/10.3390/ijms22031397
15. Das P, Ingole N. Lipoproteins and their effects on the cardiovascular system. *Cureus*. 2023 Nov;15(11):e48865. https://doi.org/10.7759/cureus.48865
16. Monteiro SS, Schnorr CE, Pasquali MA. Paraprobiotics and postbiotics—current state of scientific research and future trends toward the development of functional foods. *Foods*. 2023 Jun 16;12(12):2394. https://doi.org/10.3390/foods12122394
17. Liang B, Xing D. The current and future perspectives of postbiotics. *Probiotics and Antimicrobial Proteins*. 2023 Dec;15(6):1626–43. https://doi.org/10.1007/s12602-023-10045-x

18. Zhang D, Jian YP, Zhang YN, Li Y, Gu LT, Sun HH, Liu MD, Zhou HL, Wang YS, Xu ZX. Short-chain fatty acids in diseases. *Cell Communication and Signaling*. 2023 Aug 18;21(1):212. https://doi.org/10.1186/s12964-023-01219-9
19. Heniedy AM, Mahdy DM, Elenien WI, Mourad S, El-Kadi RA. Postbiotics as a health-promoting technique: a review article on scientific and commercial interest. *Process Biochemistry*. 2024 May 14;144:6–19. https://doi.org/10.1016/j.procbio.2024.05.010
20. Hoarau C, Lagaraine C, Martin L, Velge-Roussel F, Lebranchu Y. Supernatant of *Bifidobacterium breve* induces dendritic cell maturation, activation, and survival through a Toll-like receptor 2 pathway. *Journal of Allergy and Clinical Immunology*. 2006 Mar 1;117(3):696–702. https://doi.org/10.1016/J.JACI.2005.10.043
21. Cristofori F, Dargenio VN, Dargenio C, Miniello VL, Barone M, Francavilla R. Anti-inflammatory and immunomodulatory effects of probiotics in gut inflammation: a door to the body. *Frontiers in Immunology*. 2021 Feb 26;12:578386. https://doi.org/10.3389/fimmu.2021.578386
22. Babu AS, Guruprasath N, Adeyeye SA, Sankarganesh P, Kumar AG, Sivapriya T. A critical analysis of postbiotics: exploring their potential impact on the health and food industries. *Journal of Pure and Applied Microbiology*. 2023 Dec 1;17(4):2041–60. https://doi.org/10.22207/JPAM.17.4.18
23. Hodgkinson K, El Abbar F, Dobranowski P, Manoogian J, Butcher J, Figeys D, Mack D, Stintzi A. Butyrate's role in human health and the current progress towards its clinical application to treat gastrointestinal disease. *Clinical Nutrition*. 2023 Feb 1;42(2):61–75. https://doi.org/10.1016/j.clnu.2022.10.024
24. Kumar M, Nagpal R, Kumar R, Hemalatha R, Verma V, Kumar A, Chakraborty C, Singh B, Marotta F, Jain S, Yadav H. Cholesterol-lowering probiotics as potential biotherapeutics for metabolic diseases. *Journal of Diabetes Research*. 2012 Jan 10;2012(1):902917. https://doi.org/10.1155/2012/902917
25. Xiong RG, Zhou DD, Wu SX, Huang SY, Saimaiti A, Yang ZJ, Shang A, Zhao CN, Gan RY, Li HB. Health benefits and side effects of short-chain fatty acids. *Foods*. 2022 Sep 15;11(18):2863. https://doi.org/10.3390/foods11182863
26. Stadler JT, Marsche G. Dietary strategies to improve cardiovascular health: focus on increasing high-density lipoprotein functionality. *Frontiers in Nutrition*. 2021 Nov 22;8:761170. https://doi.org/10.3389/fnut.2021.761170
27. Ouimet M, Barrett TJ, Fisher EA. HDL and reverse cholesterol transport: basic mechanisms and their roles in vascular health and disease. *Circulation Research*. 2019 May 10;124(10):1505–18. https://doi.org/10.1161/CIRCRESAHA.119.312617
28. Yu XH, Tang CK. ABCA1, ABCG1, and cholesterol homeostasis. In *HDL metabolism and diseases*, 2022 May 17: 95–107. Singapore: Springer Nature Singapore. https://doi.org/10.1007/978-981-19-1592-5_7
29. Ikram A, Safdar SZ, Arshad MT, Rasheed A, Gnedeka KT. An overview of postbiotics: Unveiling their distinct role in gut health. *Food and Agricultural Immunology*. 2024 Dec 31;35(1):2434463. https://doi.org/10.1080/09540105.2024.2434463
30. Ji J, Jin W, Liu SJ, Jiao Z, Li X. Probiotics, prebiotics, and postbiotics in health and disease. *MedComm*. 2023 Dec;4(6):e420. https://doi.org/10.1002/mco2.420
31. Park M, Joung M, Park JH, Ha SK, Park HY. Role of postbiotics in diet-induced metabolic disorders. *Nutrients*. 2022 Sep 7;14(18):3701. https://doi.org/10.3390/nu14183701
32. Koh A, De Vadder F, Kovatcheva-Datchary P, Bäckhed F. From dietary fiber to host physiology: short-chain fatty acids as key bacterial metabolites. *Cell*. 2016 Jun 2;165(6):1332–45. https://doi.org/10.1016/j.cell.2016.05.041
33. Mishra N, Garg A, Ashique S, Bhatt S. Potential of postbiotics for the treatment of metabolic disorders. *Drug Discovery Today*. 2024 Feb 20:103921. https://doi.org/10.1016/j.drudis.2024.103921
34. Fang H, Rodrigues e-Lacerda R, Barra NG, Kukje Zada D, Robin N, Mehra A, Schertzer JD. Postbiotic impact on host metabolism and immunity provides therapeutic potential in metabolic disease. *Endocrine Reviews*. 2024 Sep 5:bnae025. https://doi.org/10.1210/endrev/bnae025
35. Cyr AR, Huckaby LV, Shiva SS, Zuckerbraun BS. Nitric oxide and endothelial dysfunction. *Critical Care Clinics*. 2020 Apr 1;36(2):307–21. https://doi.org/10.1016/j.ccc.2019.12.009
36. Om H, Chand U, Kushawaha PK. Postbiotics: an alternative and innovative intervention for the therapy of inflammatory bowel disease. *Microbiological Research*. 2024 Feb 1;279:127550. https://doi.org/10.1016/j.micres.2023.127550

37. Chambers ES, Preston T, Frost G, Morrison DJ. Role of gut microbiota-generated short-chain fatty acids in metabolic and cardiovascular health. *Current Nutrition Reports*. 2018 Dec;7:198–206. https://doi.org/10.1007/s13668-018-0248-8
38. Campos-Perez W, Martinez-Lopez E. Effects of short-chain fatty acids on metabolic and inflammatory processes in human health. *Biochimica et Biophysica Acta (BBA)-Molecular and Cell Biology of Lipids*. 2021 May 1;1866(5):158900. https://doi.org/10.1016/j.bbalip.2021.158900
39. Canani RB, Di Costanzo M, Leone L, Pedata M, Meli R, Calignano A. Potential beneficial effects of butyrate in intestinal and extraintestinal diseases. *World Journal of Gastroenterology: WJG*. 2011 Mar 3;17(12):1519. https://doi.org/10.3748/wjg.v17.i12.1519
40. Prajapati N, Patel J, Singh S, Yadav VK, Joshi C, Patani A, Prajapati D, Sahoo DK, Patel A. Postbiotic production: harnessing the power of microbial metabolites for health applications. *Frontiers in Microbiology*. 2023 Dec 19;14:1306192. https://doi.org/10.3389/fmicb.2023.1306192
41. Rahimi A, Qaisar SA, Janet T, Karimpour H, Darbandi M, Moludi J. Clinical trial of the effects of postbiotic supplementation on inflammation, oxidative stress, and clinical outcomes in patients with CVA. *Scientific Reports*. 2024 Oct 14;14(1):24021. https://doi.org/10.1038/s41598-024-76153-y
42. Guo S, Ma T, Kwok LY, Quan K, Li B, Wang H, Zhang H, Menghe B, Chen Y. Effects of postbiotics on chronic diarrhea in young adults: a randomized, double-blind, placebo-controlled crossover trial assessing clinical symptoms, gut microbiota, and metabolite profiles. *Gut Microbes*. 2024 Dec 31;16(1):2395092. https://doi.org/10.1080/19490976.2024.2395092
43. Schluter J, Matheis F, Ebina W, Jogia W, Sullivan AP, Gordon K, Fanega de la Cruz E, Victory-Hays ME, Heinly MJ, Diefenbach CS, Peled JU. A randomized controlled trial of postbiotic administration during antibiotic treatment increases microbiome diversity and enriches health-associated taxa. *medRxiv*. 2024 July 26:2024-07. https://doi.org/10.1101/2024.07.25.24311015
44. Ivashkin V, Maev I, Poluektova E, Sinitsa A, Avalueva E, Mnatsakanyan M, Simanenkov V, Karpeeva J, Kopylova D, Kuprina I, Kucheryavyy Y. Efficacy and safety of postbiotic containing inactivated *Lactobacillus reuteri* (*Limosilactobacillus reuteri*) DSM 17648 as adjuvant therapy for *Helicobacter pylori* eradication in adults with functional dyspepsia: A randomized double-blind placebo-controlled trial. *Clinical and Translational Gastroenterology*. 2022 May 13; 13(9):10–4309. https://doi.org/10.14309/ctg.0000000000000750
45. Butera A, Gallo S, Pascadopoli M, Taccardi D, Scribante A. Home oral care of periodontal patients using antimicrobial gel with postbiotics, lactoferrin, and aloe barbadensis leaf juice powder vs. conventional chlorhexidine gel: A split-mouth randomized clinical trial. *Antibiotics*. 2022 Jan 17;11(1):118. https://doi.org/10.3390/antibiotics11010118
46. Lin CW, Chen YT, Ho HH, Kuo YW, Lin WY, Chen JF, Lin JH, Liu CR, Lin CH, Yeh YT, Chen CW. Impact of the food grade heat-killed probiotic and postbiotic oral lozenges in oral hygiene. *Aging (Albany NY)*. 2022 Mar 3;14(5):2221. https://doi.org/10.18632/aging.203923
47. Singh A, D'Amico D, Andreux PA, Fouassier AM, Blanco-Bose W, Evans M, Aebischer P, Auwerx J, Rinsch C. Urolithin A improves muscle strength, exercise performance, and biomarkers of mitochondrial health in a randomized trial in middle-aged adults. *Cell Reports Medicine*. 2022 May 17;3(5):100633. https://doi.org/10.1016/j.xcrm.2022.100633
48. Catic T, Pehlivanovic B, Pljakic N, Balicevac A. The moisturizing efficacy of a proprietary dermo-cosmetic product (cls02021) versus placebo in a 4-week application period. *Medical Archives*. 2022 Apr;76(2):108. https://doi.org/10.5455/medarh.2022.76.108-114
49. Jeong K, Kim M, Jeon SA, Kim YH, Lee S. A randomized trial of *Lactobacillus rhamnosus* IDCC 3201 tyndallizate (RHT3201) for treating atopic dermatitis. *Pediatric Allergy and Immunology*. 2020 Oct;31(7):783–92. https://doi.org/10.1111/pai.13269

16 Postbiotics
Safety, Efficacy and Regulatory Considerations

Houshmand Sharafi, Seyede Hanieh Hosseini, Hanie Marhamati, Sima Valipour and Mehran Moradi
Urmia University, Urmia, Iran

16.1 INTRODUCTION

The application of biotics, primarily probiotics, in the food and beverage industry, animal feed, pharmaceuticals, personal cases, and cosmetics has expanded significantly in recent years (Castro-López et al., 2024; Duarte et al., 2022; Prajapati et al., 2023; Requena & Pérez Martínez, 2022; Urban et al., 2024). However, living microorganisms pose concerns, such as infection, antibiotic resistance transfer, and inherently higher susceptibility to storage conditions. Factors such as temperature, humidity, and water pH sensitivity can affect probiotic functionality, which may reduce their effectiveness (Moradi et al., 2021). As an alternative, a new biotic product called postbiotics offers several advantages over probiotics for medicinal use (Vinderola et al., 2022). Postbiotics, compared to probiotics, offer advantages such as inherent stability, wider applications, and a longer shelf life, making them a promising opportunity for applications in various pharmaceutical formulations and commercial food-based products for health promotion, disease prevention, and complementary treatment (Cuevas-González et al., 2020; del Valle et al., 2023; Hosseini et al., 2024; Teame et al., 2020; Vinderola et al., 2023).

The effects of postbiotics on the prevention and treatment of diseases are directly related to their chemical composition (Hernández-Granados & Franco-Robles, 2020; Moradi et al., 2021; Nasri et al., 2025). The composition of postbiotics varies and includes different compounds, based on the definition of postbiotics. The International Scientific Association of Probiotics and Prebiotics (ISAPP) defines postbiotics as inactive microorganisms and their cell fragments, including peptidoglycans, teichoic acid, polysaccharides, cell surface proteins, and beta-glucans or with or without released metabolites, including short-chain fatty acids (SCFAs), organic acids, bacteriocins, peptides, and exopolysaccharides (EPS) (Salminen et al., 2021). Inactive or nonliving forms of probiotic microorganisms are referred to as paraprobiotics or parabiotics (Barros et al., 2020; Divsalar et al., 2025; Yolmeh et al., 2025). Many researchers worldwide (Erginkaya & Konuray-Altun, 2022; Monteiro et al., 2023), including the authors of this book chapter, still rely on the definition of postbiotics as microbial metabolites, not microbial cells and cell-derived compounds (Divsalar et al., 2025; Hosseini et al., 2024; Sharafi et al., 2024).

Due to their pharmacological, immunological, and metabolic effects (Vinderola et al., 2023), postbiotics have various medical applications, including mycobiome modulation, modulation of the immune response and systemic metabolism, treatment of common infectious diseases, treatment of allergies, and wound healing (Bazjou et al., 2022; Vinderola et al., 2022). Additionally, the use of these compounds for the treatment of various internal diseases in humans and animals has been investigated (Hernández-Granados & Franco-Robles, 2020; Mishra et al., 2024; Zhong et al., 2022). Specifically, the application of postbiotics in managing certain diseases, such as nonalcoholic fatty liver disease and diabetes (Arellano-García et al., 2024), has gained attention. The ability of

DOI: 10.1201/9781003607779-16

postbiotics to modulate gut microbiota composition, regulate gut microbial metabolites, and improve the function of the intestinal barrier has attracted a lot of attention, providing a basis for the prevention and treatment of metabolic diseases (Rezaie et al., 2024). Recent studies have shown that postbiotics can help manage metabolic disorders through several potential mechanisms. These include anti-inflammatory, antioxidant, and immunomodulatory effects that improve immune system function and strengthen the intestinal barrier by directly targeting specific tissues in the intestinal epithelium as well as affecting various other organs (Bourebaba et al., 2022). These effects are attributed to the presence of a wide range of bioactive compounds in postbiotics, and any alteration in the postbiotic preparation process, even at the level of a single bacterium, can influence the type and quantity of postbiotic compounds, thereby affecting their efficacy (Hernández-Granados & Franco-Robles, 2020; Nataraj et al., 2020). The focus of this chapter is on the safety and efficacy of postbiotics, specifically addressing the regulatory aspects that pose significant challenges for postbiotics' commercial use.

16.2 SAFETY ASPECTS OF POSTBIOTICS

To qualify as a safe product, postbiotics must originate from a well-defined progenitor, be produced using a clearly described method, demonstrate confirmed health benefits in both preclinical and clinical trials, possess a known chemical composition, and have prior safety evaluations in the target host (Anonymous, 2023). Postbiotics can possess distinct characteristics, biological activities, and safety issues and precautions compared with the microbial strains from which they are derived, but the safety of postbiotics cannot be assumed solely based on the safety record of the originating microorganisms (Thorakkattu et al., 2022). Generally, products derived from food-grade strains can undergo safety procedures more easily. Postbiotics, which are composed of inactivated microorganisms or metabolites, avoid risks related to colonization, infection, or overgrowth associated with live cells (Divsalar et al., 2025).

Additionally, postbiotics offer greater stability than live probiotics and remain unaffected by storage conditions (Divsalar et al., 2025; Sharafi et al., 2024). This means that postbiotics can be used more safely in immunocompromised individuals, neonates, infants, and the elderly population than live microorganisms. However, this assumption does not eliminate the need for safety assessments of postbiotics, as their individual components or the complete formulation could pose potential risks such as gastrointestinal disorders, allergic reactions, or overstimulation of the immune system in certain cases. The production of hydroxymethylfurfural (HMF), biogenic amines, and allergic peptides during LAB propagation can pose a safety concern for prepared postbiotics (de Toledo Guimarães et al., 2023; Divsalar et al., 2025). İncili et al. (2023) conducted a study using a thermal technique to inactivate *Pediococcus acidilactici*, and HMF at a low concentration was identified in the volatile profiles of the resulting postbiotics. Therefore, comprehensive safety studies are essential to include postbiotics in the list of safe substances, such as the European Food Safety Authority (EFSA) Qualitative Presumption of Safety (QPS) database (Salminen et al., 2021). This safety concern becomes more significant when genetically engineered microbial strains are involved in the production of postbiotics (Divsalar et al., 2025).

16.3 EFFICACY OF POSTBIOTICS IN METABOLIC DISEASES

Although legal requirements for medicinal products labeled as postbiotics are not yet established, bacterial lysates (a term similar to the definition of postbiotics by ISAPP) seem to have the closest similarity to postbiotics. These products have been produced and used for human consumption worldwide under various names, mainly as immune-stimulating agents under the supervision of the European Medicines Agency (EMA) (Table 16.1). This section describes the effects of specific postbiotic compounds on metabolic diseases. More detailed information can be found in relevant chapters of this book.

TABLE 16.1
Some Bacterial Lysate-Based Products (Buckley et al., 2018; European Medicines Agency, 2019a, 2019b; Xiao et al., 2003)

Drug Name/ Form	Active Ingredients	Therapeutic Effects	Producer Company
Lacteol Capsule	Lyophilized, heat-treated *Lb. acidophilus* LB (5 billion) of human origin	Improve digestive function by reducing frequent bowel movements, abdominal discomfort, bloating, and promoting regular stool consistency	Carnot Laboratories
Pylopass Capsule	Nonviable, spray-dried *L. reuteri* DSMZ 17648	To control *Helicobacter pylori* infection	Novozymes
Luivac Tablet	Bacterial lysate containing seven bacterial strains	Prevention of recurrent respiratory tract infections (RRTI) in adults and children over 4 years old.	Daiichi Sankyo Austria Gmbh
Respivax Tablet	Bacterial lysate containing six bacterial strains	Immunotherapy and immunoprophylaxis for respiratory diseases in RRTI adults and children over 3 years old	Bb—Neipd Ltd.
Del-Immune V Powder	A lysed, lyophilized powder derived from the *Lb. rhamnosus* V strain	To reduce severity of diarrhea and constipation in children with autism	Stellar Biotics
Lantigen B Oral drops, suspension	Bacterial lysate containing six bacterial strains	Prevention and treatment of RURTI in adults and children over 3	Bruschettini S.R.L.
Buccalin Tablet	Bacterial lysate containing four bacterial strains	Prevention of RRTI in adults and children, authorized without age restrictions	Laboratorio Farmaceutico S.I.T. S.R.L.
Ismigen Sublingual, tablet	Bacterial lysate containing eight bacterial strains	Prevention and treatment of RRTI in adults and children over 3	Lallemand Pharma Europe
Broncho-Vaxom Capsule, granules in sachet	Lyophilized bacterial lysates in different dose	Prevention and treatment of RRTI in adults and children, authorized for various age groups in some countries.	Om Pharma S.A.
Ribomunyl Tablet, capsule, and granules in sachet	Membranes fractions of *Klebsiella Pneumoniae* and ribosomes of four bacterial strains	Prevention of RURTI in children over 2 or 6 years, and recurrent infections in chronic bronchitis	Pierre Fabre Medicament
Polyvaccinum Forte Suspension for injection, nasal drops	Bacterial lysate containing 10 bacterial strains	Prevention and treatment of RURTI in adults and children over 6 months (nasal drops) and recurrent infections (injectable)	Instytut Biotechnologii Surowic I SzczepionekBiomed Spółka Akcyjna

16.3.1 Short-Chain Fatty Acids

Among all microbial metabolites, SCFAs have been demonstrated to have the most effective effects in limiting inflammation, enhancing gut barrier function, and modulating gut microbiota and immune responses (Luzzi et al., 2024). SCFAs can convert the intestinal milieu to be more acidic to promote the growth of beneficial bacteria in the gut (Sanna et al., 2019). SCFAs can activate G protein-coupled receptors (GPCRs) and hinder histone deacetylase, indicating their ability to control the immune system via various mechanisms (van der Hee & Wells, 2021). They can also stimulate GPCR43, which can be used to treat metabolic disorders such as obesity and diabetes. SCFAs can also suppress hormone secretion, such as peptide YY and glucagon-like peptide-1 (GPL-1), in intestinal L cells, while also contributing to glucose homeostasis and exhibiting antidiabetic properties (Yao et al., 2020; Zhang et al., 2021).

SCFAs exhibit immunomodulatory effects, as they have been proven to restrain TNFα, IL6, and nitric oxide in macrophages and improve the release of IL-10 but hinder IL-10 in mononuclear cells (Park et al., 2022). SCFA extracted from *Lactobacillus* spp. have been shown to be effective in reducing inflammatory risk, obesity, diabetes, or other diseases (Bourebaba et al., 2022). Current data indicate that butyric acid, an important SCFA, may protect against obesity through multifaceted mechanisms, including the regulation of circulatory and nervous system functions, enhancement of intestinal barrier integrity, improvement of carbohydrate metabolism, modulation of immune responses, and regulation of appetite (Banasiewicz et al., 2020). Butyric acid, in a preclinical model of nonalcoholic fatty liver disease (NAFLD), has demonstrated the ability to repair intestinal barrier function by increasing tight junction protein expression in the small intestine and decreasing the levels of gut endotoxin in serum and liver (Ye et al., 2018; Zhou et al., 2017). SCFAs, such as butyrate, can also activate AMP-activated protein kinase (AMPK) in hepatocytes to reduce lipid accumulation and alleviate hepatic steatosis and steatohepatitis in metabolic dysfunction-associated steatotic liver diseases (MASLD) (Yilmaz, 2024). Furthermore, butyrate was able to alleviate metabolic disorders and intestinal epithelial impairment in type 2 diabetic mice by promoting insulin secretion without compensatory hyperplasia in pancreatic B cells (Matheus et al., 2017).

SCFAs, including butyrate, propionate, and acetate, play crucial roles in regulating energy homeostasis and glucose and lipid metabolism (Canfora et al., 2017). Butyrate enhances energy expenditure, fat oxidation, and metabolic health, while reducing weight and inflammation (Canfora et al., 2017; Kushwaha et al., 2022). Propionates decrease food intake, protect hepatocytes, and mitigate weight gain and liver fat (Park et al., 2022). However, acetate boosts energy expenditure, triggers intestinal hormone release, and offers anti-inflammatory benefits (Deghima et al., 2021).

16.3.2 Exopolysaccharides

EPSs have an important role in the prevention and treatment of disease. EPSa from *Lactobacillus* spp. decreased TNFα production in cervical cancer HeLa cells and showed noteworthy anti-inflammatory effects by increasing the production of IL-10 (Salva et al., 2021). EPSs derived from *L. plantarum* L-14 have been shown to inhibit the differentiation of immature cells into mature adipocytes, while also regulating body weight gain and lipid profiles in mouse models through upregulation of the TLR2-AMPK signaling pathway. Additionally, long-chain polyphosphates sourced from *Lb. brevis* enhance intestinal inflammation and bolster intestinal barrier function by activating the extracellular regulated protein kinase (ERK) signaling pathway (Lee et al., 2021). A study on rats fed a high-fat diet investigated the antidiabetic effects of the oral administration of EPSs derived from *B. animalis* IPLA R1 (Salazar et al., 2019). The results showed a decrease in serum insulin content without considerable fluctuations in fasting glucose and homeostasis model assessment indices, indicating the positive effect of EPSs on insulin sensitivity. EPS from *Lb. plantarum* H31-2 has been suggested to be effective in suppressing pancreatic α-amylase activity, resulting in a reduction in blood glucose levels in T2DM patients (Huang et al., 2020). Recent studies on EPSs have suggested their ability to reduce bacterial translocation to the liver and prevent inflammatory responses (Yilmaz, 2024).

16.3.3 Bacteriocins

Bacteriocins may play a role in managing metabolic conditions by promoting a healthy gut microbiome. Plantaricin EF is a bacteriocin produced by *Lb. plantarum* that has been shown to ameliorate inflammatory bowel disease via anti-inflammatory activity against 2,4,6-trinitro benzene sulfonic acid (TNBS)–induced inflammatory bowel disease in mouse models (Yin et al., 2018).

16.3.4 Polyamines

Polyamines, such as spermidine, produced by *Bacteroides* spp. and *Fusobacterium* spp. can reduce hepatic lipid accumulation, insulin resistance, and inflammation. They can also help manage MASLD by activating AMPK and autophagy (Yilmaz, 2024).

16.3.5 Peptidoglycan

Peptidoglycan and teichoic acid are the major components of the cell wall and have been shown to possess immunomodulatory properties. Peptidoglycan has also been shown to inhibit proinflammatory cytokines such as TNFα, IL-1β, and IL-6 (Salva et al., 2021). Different *Lactobacillus*-derived peptidoglycans prevent the release of inflammatory cytokines in LPS-stimulated RAW 264.7 cell models (Matsuguchi et al., 2003; Salva et al., 2021). The results of this study showed that peptidoglycans derived from *Lb. rhamnosus* enhances the innate immune response in *Streptococcus*-infected immunodeficient mice (Kolling et al., 2018). Lipoteichoic acid can bind to the bacterial membrane through a glycolipid, whereas wall teichoic acid can bind to peptidoglycan (Kolling et al., 2018). Teichoic acid derived from *Lactobacillus* spp. has been proven to have anti-inflammatory and anti-obesogenic effects (Bourebaba et al., 2022). Teichoic acid has also positive effects on the management of metabolic disfunction-associated steatotic liver disease due to its ability to reduce intestinal permeability and its positive effects on lipid metabolism (Yilmaz, 2024).

16.4 REGULATORY ISSUES

Currently, postbiotics lack dedicated guidelines at the government or international agency levels (Vinderola et al., 2022). A key consideration in regulating manufacture and distribution of postbiotics for medical purposes is determining the appropriate category of medical products or devices. Defining this classification system can greatly facilitate the development of national and international regulations for postbiotics. Similar to certain biological substances, postbiotics may be categorized differently for medical application in various countries.

According to Directive 2004/27/CE of the EU, medicinal products are substances or combinations presented for treating or preventing diseases in humans, or for restoring or modifying physiological functions. In addition, according to Directive 2001/83/CE, biological medicinal products contain biological substances that are produced or extracted from biological sources and require a combination of testing for quality determination. Postbiotics derived from living organisms can be considered active substances in biological medicinal products (Vinderola et al., 2023). Then developers can follow the established guidelines for biological medicinal products when developing postbiotic-based medicinal products. The Food and Drug Administration (FDA) has established a classification for living microorganisms known as live biotherapeutic products (LBPs) (Cordaillat-Simmons et al., 2020). Although LBPs are designed to prevent or treat various human diseases and health issues, this classification may not be suitable for postbiotics, which consist of nonliving microorganisms or their byproducts. In Canada, based on the Health Canada proposal at Probiota 2023 in Chicago, and by considering the ISAPP definition, postbiotics can be placed under the Natural and Non-Prescription Health Products Directorate (NNHPD) (Vinderola, 2024). The selection of microorganisms is of critical importance. The chosen microorganism(s) must possess characteristics similar to those required for the use of live microorganisms (de Melo Pereira et al., 2018). In addition, the chemical composition of postbiotic products is crucial for the regulatory authorities. Therefore, manufacturers must provide detailed composition profiles for each postbiotic product (Divsalar et al., 2025).

The regulation of postbiotic-based products should cover a range of clinical safety assessments that must be considered before their introduction to the market. Preclinical toxicological investigations of allergenicity, carcinogenicity, subchronic and acute toxicity, reproductive toxicity, and genotoxicity should be performed to evaluate the negative effects of antibiotics on various internal organs of the host. Therefore, this requires clear regulations for the safety evaluation of postbiotic products in preclinical and clinical stages. The use of omics techniques, such as genomics/proteomics (Odorskaya et al., 2024) and metabolomics approaches (Pang et al., 2025), is essential, and phenotypic investigation of probiotic cells can be applied to ensure the safety of the postbiotic profile. Although numerous in vitro studies have been conducted on postbiotics, randomized controlled trials remain scarce. Multicenter randomized controlled trials are the most reliable method to evaluate the safety of postbiotics. The clinical evaluation process followed a standardized protocol consisting of four phases: safety evaluation (phase 1), efficacy testing (phase 2), effectiveness validation (phase 3), and post-market surveillance (phase 4) (Zielińska et al., 2018). In the initial phase, the safety of postbiotics derived from probiotics or beneficial microorganisms should be assessed and their effectiveness should be evaluated through animal experiments (Smith et al., 2018). The results of animal experiments should be analyzed with consideration of human characteristics, while in the critical third stage, implementation is influenced by study requirements, regulations, and population, with placebo-controlled randomized trials and long-term follow-up being the gold standard. It has been suggested that the participation of microbiologists, clinicians, regulators, and food and drug industry representatives can improve clinical trial quality, field research, and finally human health (Shane et al., 2010). Meta-analysis is another method to assess the effectiveness of postbiotics on human health.

A key issue in the labeling of postbiotics is related to their definition. As discussed previously, the challenge with the definition of postbiotics lies in the absence of a universally agreed-upon definition, leading to confusion and potential misinterpretations in both research and product manufacturing. Consequently, the definition of postbiotics adopted by regulatory organizations may differ, which appears to be the most significant challenge regarding postbiotics labeling. This discrepancy can lead to variations in the composition of postbiotics based on different definitions, potentially affecting their usage and efficacy. For example, according to Anonymous (European Medicines Agency, 2019a), a list of EU-authorized medicinal products based on bacterial lysates includes various names, such as bacterial lysate, freeze-dried (lyophilized), standardized bacterial lysate, standardized lyophilisate, proteoglycans, ribosomal fractions, membrane fractions, standardized bacterial extracts, and inactivated cells. These varying terms may confuse and mislead consumers. A similar situation could arise if regulatory bodies fail to establish a clear definition of postbiotics for pharmaceutical and food applications.

The labeling of postbiotics must include details about the microorganisms used, postbiotics chemical composition, postbiotics formulation, health benefits and possible contraindications, recommended dosage, target human population, and method of application (Divsalar et al., 2025). A review on bacterial lysates for the prevention and treatment of respiratory tract infections did not clarify the mechanism of action of these products (European Medicines Agency, 2019b). The composition, production process, formulation, treatment dose, and administration routes of various postbiotics used in human therapy vary significantly. It is still uncertain whether these variations lead to different clinical outcomes, but it seems that they can affect the functionality of these products (European Medicines Agency, 2019b). As postbiotic products are relatively new, there have been no documented cases of mislabeling to date. However, several instances of mislabeling have been reported for probiotic products and supplements. For example, labels for 16 *Bifidobacterium* probiotic products have been evaluated (Lewis et al., 2016). The result indicates that the probiotic content of products is different from the list of ingredients on the label, and only one product's contents matched the label claims.

16.5 CONCLUSION

Postbiotics is a promising area in medicine and supplement formulation, offering several advantages over traditional probiotics. The therapeutic efficacy of postbiotics is closely linked to their chemical composition, including metabolites such as SCFAs, EPSs, bacteriocins, and other bioactive compounds. These components contribute to anti-inflammatory, antioxidant, and immunomodulatory effects, which have shown promise in managing metabolic diseases, improving gut health, and supporting immune responses. Despite these advancements, significant challenges remain, particularly regarding safety and regulatory frameworks. Comprehensive safety assessments and clinical trials are essential to address potential risks, such as gastrointestinal disorders and allergic reactions. Additionally, clear labeling, harmonizing definitions and regulatory standards across global and national agencies is critical to ensuring the consistency, efficacy, and safety of postbiotics-based products. Continued research, multidisciplinary collaboration, and robust regulatory frameworks are imperative to realize their full potential in medicine.

ACKNOWLEDGMENTS

The authors acknowledge Urmia University for financial support.

USE OF GENERATIVE AI

During the preparation of this work, the authors used ChatGPT to proofread the text for errors and clarity. After using this tool/service, the authors reviewed and edited the content as needed and take full responsibility for the content of the publication.

REFERENCES

Anonymous. (2023). *Harness the power of postbiotics: Trends and innovations reinventing the biotics market. Dietary supplements*. Whitepaper. DSM-Firmenich. https://www.dsm-firmenich.com/content/dam/dsm-firmenich/health-nutrition-care/human-nutrition/products/humiome/dsm-firmenich-whitepaper-postbiotics.pdf

Arellano-García, L. I., Portillo, M. P., Martínez, J. A., Courtois, A., & Milton-Laskibar, I. (2024). Postbiotics for the management of obesity, insulin resistance/type 2 diabetes and NAFLD: Beyond microbial viability. *Critical Reviews in Food Science and Nutrition*, In press. https://doi.org/10.1080/10408398.2024.2437143

Banasiewicz, T., Domagalska, D., Borycka-Kiciak, K., & Rydzewska, G. (2020). Determination of butyric acid dosage based on clinical and experimental studies—a literature review. *Gastroenterology Review*, *15*(2), 119–125. https://doi.org/10.5114/pg.2020.95556

Barros, C. P., Guimarães, J. T., Esmerino, E. A., Duarte, M. C. K. H., Silva, M. C., Silva, R., Ferreira, B. M., Sant'Ana, A. S., Freitas, M. Q., & Cruz, A. G. (2020). Paraprobiotics and postbiotics: Concepts and potential applications in dairy products. *Current Opinion in Food Science*, *32*, 1–8. https://doi.org/10.1016/j.cofs.2019.12.003

Bazjou, A., Jafari, P., Marjani, A., & Akbari, N. (2022). Effect of cell-free supernatant of *Bifidobacterium bifidum* combined with chitosan biodegradable film on full thickness wound healing in rats. *Physiology-Pharmacology*, *26*(4), 468–479. https://doi.org/10.52547/phypha.26.4.2

Bourebaba, Y., Marycz, K., Mularczyk, M., & Bourebaba, L. (2022). Postbiotics as potential new therapeutic agents for metabolic disorders management. *Biomedicine & Pharmacotherapy*, *153*, 113138. https://doi.org/10.1016/j.biopha.2022.113138

Buckley, M., Lacey, S., Doolan, A., Goodbody, E., & Seamans, K. (2018). The effect of *Lactobacillus reuteri* supplementation in *Helicobacter pylori* infection: A placebo-controlled, single-blind study. *BMC Nutrition*, *4*(1), 48. https://doi.org/10.1186/s40795-018-0257-4

Canfora, E. E., van der Beek, C. M., Jocken, J. W. E., Goossens, G. H., Holst, J. J., Olde Damink, S. W. M., Lenaerts, K., Dejong, C. H. C., & Blaak, E. E. (2017). Colonic infusions of short-chain fatty acid mixtures promote energy metabolism in overweight/obese men: A randomized crossover trial. *Scientific Reports*, *7*(1), 2360. https://doi.org/10.1038/s41598-017-02546-x

Castro-López, C., Romero-Luna, H. E., García, H. S., Vallejo-Cordoba, B., González-Córdova, A. F., & Hernández-Mendoza, A. (2024). Postbiotics: Perspectives on innovative applications. In J. M. Lorenzo (Ed.), *Strategies to improve the quality of foods* (pp. 235–257). Academic Press. https://doi.org/10.1016/B978-0-443-15346-4.00009-4

Cordaillat-Simmons, M., Rouanet, A., & Pot, B. (2020). Live biotherapeutic products: The importance of a defined regulatory framework. *Experimental & Molecular Medicine*, *52*(9), 1397–1406. https://doi.org/10.1038/s12276-020-0437-6

Cuevas-González, P. F., Liceaga, A. M., & Aguilar-Toalá, J. E. (2020). Postbiotics and paraprobiotics: From concepts to applications. *Food Research International*, *136*, 109502. https://doi.org/10.1016/j.foodres.2020.109502

de Melo Pereira, G. V., de Oliveira Coelho, B., Magalhães Júnior, A. I., Thomaz-Soccol, V., & Soccol, C. R. (2018). How to select a probiotic? A review and update of methods and criteria. *Biotechnology Advances*, *36*(8), 2060–2076. https://doi.org/10.1016/j.biotechadv.2018.09.003

de Toledo Guimarães, J., Barros, C., Sharafi, H., Moradi, M., Esmerino, E. A., & da Cruz, A. G. (2023). Postbiotics preparation for use in food and beverages. In A. Gomes da Cruz, M. C. Silva, T. Colombo Pimentel, E. A. Esmerino, & S. Verruck (Eds.), *Probiotic foods and beverages: technologies and protocols* (pp. 223–242). Springer US. https://doi.org/10.1007/978-1-0716-3187-4_16

Deghima, A., Righi, N., Rosales-Conrado, N., León-González, M. E., Baali, F., Gómez-Mejía, E., Madrid, Y., & Bedjou, F. (2021). Anti-inflammatory activity of ethyl acetate and n-butanol extracts from *Ranunculus macrophyllus* Desf. and their phenolic profile. *Journal of Ethnopharmacology*, *265*, 113347. https://doi.org/10.1016/j.jep.2020.113347

del Valle, J. C., Bonadero, M. C., & Fernández-Gimenez, A. V. (2023). *Saccharomyces cerevisiae* as probiotic, prebiotic, synbiotic, postbiotics and parabiotics in aquaculture: An overview. *Aquaculture*, *569*, 739342. https://doi.org/10.1016/j.aquaculture.2023.739342

Divsalar, E., İncili, G. K., Shi, C., Semsari, E., Hosseini, S. H., Ebrahimi Tirtashi, F., Toker, O. S., Mojgani, N., Liu, S.-Q., & Moradi, M. (2025). Challenges with the use of postbiotics/parabiotics in food industry. *Critical Reviews in Food Science and Nutrition*, In press.

Duarte, M., Oliveira, A. L., Oliveira, C., Pintado, M., Amaro, A., & Madureira, A. R. (2022). Current postbiotics in the cosmetic market—an update and development opportunities. *Applied Microbiology and Biotechnology*, *106*(18), 5879–5891. https://doi.org/10.1007/s00253-022-12116-5

Erginkaya, Z., & Konuray-Altun, G. (2022). Potential biotherapeutic properties of lactic acid bacteria in foods. *Food Bioscience*, *46*, 101544. https://doi.org/10.1016/j.fbio.2022.101544

European Medicines Agency. (2019a). Bacterial lysate medicines Article 31 referral—Annex I. The European Medicines Agency. Committee for Medicinal Products for Human Use. https://www.ema.europa.eu/en/documents/referral/bacterial-lysate-medicines-article-31-referral-annex-i_en.pdf

European Medicines Agency. (2019b). Bacterial lysate medicines Article 31 referral—CHMP assessment report. EMA/457345/2019—The European Medicines Agency. Committee for Medicinal Products for Human Use. https://www.ema.europa.eu/en/documents/referral/bacterial-lysate-medicines-article-31-referral-chmp-assessment-report_en.pdf

Hernández-Granados, M. J., & Franco-Robles, E. (2020). Postbiotics in human health: Possible new functional ingredients? *Food Research International*, *137*, 109660. https://doi.org/10.1016/j.foodres.2020.109660

Hosseini, S. H., Farhangfar, A., Moradi, M., & Dalir-Naghadeh, B. (2024). Beyond probiotics: Exploring the potential of postbiotics and parabiotics in veterinary medicine. *Research in Veterinary Science*, *167*, 105133. https://doi.org/10.1016/j.rvsc.2023.105133

Huang, Z., Lin, F., Zhu, X., Zhang, C., Jiang, M., & Lu, Z. (2020). An exopolysaccharide from *Lactobacillus plantarum* H31 in pickled cabbage inhibits pancreas α-amylase and regulating metabolic markers in HepG2 cells by AMPK/PI3K/Akt pathway. *International Journal of Biological Macromolecules*, *143*, 775–784. https://doi.org/10.1016/j.ijbiomac.2019.09.137

İncili, G. K., Akgöl, M., Karatepe, P., Tekin, A., Kanmaz, H., Kaya, B., & Hayaloğlu, A. A. (2023). Whole-cell postbiotics: An innovative approach for extending the shelf life and controlling major foodborne pathogens in chicken breast fillets. *Food and Bioprocess Technology*, *16*, 1502–1524. https://doi.org/10.1007/s11947-023-03009-0

Kolling, Y., Salva, S., Villena, J., & Alvarez, S. (2018). Are the immunomodulatory properties of *Lactobacillus rhamnosus* CRL1505 peptidoglycan common for all *Lactobacilli* during respiratory infection in malnourished mice? *PLOS ONE*, *13*(3), e0194034. https://doi.org/10.1371/journal.pone.0194034

Kushwaha, V., Rai, P., Varshney, S., Gupta, S., Khandelwal, N., Kumar, D., & Nilkanth Gaikwad, A. (2022). Sodium butyrate reduces endoplasmic reticulum stress by modulating CHOP and empowers favorable anti-inflammatory adipose tissue immune-metabolism in HFD fed mice model of obesity. *Food Chemistry: Molecular Sciences*, *4*, 100079. https://doi.org/10.1016/j.fochms.2022.100079

Lee, J., Park, S., Oh, N., Park, J., Kwon, M., Seo, J., & Roh, S. (2021). Oral intake of *Lactobacillus plantarum* L-14 extract alleviates TLR2- and AMPK-mediated obesity-associated disorders in high-fat-diet-induced obese C57BL/6J mice. *Cell Proliferation*, *54*(6), e13039. https://doi.org/10.1111/cpr.13039

Lewis, Z. T., Shani, G., Masarweh, C. F., Popovic, M., Frese, S. A., Sela, D. A., Underwood, M. A., & Mills, D. A. (2016). Validating bifidobacterial species and subspecies identity in commercial probiotic products. *Pediatric Research*, *79*(3), 445–452. https://doi.org/10.1038/pr.2015.244

Luzzi, A., Briata, I. M., Di Napoli, I., Giugliano, S., Di Sabatino, A., Rescigno, M., & Cena, H. (2024). Prebiotics, probiotics, synbiotics and postbiotics to adolescents in metabolic syndrome. *Clinical Nutrition*, *43*(6), 1433–1446. https://doi.org/10.1016/j.clnu.2024.04.032

Matheus, V., Monteiro, L., Oliveira, R., Maschio, D., & Collares-Buzato, C. (2017). Butyrate reduces high-fat diet-induced metabolic alterations, hepatic steatosis and pancreatic beta cell and intestinal barrier dysfunctions in prediabetic mice. *Experimental Biology and Medicine*, *242*(12), 1214–1226. https://doi.org/10.1177/1535370217708188

Matsuguchi, T., Takagi, A., Matsuzaki, T., Nagaoka, M., Ishikawa, K., Yokokura, T., & Yoshikai, Y. (2003). Lipoteichoic acids from *Lactobacillus strains* elicit strong tumor necrosis factor alpha-inducing activities in macrophages through toll-like receptor 2. *Clinical and Vaccine Immunology*, *10*(2), 259–266. https://doi.org/10.1128/CDLI.10.2.259-266.2003

Mishra, N., Garg, A., Ashique, S., & Bhatt, S. (2024). Potential of postbiotics for the treatment of metabolic disorders. *Drug Discovery Today*, *29*(4), 103921. https://doi.org/10.1016/j.drudis.2024.103921

Monteiro, S. S., Schnorr, C. E., & Pasquali, M. A. (2023). Paraprobiotics and postbiotics—current state of scientific research and future trends toward the development of functional foods. *Foods*, *12*(12), 2394. https://doi.org/10.3390/foods12122394

Moradi, M., Molaei, R., & Guimarães, J. T. (2021). A review on preparation and chemical analysis of postbiotics from lactic acid bacteria. *Enzyme and Microbial Technology*, *143*, 109722. https://doi.org/10.1016/j.enzmictec.2020.109722

Nasri, F., Alizadeh, A., İncili, G. K., Hayaloğlu, A. A., & Moradi, M. (2025). Investigating chemical composition and functionality of *Lactobacillus acidophilus* LA-5 postbiotics prepared in classic and cheese whey media. *Probiotics and Antimicrobial Proteins*, In press. https://doi.org/10.1007/s12602-024-10435-9

Nataraj, B. H., Ali, S. A., Behare, P. V. & Yadav, H. (2020). Postbiotics-parabiotics: The new horizons in microbial biotherapy and functional foods. *Microbial Cell Factories*, *19*(1), 168. https://doi.org/10.1186/s12934-020-01426-w

Odorskaya, M. V., Mavletova, D. A., Nesterov, A. A., Tikhonova, O. V., Soloveva, N. A., Reznikova, D. A., Galanova, O. O., Vatlin, A. A., Slynko, N. M., Vasilieva, A. R., Peltek, S. E., & Danilenko, V. N. (2024). The use of omics technologies in creating LBP and postbiotics based on the *Limosilactobacillus fermentum* U-21. *Frontiers in Microbiology*, *15*, 1416688. https://doi.org/10.3389/fmicb.2024.1416688

Pang, L., Huang, Y., Li, R., Guo, L., Man, C., Yang, X., & Jiang, Y. (2025). Effects of postbiotics produced by *Lactobacillus plantarum* JM015 isolated from traditional fermented dairy products on *Salmonella*-induced intestinal inflammation: A preventive strategy. *Food Chemistry*, *469*, 142549. https://doi.org/10.1016/j.foodchem.2024.142549

Park, M., Joung, M., Park, J. H., Ha, S. K., & Park, H. Y. (2022). Role of postbiotics in diet-induced metabolic disorders. *Nutrients*, *14*(18), 3701. https://doi.org/10.3390/nu14183701

Prajapati, N., Patel, J., Singh, S., Yadav, V. K., Joshi, C., Patani, A., Prajapati, D., Sahoo, D. K., & Patel, A. (2023). Postbiotic production: Harnessing the power of microbial metabolites for health applications. *Frontiers in Microbiology*, *14*, 1306192. https://doi.org/10.3389/fmicb.2023.1306192

Requena, T., & Pérez Martínez, G. (2022). Probiotics, prebiotics, synbiotics, postbiotics and other biotics. What's next? In M. Glibetic (Ed.), *Comprehensive gut microbiota* (pp. 197–210). Elsevier. https://doi.org/10.1016/B978-0-12-819265-8.00094-2

Rezaie, N., Aghamohammad, S., Haj, E., Gholizadeh, A., Khatami, S., Sohrabi, A., & Rohani, M. (2024). The preventive effects of native probiotic and postbiotic on inflammation and oxidative stress in DSS-induced colitis with normal diet: Which of these agents may offer greater advantages? *Heliyon*, *10*(17), e37279. https://doi.org/10.1016/j.heliyon.2024.e37279

Salazar, N., Neyrinck, A. M., Bindels, L. B., Druart, C., Ruas-Madiedo, P., Cani, P. D., de los Reyes-Gavilán, C. G., & Delzenne, N. M. (2019). Functional effects of EPS-producing *Bifidobacterium* administration on energy metabolic alterations of diet-induced obese mice. *Frontiers in Microbiology*, *10*, 1809. https://doi.org/10.3389/fmicb.2019.01809

Salminen, S., Collado, M. C., Endo, A., Hill, C., Lebeer, S., Quigley, E. M. M., Sanders, M. E., Shamir, R., Swann, J. R., Szajewska, H., & Vinderola, G. (2021). The International Scientific Association of

Probiotics and Prebiotics (ISAPP) consensus statement on the definition and scope of postbiotics. *Nature Reviews. Gastroenterology & Hepatology*, *18*(9), 649–667. https://doi.org/10.1038/s41575-021-00440-6

Salva, S., Tiscornia, I., Gutiérrez, F., Alvarez, S., & Bollati-Fogolín, M. (2021). *Lactobacillus rhamnosus* postbiotic-induced immunomodulation as safer alternative to the use of live bacteria. *Cytokine*, *146*, 155631. https://doi.org/10.1016/j.cyto.2021.155631

Sanna, S., van Zuydam, N. R., Mahajan, A., Kurilshikov, A., Vich Vila, A., Võsa, U., Mujagic, Z., Masclee, A. A. M., Jonkers, D. M. A. E., Oosting, M., Joosten, L. A. B., Netea, M. G., Franke, L., Zhernakova, A., Fu, J., Wijmenga, C., & McCarthy, M. I. (2019). Causal relationships among the gut microbiome, short-chain fatty acids and metabolic diseases. *Nature Genetics*, *51*(4), 600–605. https://doi.org/10.1038/s41588-019-0350-x

Shane, A. L., Cabana, M. D., Ellis, C. L., Heimbach, J. T., Hempel, S., Hummelen, R., Lynch, S., Merenstein, D. J., Sanders, M. E., Tancredi, D. J., & Vidry, S. (2010). Guide to designing, conducting, publishing, and communicating results of clinical studies involving probiotic applications in human participants. *Gut Microbes*, *1*(4), 243–253. https://doi.org/10.4161/gmic.1.4.12707

Sharafi, H., Divsalar, E., Rezaei, Z., Liu, S. Q., & Moradi, M. (2024). The potential of postbiotics as a novel approach in food packaging and biopreservation: A systematic review of the latest developments. *Critical Reviews in Food Science and Nutrition*, *64*(33), 12524–12554. https://doi.org/10.1080/10408398.2023.2253909

Smith, D., Anderson, D., Degryse, A.-D., Bol, C., Criado, A., Ferrara, A., Franco, N. H., Gyertyan, I., Orellana, J. M., Ostergaard, G., Varga, O., & Voipio, H.-M. (2018). Classification and reporting of severity experienced by animals used in scientific procedures: FELASA/ECLAM/ESLAV Working Group report. *Laboratory Animals*, *52*(1_suppl), 5–57. https://doi.org/10.1177/0023677217744587

Teame, T., Wang, A., Xie, M., Zhang, Z., Yang, Y., Ding, Q., Gao, C., Olsen, R. E., Ran, C., & Zhou, Z. (2020). Paraprobiotics and postbiotics of probiotic *Lactobacilli*, their positive effects on the host and action mechanisms: A review. *Frontiers in Nutrition*, *7*, 191. https://doi.org/10.3389/fnut.2020.570344

Thorakkattu, P., Khanashyam, A. C., Shah, K., Babu, K. S., Mundanat, A. S., Deliephan, A., Deokar, G. S., Santivarangkna, C., & Nirmal, N. P. (2022). Postbiotics: Current trends in food and pharmaceutical industry. *Foods*, *11*(19), 1–29. https://doi.org/10.3390/foods11193094

Urban, J., Kareem, K. Y., Atanasov, A. G., Matuszewski, A., Bień, D., Ciborowska, P., Rygało-Galewska, A., & Michalczuk, M. (2024). Postbiotics, a natural feed additive for growth performance, gut microbiota and quality of poultry products. *Current Research in Biotechnology*, *8*, 100247. https://doi.org/10.1016/j.crbiot.2024.100247

van der Hee, B., & Wells, J. M. (2021). Microbial regulation of host physiology by short-chain fatty acids. *Trends in Microbiology*, *29*(8), 700–712. https://doi.org/10.1016/j.tim.2021.02.001

Vinderola, G. (2024). Postbiotics: A global perspective on regulatory progress. *ISAPP Science Blog*. https://isappscience.org/postbiotics-a-global-perspective-on-regulatory-progress/

Vinderola, G., Druart, C., Gosálbez, L., Salminen, S., Vinot, N., & Lebeer, S. (2023). Postbiotics in the medical field under the perspective of the ISAPP definition: Scientific, regulatory, and marketing considerations. *Frontiers in Pharmacology*, *14*(1), 1–23. https://doi.org/10.3389/fphar.2023.1239745

Vinderola, G., Sanders, M. E., Salminen, S., & Szajewska, H. (2022). Postbiotics: The concept and their use in healthy populations. *Frontiers in Nutrition*, *9*, 1002213. https://doi.org/10.3389/fnut.2022.1002213

Xiao, S.-D., De Zhang, Z., Lu, H., Jiang, S. H., Liu, H. Y., Wang, G. S., Xu, G. M., Zhang, Z. B., Lin, G. J., & Wang, G. L. (2003). Multicenter, randomized, controlled trial of heat-killed *Lactobacillus acidophilus* LB in patients with chronic diarrhea. *Advances in Therapy*, *20*(5), 253–260. https://doi.org/10.1007/BF02849854

Yao, Y., Yan, L., Chen, H., Wu, N., Wang, W., & Wang, D. (2020). *Cyclocarya paliurus* polysaccharides alleviate type 2 diabetic symptoms by modulating gut microbiota and short-chain fatty acids. *Phytomedicine*, *77*, 153268. https://doi.org/10.1016/j.phymed.2020.153268

Ye, J., Lv, L., Wu, W., Li, Y., Shi, D., Fang, D., Guo, F., Jiang, H., Yan, R., Ye, W., & Li, L. (2018). Butyrate protects mice against methionine–choline-deficient diet-induced non-alcoholic steatohepatitis by improving gut barrier function, attenuating inflammation and reducing endotoxin levels. *Frontiers in Microbiology*, *9*, 1967. https://doi.org/10.3389/fmicb.2018.01967

Yilmaz, Y. (2024). Postbiotics as antiinflammatory and immune-modulating bioactive compounds in metabolic dysfunction-associated steatotic liver disease. *Molecular Nutrition & Food Research*, *68*, 2400754. https://doi.org/10.1002/mnfr.202400754

Yin, X., Heeney, D., Srisengfa, Y., Golomb, B., Griffey, S., & Marco, M. (2018). Bacteriocin biosynthesis contributes to the anti-inflammatory capacities of probiotic *Lactobacillus plantarum*. *Beneficial Microbes*, *9*(2), 333–344. https://doi.org/10.3920/BM2017.0096

Yolmeh, M., Sforça, M. L., & Sant'Ana, A. S. (2025). Antimicrobial properties of *Bifidobacterium animalis* subsp. lactis Bb-12 paraprobiotic obtained using ohmic heating against *Salmonella enterica*: A metabolomic approach. *International Journal of Food Microbiology*, *429*, 111016. https://doi.org/10.1016/j.ijfoodmicro.2024.111016

Zhang, S., Zhao, J., Xie, F., He, H., Johnston, L. J., Dai, X., Wu, C., & Ma, X. (2021). Dietary fiber-derived short-chain fatty acids: A potential therapeutic target to alleviate obesity-related nonalcoholic fatty liver disease. *Obesity Reviews*, *22*(11), e13316. https://doi.org/10.1111/obr.13316

Zhong, Y., Wang, S., Di, H., Deng, Z., Liu, J., & Wang, H. (2022). Gut health benefit and application of postbiotics in animal production. *Journal of Animal Science and Biotechnology*, *13*(1), 38. https://doi.org/10.1186/s40104-022-00688-1

Zhou, D., Pan, Q., Xin, F.-Z., Zhang, R.-N., He, C.-X., Chen, G.-Y., Liu, C., Chen, Y.-W., & Fan, J.-G. (2017). Sodium butyrate attenuates high-fat diet-induced steatohepatitis in mice by improving gut microbiota and gastrointestinal barrier. *World Journal of Gastroenterology*, *23*(1), 60. https://doi.org/10.3748/wjg.v23.i1.60

Zielińska, D., Sionek, B., & Kołożyn-Krajewska, D. (2018). Safety of probiotics. In A. M. Holban & G. A. Mihai (Eds.), *Diet, microbiome and health* (pp. 131–161). Elsevier. https://doi.org/10.1016/B978-0-12-811440-7.00006-5

Index

Pages in **bold** refer to tables.

T

W